Stellar photometry from space, automatic photometric telescopes and CCD photometers, these are just some of the exciting areas of current interest and future developments in stellar photometry covered in this timely review. Articles from international experts – drawn together at the IAU Colloquium 136, in Dublin, 1992 – are gathered here to cover all aspects of this fundamental technique.

In this survey, professionals d[illegible]-art and future technology including photometry with milli[illegible]nnel arrays used in the optical and IR, a global net[illegible]lescopes, time-series photometry of faint sour[illegible]rom space.

These articles provide an u[illegible] photometry and a guide to future developments – an [illegible]nals involved in the design and use of such instruments.

Stellar Photometry –
Current Techniques and Future Developments

IAU Astronomical Union
Union Astronomique International

The following Colloquia of the International Astronomical Union are published for the Union by Cambridge University Press.

82. Cepheids. *Edited by Barry F. Madore.* 0 521 30091 6. 1985

91. History of Oriental Astronomy. *Edited by G. Swarup, A. K. Bag and K. S. Shukla.* 0 521 34659 2. 1987

92. Physics of Be Stars. *Edited by A. Slettebak and T. P. Snow.* 0 521 33078 5. 1987

101. Supernova Remnants and the Interstellar Medium. *Edited by R. S. Roger and T. L. Landecker.* 0 521 35062 X. 1988

105. The Teaching of Astronomy. *Edited by Jay M. Pasachoff and John R. Percy.* 0 521 35331 8. 1989

106. Evolution of Peculiar Red Giant Stars. *Edited by Hollis Johnson and Ben Zuckerman.* 0 521 36617 8. 1989

111. The Use of Pulsating Stars in Fundamental Problems of Astronomy. *Edited by Edward G. Schmidt.* 0 521 37023 X. 1989

136. Stella Photometry – Current Techniques and Future Developments. *Edited by C. J. Butler and I. Elliott.* 0 521 41866 6. 1993

139. Stellar Pulsation and Pulsating Variable Stars. *Edited by J. Nemec and J. Matthews.* 0 521 44382 2. 1993

Stellar Photometry – Current Techniques and Future Developments

Proceedings of the IAU Colloquium No. 136 held in Dublin, Ireland
4–7 August 1992

Edited by
C. J. BUTLER
Armagh Observatory

and

I. ELLIOTT
Dunsink Observatory

Published by the Press Syndicate of the University of Cambridge
The Pitt Building, Trumpington Street, Cambridge CB2 1RP
40 West 20th Street, New York, NY 10011–4211, USA
10 Stamford Road, Oakleigh, Melbourne 3166, Australia

First published 1993

Printed in Great Britain at the University Press, Cambridge

A catalogue record for this book is available from the British Library

Library of Congress cataloguing in publication data available

ISBN 0 521 41866 6 hardback

Contents

The History of Stellar Photometry

Session 1: Photometric Systems

Session 2: High Precision Photometry

Session 3: New Techniques

Session 4: Automatic Photoelectric Telescopes, and Extinction

Session 5: Global Networks

Session 6: Photometry with CCDs

Session 7: Photometry from Space

Participants

B.J. Anthony-Twarog	University of Kansas, Lawrence, KS 66044-2151, U.S.A.
E. Antonello	Osserv. Astronomico di Brera, I-22055 Merate(CO), Italy
A.P. Antov	Dept. of Astr., Belogradchik Ast. Obs., 1784 Sofia, Bulgaria
K. Arai	Kuki-Hokuyo Senior High School, Kuki City, Japan
J. Baruch	University of Bradford, Bradford BD7-1DP, U.K.
M.S. Bessell	Mount Stromlo & Siding Spring Obs., Australia
R.P. Boyle	Vatican Obs., Univ. of Arizona, Tucson, AZ 85721, U.S.A.
M. Breger	Institute for Astronomy, Tuerkenschanzstr. 17, Wien, Austria
G.E. Bromage	Rutherford Appleton Lab., Chilton, Didcot OX11 0QX, U.K.
E. Budding	Carter Observatory, P.O. Box 2909, Wellington, New Zealand
C.J. Butler	Armagh Observatory, College Hill, Armagh, N. Ireland
M. Chevreton	Observatoire de Paris, Meudon, 92195 Meudon Cedex, France
M. Cohen	Radio Ast. Lab., Univ. of Calif., Berkeley, CA 94720, U.S.A.
D.L. Crawford	Kitt Peak National Observatory, Tucson, AZ 85726, U.S.A.
G. Cutispoto	Osservatorio Astrofisico, I-95125 Catania, Italy
A.J. Delgado	Institute de Astrofisica de Andalucia, 18080- Granada, Spain
D. Dravins	Lund Observatory, Box 43, S-22100 Lund, Sweden
R.P. Edwin	Univ. of St. Andrews, Fife, KY16 9SS, Scotland
I. Elliott	Dunsink Observatory, Dublin 15, Ireland
J.D. Fernie	Univ. of Toronto, Richmond Hill, Ontario L4C 4Y6, Canada
F. Figueras	Astr. i Met., Univ. de Barcelona, 08028 Barcelona, Spain
R. Florentin-Nielsen	Copenhagen Univ. Obs., Brorfelde, DK-4340 Toelloese, Denmark
E.L. Folgheraiter	University, Leeds LS2 9JT, W. Yorks., U.K.
M. Fulle	Osservatorio Astronomico, via Tiepolo 11, I-34131 Trieste, Italy
J.R. Garcia	Instituto Copernico, 1448 Buenos Aires, Argentina
R.F. Garrison	David Dunlap Obs., Richmond Hill, ONT L4C 4Y6, Canada
R.M. Genet	Fairborn Obs., 3435 E. Edgewood Ave., Mesa, Arizona, U.S.A.
I.S. Glass	S.A.A.O., P.O. Box 9, Observatory 7935, S. Africa
I.V. Glushneva	Sternberg Inst., Univ. of Moscow, 119899 Moscow, Russia
W.K. Griffiths	Physics Dept., The University of Leeds, Leeds LS2 9JT, U.K.
V. Grossmann	Astronomiches Institut, Tuebingen, D 7400, Germany
J. Guarinos	Obs. de Geneve, 51, CH-1290 Sauverny, Switzerland
M.D. Guarnieri	c/o Oss. Astr. , 10025 Pino Torinese, Torino, Italy
D.S. Hall	Dyer Obs., Vanderbilt Univ., Nashville, TN 37235, U.S.A.
B. Hauck	l'Univ. de Lausanne, CH-1290 Chavannes-des-Bois, Switzerland
R.L. Hawkins	Whitin Obs., Wellesley College, Wellesley, MA 02181, U.S.A.
J. Hearnshaw	University of Canterbury, Christchurch, New Zealand
P.W. Hill	Univ. of St. Andrews, Fife KY16 9SS, Scotland
J. Hilton	Math. Dept., Goldsmiths' College, London SE14 6NW, U.K.
S.B. Howell	Planetary Science Institute, Tucson, AZ 85719, U.S.A.
S.M.G. Hughes	California Inst. of Tech., Pasadena, CA 91125, U.S.A.
A. Imadache-Guarinos	Obs. de Geneve, 51, CH-1290 Sauverny, Switzerland
Y. Itoh	Sendai West Senior High School, Sendai-City, Miyagi, Japan

I. Jankovics	Gothard Astrophysical Obs., Sombathely, 9707, Hungary
H. Jonch-Sorensen	Copenhagen Univ. Obs., DK-1350 Copenhagen-K, Denmark
D.H.P. Jones	Royal Greenwich Observatory, Cambridge CB3 0EZ, U.K.
B.D. Jordan	Dunsink Observatory, Dublin 15, Ireland
C. Jordi	Univ. de Barcelona, E-08028 Barcelona, Spain
I.D. Karachentsev	Special Astrophysical Observatory, Zelenchukskaja, Russia
D. Kilkenny	S.A.A.O., P.O. Box 9, Observatory 7935, S. Africa
H.S. Kim	Kongju National Univ. , Choongnam, 314-901, Korea
M. Kitamura	National Astronomical Obs., Mitaka, Tokyo 181, Japan
T.J. Kreidl	Lowell Observatory, Flagstaff, AZ 86001, U.S.A.
J. Kreiner	Pedagogical Univ., Inst. of Physics, Krakow, Poland
K. Kuratov	Astrophys. Inst., Kazakh Academy of Science, Kazakhstan
M. Kurpinska-Winiarska	Astr. Obs. of Jagiellonian Univ., 30-244 Krakow, Poland
S.K. Leggett	U.S. Naval Observatory, Flagstaff, AZ 86002, U.S.A.
C. Lloyd	Rutherford Appleton Lab., Chilton, Didcot, OX11 0QX, U.K.
G.W. Lockwood	Lowell Observatory, Flagstaff, AZ 86001, U.S.A.
K.-H. Mantel	Universitaets-Sternwarte, D-8000 Muenchen 80, Germany
M.F. McCarthy S.J.	Vatican Obs. Research Group, Univ. of Arizona, Tuscon, U.S.A.
J.W. Menzies	S.A.A.O., P.O. Box 9, Observatory 7935, S. Africa
E.F. Milone	Univ. of Calgary, Calgary AB T2N 1N4, Canada
C. Morossi	Osservatorio Astronomico, I-34131 Trieste, Italy
B. Nicolet	Observatoire de Geneve, Ch-1290 Sauverny, Switzerland
P. North	Univ. de Lausanne, CH-1290 Chavannes-des-Bois, Switzerland
V.V. Novikov	Main Astr. Obs., Pulkovo, St. Petersburg, 196140, Russia
C. O'Byrne	Physics Department, University College, Galway, Ireland
D. O'Donoghue	Univ. of Cape Town, Rondebosch 7700, South Africa
E. O'Mongain	Physics Dept., University College, Belfield, Dublin 4, Ireland
T. O'Sullivan	10 Mountjoy Parade, Dublin 1, Ireland
E. Oblak	Obs. de Besançon, BP 1615, 25010 Besançon Cedex, France
S. Ohmori	Science Museum, Kawasaki City, Kanagwa-Ken, Japan
T. Oja	Kvistaberg Observatory, S-197H91 BRO, Sweden
K. Oláh	Konkoly Observatory, P.O. Box 67, 1525 Budapest, Hungary
R. Oriol i Palarea	Escola Univ. Pol., 08800 - Vilanova i la Geltru, Spain
M. Othman	Planetarium Div., 50480 Kuala Lumpur, Malayasia
H.S. Park	Korea Astronomy Observatory, Taejon 305-348, Korea
J.H. Peña	Inst. de Astr. UNAM & INAOE, 04510 Mexico D.F., Mexico
R. Peniche	Inst. de Astr. UNAM & INAOE, 04510 Mexico D.F., Mexico
A.J. Penny	Rutherford Appleton Lab., Chilton, Didcot, OX11 0QX, U.K.
W. Pfau	University Obs., Schillergasschen 2, D-(O)-6900 Jena, Germany
A.G. Davis Philip	1125 Oxford Place, Schnectady, New York 12308, U.S.A.
A. Piccioni	Dipartimento di Astronomia, 40 Bologna, Italy
E. Poretti	Osservatorio Astr. di Brera, I-22055 Merate (CO), Italy
L. Pulone	Osservatorio Astr. di Trieste, I-34131 Trieste, Italy
D.P. Pyper Smith	University of Nevada, Las Vegas, 89154, U.S.A.
F.R. Querçi	Obs. Midi-Pyrenees, 14 Ave. E. Belin, 31400 Toulouse, France
M. Rabbette	Physics Dept., University College, Belfield, Dublin 4, Ireland

T.P. Ray	School of Cosmic Physics, 5 Merrion Square, Dublin 2, Ireland
R.M. Redfern	Physics Department, University College, Galway, Ireland
H.-G. Reimann	Univ.-Sternwarte und Astro. Inst., D(O)-6900 Jena, Germany
A. Ruelas	Apartado Postal 70-264, CP 04510 Mexico D.F., Mexico
S.C. Russell	School of Cosmic Physics, 5 Merrion Square, Dublin 2, Ireland
H.J. Schöber	Institut fuer Astronomie, Universitatplatz 5, A-8010 Graz, Austria
R.R. Shobbrook	University of Sydney, New South Wales 2006, Australia
N. Smith	Regional Technical College, Rossa Avenue, Cork, Ireland
M.A.J. Snijders	Astronomisches Inst. Tuebingen, D 7400, Tuebingen 1, Germany
C.L. Sterken	Astrophysical Institut, Vrije Univ. Brussel, 1050 Brussels, Belgium
P.B. Stetson	Dominion Astrophysical Obs., Victoria, BC V8X 4M6, Canada
M.J. Stift	Institut fur Astronomie, Turkenschanstr. 17, A-1180 Wien, Austria
V. Straižys	1125 Oxford Place, Schenectady, N.Y. 12308, U.S.A.
D.J. Sullivan	Victoria University, P.O. Box 600 Wellington, New Zealand
G. Ṡzécsényi-Nagy	Eötvös University, Ludovika ter 2, Budapest VIII, H-1083, Hungary
M. Taylor	Univ. of Wisconsin, 475 N. Charter, Madison, WI 53706, U.S.A.
J. Tinbergen	Kapteyn Observatory, NL-9310 KA Roden, The Netherlands
W. Tobin	Physics Dept., Univ. of Canterbury, Christchurch 1, New Zealand
J. Torra	Univ. de Barcelona, Avda. Diagonal, 647, E-08028 Barcelona, Spain
J.P. Toutonghi	Seattle University, Seattle, WA 98122, U.S.A.
J.L. Trudel	Univ. of Toronto, Toronto, Ontario, M5S 1A1, Canada
M.K. Tzvetkov	Bul. Acad. of Sciences, BG-1784 Sofia, Bulgaria
A.R. Upgren	Van Vleck Obs., Wesleyan Univ., Middleton, CT 06457, U.S.A.
J.C. Valtier	Obs. de la Côte d'Azur, B.P. 229, 06304 Nice Cedex 4, France
WANG Chuan-jin	Purple Mountain Observatory, Nanjing, 210008, China
A. Walker	Cerro Tololo Inter-American Obs., Casilla 603, La Serena, Chile
B. Warner	University of Capetown, Private Bag, Rondebosch 7700, S. Africa
P.A. Wayman	Dunsink Observatory, Dublin 15, Ireland
D.L. Weaire	Trinity College, Dublin 2, Ireland
R.F. Wing	Ohio State University, Columbus, OH 43210, U.S.A.
W.Z. Wisniewski	Space Sciences Bld., Univ. of Arizona, Tucson, AZ 85721, U.S.A.
G. Wlérick	Obs. de Paris, Section d'Astrophysique, F-92195 Meudon, France
A.T. Young	ESO Headquarters, D-8046 Garching bei Munchen, Germany
R. Yudin	Central Astronomical Observatory, Pulkovo, St. Petersburg, Russia
M. Zeilik	University of New Mexico, Albuquerque, NM 87131-1156, U.S.A.
M. de Groot	Armagh Observatory, College Hill, Armagh BT61 9DG, N. Ireland
I.G. van Breda	Dunsink Observatory, Dublin 15, Ireland
N.S. van der Bliek	Sterrewacht Leiden, 2300 RA Leiden, The Netherlands
T. von Hippel	Univ. of Cambridge, Madingley Road, Cambridge, CB3 OHA, U.K.

1 A.G. Davis Philip
2 Michael Zeilik
3 Martin McCarthy
4 MaryJane Taylor
5 Dainis Dravins
6 Denis Weaire
7 Richard Boyle
8 Hee Soo Kim
9 Yoshiharu Itoh
10 Shigeo Ohmori
11 Mart de Groot
12 Ian van Breda
13 Hong Suh Park
14 Sheila Hill
15 Gérard Wlérick
16 Tarmo Oja
17 Bill Griffiths
18 Kikuichi Arai
19 Hans-Georg Reimann
20 Istvan Jankovics
21 Masatoshi Kitamura
22 Pierre North
23 Alexandar Antov
24 Roger Edwin
25 Patrick Wayman
26 Emilio Folgheraiter
27 Werner Pfau
28 Alistair Walker
29 Shaun Hughes
30 Hans Schöber
31 Don Fernie
32 Jacques Guarinos
33 Victor Novikov
34 Lee Hawkins
35 Irina Glushneva
36 Ralph Florentin-Nielsen
37 Jaap Tinbergen
38 Ruslan Yudin
39 Martin Stift
40 Aiche Imadache-Guarinos
41 Bernard Nicolet
42 Tobias Kreidl
43 Steve Howell
44 Mazlan Othman
45 Russ Genet
46 Ed Budding
47 Kenesken Kuratov
48 Chris Lloyd
49 Edouard Oblak
50 John Baruch
51 Ian Glass
52 Karl-Heinz Mantel
53 Helen Milone
54 Douglas Hall
55 John Hearnshaw
56 Chris O'Byrne
57 Giuseppe Cutispoto
58 Barbara Anthony-Twarog
59 Andrew Young
60 John Toutonghi
61 Alan Penny
62 Chris Sterken
63 Denis Sullivan
64 Jaime Garcia
65 Peter Stetson
66 Martin Cohen
67 Alex Ruelas
68 Bob Garrison
69 Dave Crawford
70 Eon O'Mongain
71 Jean Claude Valtier
72 Francesca Figueras
73 Michael Bessell
74 William Tobin
75 Ted von Hippel
76 Gabor Ŝzécsényi-Nagy
77 Roser Oriol
78 Sandy Leggett
79 Francois Querçi
80 Bob Wing
81 Brian Warner
82 Jordi Torra
83 Carme Jordi
84 Toon Snijders
85 Gene Milone
86 Nicole van der Bliek
87 Brendan Jordan
88 Carlo Morossi
89 Michel Chevreton
90 Darragh O'Donoghue
91 Marco Fulle
92 Diane Pyper Smith
93 Rosario Peniche
94 Vytautas Straizys
95 Elio Antonello
96 Ian Elliott
97 Gordon Bromage
98 Volkmar Grossmann
99 José Peña
100 Luigi Pulone
101 Arthur Upgren
102 Michel Breger
103 Bernard Hauck
104 Ennio Poretti
105 Steven Russell
106 Helge Jonch-Sorensen
107 John Hilton
108 Bob Shobbrook
109 Katalin Oláh
110 Milcho Tzvetkov
111 Adalberto Piccioni
112 Wes Lockwood
113 John Menzies
114 Jean-Louis Trudel
115 John Butler
116 Derek Jones
117 Phil Hill
118 Wieslaw Wisniewski
119 Maria Kurpinska-Winiarska
120 WANG Chuan-jin
121 Maura Rabbette
122 Jerzy Kreiner

Foreword

I have been asked to summarize the goals and accomplishments of the Colloquium, and provide a guide to the reader — a tall order, and one that cannot be filled impartially, for different participants had different goals, and will regard different accomplishments as most important. Because photometry is a vigorous and active field, anyone's choices will be disputed by someone else.

Some of the goals were certainly to bring together as many leading photometrists as possible; to review the current state of our art; and to discuss where we should go from here, and how we can get there. We wanted to have an interesting meeting that would stimulate lots of discussion, in greater depth than is possible at IAU General Assemblies. Moreover, the IAU encouraged wide geographic participation, and wanted to ensure that astronomers far removed from the leading edge of photometry would be brought closer to it. Finally, we wanted to honor the pioneers of photometry, and to develop some perspective from the history of the subject.

To meet these goals, the Scientific Organizing Committee selected seven major topics, and devoted one session to each: Photometric Systems; High-Precision Photometry; New Techniques; Automatic Photoelectric Telescopes; Global Networks; Photometry with CCDs; and Photometry from Space. The papers in this volume are arranged in this order. We invited a number of people to cover each area, trying to obtain speakers who could give good talks as well as provide technical expertise. In addition, we were fortunate in having an excellent historical review, presented by John Hearnshaw.

As the abstracts poured in, our plans changed a little. Some invited speakers were unable to come, and were replaced by contributed talks selected by the SOC. Many more excellent CCD papers were submitted than could fit into a single session, so some appeared in other sessions. A group of papers on extinction elbowed their way into the APT session. Members of the SOC wanted more papers presented orally than would fit in the program, so many interesting papers had to become poster talks.

I believe the quality of the papers is illustrated by complaints I received from two participants. "I had expected to skip a couple of sessions to explore the town," one said. "But the talks have been so interesting, I haven't had a chance to see Dublin." Another astronomer brought his family, expecting to do some sightseeing with them. But he spent so much time in the sessions that they vowed never to go to a scientific meeting with him again. I hope the reader finds the printed papers as interesting as the participants found the talks.

Rather than picking my personal favorites, I would like to call attention to some recurrent themes and ideas, which seem to characterize photometry today. We could have picked these, instead of the topics we chose, as the basis for organizing the Colloquium.

First, as Lockwood et al., Breger, Grossmann, and others pointed out, a short-term precision of better than a millimagnitude has been reached repeatedly in the best work. This seems relatively straightforward to achieve by conventional means if the well-known problems such as temperature effects, centering errors, and filter nonuniformities are properly attended to. Millimagnitude precision has been reached with photomultipliers, CCDs, and silicon photodiodes. As the best ground-based observations are limited primarily by scintillation noise, and the best space-based observations by photon noise, the current limits in precision seem to be set more by telescope size than by detectors. Thus, despite the tantalizing prospects of new detectors and instruments offered in the session on new techniques, it appears that one can do very well indeed with the old techniques, simply by being careful and consistent. Wes Lockwood reminded us how important it is to have reliable funding in such work, as well as reliable equipment.

Second, CCDs and IR arrays continue to improve, but still have severe calibration problems at the 1% level. Walker, Stetson, Tobin, Kreidl, and Zeilik, among others, emphasized that each diode in these arrays is a different detector, with its own characteristics that must be calibrated. The best precision with CCDs still requires keeping the stars fixed on the same diodes, frame after frame, because of uncontrolled calibration errors. Both Walker and Tobin pointed out that at least some of the problem with flat-fielding is due to violating the basic rule of photometry: the calibration and program exposures *must* illuminate the detector identically. In many cases, stray light, often (according to Walker and Tobin) due to improper telescope baffling, causes large-scale flat-fielding errors. This is clearly an area where more effort needs to be invested, if CCDs are to compete more effectively with photomultipliers.

The areas where each of these two major detectors excels are now clearly defined. CCDs have completely displaced other methods for work in crowded fields, and for the faintest stars, where sky-brightness fluctuations demand simultaneous measurements of star and sky. For bright, isolated stars — especially those limited by scintillation rather than photon noise — the real-time readout and superior dynamic range of photomultipliers, as well as their simpler data-handling, make CCD methods "completely uninteresting" (according to Kreidl, in the discussion). We now have the major problem of joining together the systems of bright-star standards established with PMTs, and extending them to the faintest stars reachable with CCDs.

This brings me to the question of accuracy. Here, as opposed to precision, we have made little progress in recent times. This problem arose again and again: intercalibration of the channels of multichannel instruments, unifying the output of photometric networks, establishing the faint standards needed for CCD work, comparing ground- and space-based observations, and many other areas require accuracy as well as precision. In retrospect, we could well have had a session or two devoted to this topic.

Many speakers, including Bessell, Menzies, Straizys, Leggett et al., Budding, Sterken, Tinbergen, Milone, Zeilik, and myself, emphasized the importance of matching instrumental response functions to those of the standard system. Both Bessell and Zeilik insisted that one must first measure the spectral response of the detector before even designing the filters needed; and Menzies and I both pointed out that glass filters deviate significantly from nominal catalog curves. In particular, CCDs still show large variations in spectral response from one sample to the next; manufacturers' curves are *not* good enough to use as a basis for filter design. Even with the best efforts at matching response functions, transformations are necessarily nonlinear (Menzies; Young). Cubic transformation equations are now becoming standard (Dodd et al.; Stetson). It is more important to obtain small residuals from the transformation than to obtain small transformation coefficients (Bessell).

The substantial differences between Cousins's and Landolt's careful and independent realizations of the UBV system, documented by Menzies, show that Johnson's insistence on defining the system only by standard-star values, rather than by instrumental response functions, was a serious mistake. Both standard stars and standard response functions are needed. It is clear that experimental measurement and control of instrumental spectral response on the one hand, and theoretical investigation of the transformation problem on the other, are the areas in which progress must be made if we are to solve the accuracy problem. Without this progress, we will be unable to use CCDs and other new techniques, such as multichannel instruments, multitelescope networks, and observations from above the atmosphere, to their full potential.

Other points, such as the importance of improved spectrophotometry for synthetic photometry (Shobbrook; Glushneva); the measurement of extinction and its variations (Poretti & Zerbi; Reimann & Ossenkopf; Milone & Young); the need for well-documented reduction programs (Sterken; Hauck; Stetson); the lack of late-type standard stars (Bessell; Sterken), or even standard stars that are really constant (Lockwood et al.); the importance of high-quality automated telescopes for future photometric investigations (Genet; Pyper et al.; Hall; Tinbergen; Florentin-Nielsen; Crawford; O'Donoghue & Provencal); and economic considerations (Crawford; Lockwood; Budding; Taylor & Bless) all deserve careful attention; but I have no room to discuss them. The attentive reader should find something of interest in every paper.

In sum, we have here a long-needed review of the state of photometry, which suggests where future effort should be directed. We have made good progress in precision, but much work is needed elsewhere. I believe the two most pressing problems are the need for better CCD flat fielding, which seems to require attention to the entire telescope optical train; and the development of techniques that can provide accurate transformations between different instruments. The increasing use of CCDs makes these photometric problems urgent for users of the largest telescopes, as well as the smaller ones to which photometry is often confined. I hope these hurdles will be overcome by the time of the next IAU Colloquium on photometry.

Andrew T. Young
Chairman of the Scientific Organizing Committee

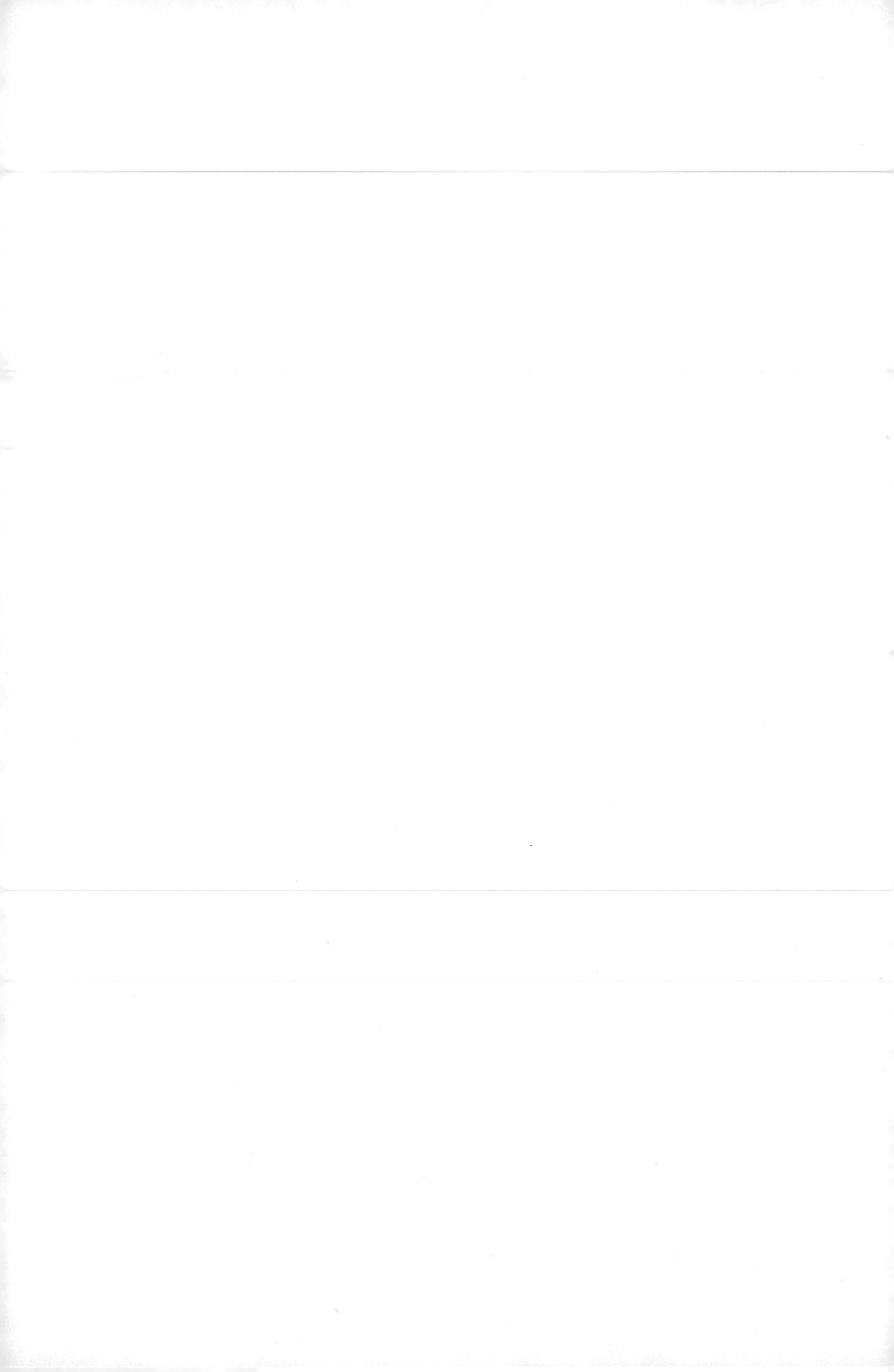

Preface

The idea of holding a colloquium on stellar photometry in Dublin had its origin in the historical interests of Russell Genet and the editors. The seed of the idea was sown at a meeting of the Astronomical Science Group of Ireland in September 1985 when John Butler gave a talk on the early electrical measurements of starlight which were carried out in Dublin in August 1892 and three years later at Daramona Observatory in County Westmeath. With Russell Genet's enthusiastic support it was agreed to mark the centenary by a meeting to assess current techniques and to study possible future developments. The meeting was supported by Commission 9 (Instruments and Techniques) and Commission 25 (Photometry and Polarimetry) of the International Astronomical Union which granted colloquium status. The Colloquium was preceded by a three-day workshop on robotic observatories which was held in Kilkenny (29–31 July) and was attended by 27 participants.

The main credit for the scientific content of the Colloquium must go to the Scientific Organising Committee and especially to Andrew T. Young who chaired it. We thank our fellow members: E. Budding, V.I. Burnashev, D.L. Crawford, R.F. Garrison, R.M. Genet, I.S. Glass, J.A. Graham, J. Tinbergen and B. Warner for their assistance in drawing up the programme. We are grateful also to the persons who chaired the sessions: M. de Groot, D.L. Crawford, J.D. Fernie, I.S. Glass, R.F. Garrison, P.A. Wayman and C.L. Sterken.

For hard work and good advice in making the local arrangements we thank the other members of the Local Organising Committee: P.A. Wayman (Chair), D.L. Weaire and E. O'Mongain. We thank Miss M. Heanue and her team of students and also Mr W. Dumpleton for practical help during the meeting.

We acknowledge the generous support of the following: Aer Lingus, Bord Failte, EOLAS, Bristol Myers Squibb, the Royal Irish Academy, Siemens Ireland Ltd., the Ulster Bank, Optronics Ireland, Fred Hanna Ltd., AGB Scientific, Mason Technology, T.S. Maharry, R.G. Tennant and T.K. Laidlaw. In particular, we are grateful to Trinity College for recognising the Colloquium as part of the College's Quatercentenary celebrations and for making available a lecture theatre and other facilities. One of the chief aims of the organisers was to bring together experts from East and West and this was facilitated by a grant from the International Astronomical Union towards the travel and subsistence expenses of thirteen participants.

The Colloquium was opened by Prof. T.D. Spearman, M.R.I.A., Vice-Provost of Trinity College who welcomed the participants on behalf of the Royal Irish Academy and Trinity College. A highlight of the Colloquium was the illustrated historical review, *Photoelectric Photometry – The First Fifty Years* , given by Dr John B. Hearnshaw which is reproduced in this volume. Most of the participants took the excursions to the megalithic monuments of the Boyne Valley or to the monastic site at Glendalough; afterwards an informal evening reception was held at Dunsink Observatory. The following evening the deputy Lord Mayor of Dublin received the participants at the Mansion House; the Hon. Mrs. W. Tirikatene-Sullivan from New Zealand replied on behalf of the guests. Later that evening there were three short talks on *Archaeoastronomy in Ireland* given by Dr Tom Ray, Mr Frank Prendergast and Prof. Patrick Wayman. The Colloquium closed with a banquet in the College Dining Hall; the company was entertained by a witty speech from Prof. Gordon Herries Davies and by a recital on the Irish harp by Ms. Fionnuala Monks. After the Colloquium, some thirty participants took a coach tour to Armagh where they visited the Observatory, the Planetarium and sites of historic interest.

This volume contains only the review papers and oral contributions; the poster papers will be published separately as *Poster Papers on Stellar Photometry, IAU Colloquium 136* by the Dublin Institute for Advanced Studies. Our thanks are due to Miss A. Brannigan of Armagh Observatory for assistance in preparing the manuscripts for publication. The proceedings of the Kilkenny Workshop on Robotic Observatories is edited by M.F. Bode and B.P. Hine and will be published by Ellis Horwood.

January 1993

C. John Butler
Armagh Observatory

Ian Elliott
Dunsink Observatory

The History of Stellar Photometry

George M. Minchin, FRS (1845–1914)

William H.S. Monck (1839–1915)

George F. Fitzgerald, FRS (1851–1901)

Stephen M. Dixon (1866–1940)

Biographical and Historical Notes on the Pioneers of Photometry in Ireland

The Editors

Introduction

As the circumstances of the early electrical measurements of starlight in Ireland are not widely known we wish to take this opportunity to set down the facts, as far as they are known to us. Corrections or new information will be welcomed.

The observations made in Dublin in 1892 were the result of a collaboration between four graduates of Trinity College: George M. Minchin, William H.S. Monck, Stephen M. Dixon and George F. Fitzgerald. The observations in 1895 were made at Daramona Observatory, Co. Westmeath with a 24-inch reflector by Minchin, Fitzgerald and the owner of the telescope, William E. Wilson. In 1875 Minchin was appointed Professor of Applied Mechanics at the Royal Indian Engineering College at Coopers Hill, near Staines in London. In 1877 he started a long series of investigations of photoelectricity using a small optical laboratory at Coopers Hill and the laboratories of University College London. His initial aim was to transmit images electrically but he became skilled in making photovoltaic cells of selenium.

By September 1891 Minchin had succeeded in making some working cells and he wished "to test them on the stars". He contacted his friend Monck who had recently set up an observatory in his back garden in Dublin. Monck in turn asked Fitzgerald for the loan of a galvanometer or electrometer and for advice in its use. As Fitzgerald did not have a suitable instrument, he ordered Clifton's form of Thomson's quadrant electrometer, which arrived near the end of the year. It seems that Minchin also sent some of his cells to Dr Boeddicker in the spring of 1892 for trials on the 72-inch telescope but no reports are available.

Minchin visited Dublin in August 1892 with some improved cells but on account of bad weather he had to return to England before a test could be made. However, the weather improved near the end of the month and a trial was carried out with Monck's telescope on the morning of the 28th. Monck was assisted by his 26-year old neighbour, Stephen Dixon and they succeeded in measuring the relative brightness of Jupiter and Venus. They failed to obtain 'certain' results from the stars on account of instrumental drift and other difficulties.

In December 1893 Minchin met Wilson in London and Wilson invited him to try his cells on the Daramona telescope. Minchin probably visited Daramona in January 1894 but there is no record of the visit. The first recorded observations at Daramona took place in April 1895. Wilson and Minchin operated the telescope and Fitzgerald attended to the electrometer in a room below the observing floor. Observations were possible on four nights and the results were published by Minchin in the *Proceedings of the Royal Society*. Minchin acknowledged that the first measurements of planets and stars were made with Monck's telescope to which he mistakenly attributed an aperture of nine inches; this error has been copied by several authors. Minchin visited Daramona again in September 1895 but Fitzgerald was not present and there is no report of observations being made. The second set of observations at Daramona was made in January 1896, despite bad weather. It seems that Minchin visited Daramona a year later but no more observations were reported.

G.M. Minchin (1845-1914)

George Minchin Minchin was born on 25 May 1845 at Valentia Island, Co. Kerry; for some reason his name on the baptismal certificate is recorded as 'George Minchin Smith'. His father George was an attorney and lived in Donnybrook, Dublin. His mother Alice died when he was about nine years old and he was put in the care of his uncle-in-law, Mr. David Bell, a schoolmaster and Shakespearean scholar.

Minchin entered Trinity College in 1862. He was awarded the first Scholarship in Mathematics in 1865 and in the following year he obtained a gold medal in mathematics and began to take an interest in experimental science. He was unsuccessful in the Fellowship examination in 1871 and 1872 but was awarded the Madden Premium, a consolation prize for unsuccessful candidates.

In 1875 he was appointed Professor of Applied Mechanics at the Royal Indian Engineering College at Coopers Hill, near Staines in London which is now part of Brunel University. The purpose of the College was to train engineers for service in India and by all accounts Minchin was a genial and inspiring teacher. He wrote several mathematical texts, the best known being his *Statics* which ran to seven editions. He was renowned for his clear reasoning and elegance of expression and it was said of him that he could never tolerate a slipshod argument or a carelessly written sentence. He was a notable correspondent, especially with Fitzgerald with whom he swapped mathematical problems. His letters bear testimony to his clear style and his logical mind. Apart from mathematical work he was a keen and skilful experimenter and he carried out investigations into wireless waves, X-rays and photoelectricity not only at Coopers Hill but also at the recently established laboratories of University College London. He invented and developed an absolute sine-electrometer for measuring voltage which was a modification of the gold-leaf electrometer. He was elected a Fellow of the Royal Society in 1895.

He maintained his links with Ireland, not only through letters but by visits to an aunt in Westmeath. In 1887 he married Emma Fawcett of Strand Hill, Co. Leitrim. While a bachelor and living in Coopers Hill he kept a collection of small birds in cages in his rooms. He was an early riser and his books were written mostly before breakfast. He was a keen athlete, being rated as one of the best lawn tennis players at Coopers Hill; in his younger days he had played cricket for the Gentlemen of Ireland.

Coopers Hill closed in 1906 and Minchin moved to New College Oxford where he still had access to telescopes and laboratories. He died on 23 March 1914 and was survived by his widow, a son and a daughter.

W.H.S. Monck (1839-1915)

William Henry Stanley Monck was born on 21 April 1839 and spent his childhood at Skeirke, near Borris-in-Ossory, Co. Laois where his father was curate from 1829 to 1850. Skeirke is about 30km from Birr (Parsonstown), the seat of the Rosse family in Co. Offally where William Parsons, the Third Earl of Rosse completed his great 72-inch (1.83m) speculum reflector in 1845. It seems likely that the boy Monck would have seen the great telescope and this may have inspired his lifelong interest in astronomy. In 1850 the family moved to Inistioge, Co. Kilkenny where William's father was rector until his death in 1858; his grandfather, the Rev. Thomas Stanley Monck, was an elder brother of Charles Stanley, the First Viscount Monck of Charleville, Enniskerry, Co. Wicklow. Charles, the Fourth Viscount Monck was Governor General of Canada from 1861 to 1868.

William never attended school but was for a short time educated at home by tutors. He distinguished himself on entry to Trinity College Dublin and in 1861 he obtained the first Scholarship in Science with a gold medal and a senior moderatorship in logic and ethics besides the Wray Prize for the encouragement of metaphysical studies. He studied divinity with distinction for several years but instead of following his father and grandfather into the Church, he turned to Law and was called to the Bar in 1873, being later appointed Chief Registrar in Bankruptcy in the High Court. In 1878 he returned to academic life and was Professor of Moral Philosophy in Trinity (the chair formerly occupied by George Berkeley) until 1882. He wrote *An Introduction to Kant's Philosophy* and a well-received *Introduction to Logic.*

Astronomical questions always interested him. In 1886 he became a Fellow of the Royal Astronomical Society and a member of the Liverpool Astronomical Society. On 12 July 1890 he wrote a letter to the *English Mechanic* advocating the formation of an association of amateur astronomers to cater for those who found the R.A.S. subscription too high, or its papers too technical, or who, being women, were excluded. Moves towards setting up such a society had already been made by E.W. Maunder and at a meeting on 24 October 1890 the British Astronomical Association was established. Monck was a prolific writer and many of his letters and articles appeared in the early volumes of the *Publications of the Astronomical Society of the Pacific*, *The Sidereal Messenger* and *Astronomy and Astro-Physics* (the forerunner of the *Astrophysical Journal*) as well as the *Journal of the British Astronomical Society* and the *English Mechanic.* In 1899 he published an *Introduction to Stellar Astronomy.*

About 1888 he took up residence at 16 Earlsfort Terrace and in 1891 he bought the $7\frac{1}{2}$-inch Alvan Clark refractor which had been owned by Dawes and Erck and set up a small observatory in his back garden. While the observations made in August 1892 are significant in retrospect, they represent only one facet of his many astronomical interests. Meteors interested him greatly and he corresponded for many years with W.F. Denning. In 1894 Monck suggested that there were probably two distinct classes of yellow stars — one being dull and near, the other being bright and remote. This clue to the existence of dwarf and giant stars was taken up by J.E. Gore of Sligo who estimated the size of Arcturus. If better data had been available to Monck, his discussion of proper motions and spectra might have led him to the relationship between luminosity and colour later discovered by Hertzsprung and Russell.

Monck was quiet and reserved in manner. He was an authority on economics, church history and legal matters. He was a formidable chess opponent but took more pleasure in solving problems than in the cut and thrust of games. He died on 24 June 1915 at the age of 76, leaving no children. His widow survived him by only a few months.

S.M. Dixon (1866-1940)

Stephen Mitchell Dixon was born in Dublin in 1866, the seventh of nine children of George Dixon and his wife Rebecca, who was the daughter of George Yeates, Dublin's leading instrument maker. His father died when he was five but his mother brought up her large family with loving care and all her seven sons had successful careers, the best known being Henry H. Dixon, F.R.S. and Andrew F. Dixon who became professors of Botany and Anatomy respectively in Trinity College (they were fondly known as 'Botany Dick' and 'Anatomy Dick'). The Dixon family lived at 17 Earlsfort Terrace from 1889 to 1894.

In Trinity College Stephen was a Senior Moderator and Gold Medalist in experimental science. For two years he was demonstrator to Dr W.A. Trail, the professor of Applied Mechanics in the Engineering School. After Trinity he spent two years as a civil engineer on railway construction in England and six months with the Portrush Electric Railway Company — a narrow gauge tramway constructed by Trail and his brother in 1883. The tramway connected Portrush with the Giant's Causeway and it was the first in the world to be powered by hydro-electricity.

In 1892 Dixon was appointed professor of civil engineering at the University of New Brunswick in Nova Scotia where he was a popular teacher and an effective administrator. He was largely responsible for the modernisation of the engineering course and for the design of a new Science and Engineering Building which was erected in 1900–1901. He directed the plays at the annual concerts of the university Glee Club. In 1894 he married Aline Harrison, the daughter of the Chancellor of the University; in order to outwit the students the marriage ceremony was held at 5 a.m. on the day before the announced date. They had one daughter, Sibyl who was born in 1899. Dixon held the chairs of civil engineering in the Universities of Dalhousie in Halifax (1902–05), and Birmingham (1905–13) and at Imperial College London (1913–33). He was Dean of the City and Guilds Engineering College (1930–33). During the First World War he had an important post in the Ministry of Munitions and he served with the Royal Engineers in France. He was awarded the O.B.E. in 1937.

After the death of his first wife in 1934 he married Josephine Jud in 1936 and lived in the south of France. He died on 25 March 1940 in Nice at the age of 74.

G.F. Fitzgerald (1851–1901)

George Francis Fitzgerald was born on 3 August 1851 and spent his childhood in Monkstown, Co. Dublin where his father was rector of Kill-o'-the-Grange parish. His mother was a sister of George Johnstone Stoney who proposed the name of 'electron' for the particle of electricity in 1891. George with his brothers and sisters were tutored at home by the sister of George Boole, the logician and Professor of Mathematics in the Queen's College at Cork.

Fitzgerald had a most distinguished academic record in Trinity and became a Fellow of the College at the age of 26. Four years later he was appointed Erasmus Smith Professor of Natural and Experimental Philosophy in the College. While the volume of his published work is comparatively modest, he had a great influence on his contemporaries and was noted for his generosity of time and ideas. He was a champion of Clerk Maxwell's electromagnetic theory and he did much to make it understood. When Hertz succeeded in generating electromagnetic waves, Fitzgerald brought his work to the attention of the British Association for the Advancement of Science at its meeting in Bath in 1888 and ensured that its significance was appreciated. By suggesting a suitable means of producing radio waves he helped to lay the foundation of wireless telegraphy.

Fitzgerald is probably best remembered for the theory which bears his name — the Fitzgerald-Lorentz Contraction. After the failure of the Michelson-Morley experiment to detect the existence of the ether, Fitzgerald proposed in 1892 that the explanation lay in the contraction of a moving body in the direction of its motion. The mathematical basis of the theory was developed by the Dutch physicist, H.A. Lorentz. Though the concept of the ether was later abandoned, the Fitzgerald-Lorentz Contraction was a significant milestone towards Einstein's Theory of Relativity.

When Fitzgerald was appointed to the chair of natural and experimental philosophy, there was no teaching of practical physics in Dublin so he obtained a disused chemical laboratory and began classes in experimental physics. He took an active part in general educational matters and was concerned with improving the general level of education in the country. It was largely through his efforts that technical education was established in Ireland, including the foundation of Kevin Street College of Technology. In 1895 he bought a Lilienthal glider and he attempted to fly it in Trinity College Park.

In 1885 he married Harriette, the second daughter of the Rev. J.H. Jellett, Provost of Trinity. They had five daughters and three sons. He served as secretary to the Royal Dublin Society from 1881 to 1889 and took an active part in its meetings and in the annual meetings of the British Association. Among distinctions awarded to him were Fellowship of the Royal Society in 1883 and the award of its Royal Medal in 1899.

His generosity and willingness brought him many tasks and continuous overwork undermined his health. Fitzgerald died after an operation for a digestive complaint on 22 February 1901, at the age of 49 and his death was a great blow to the College and to the wider scientific community. Oliver J. Lodge wrote: "...it may be doubted whether there ever was a man of equal scientific power, agility of thought and selflessness combined".

W.E. Wilson (1851-1908)

William Edward Wilson was born on 19 July 1851, just two weeks before G.F. Fitzgerald. Wilson's father owned a large estate in County Westmeath at Daramona which lies north-east of the road between Mullingar and Longford. Due to delicate health he was educated at home. He showed a keen interest in astronomy and in 1870, at the age of nineteen, he took part in a total eclipse expedition to Iran. The following year he set up an observatory equipped with a 12-inch Grubb reflector in the garden at Daramona.

In 1881 Wilson bought a 24-inch mirror and tube from Grubb and installed them in a new dome which adjoined Daramona House. Wilson carried out three main projects: a determination of the temperature of the solar photosphere, the electrical measurement of the brightnesses of stars and photography of the Sun and stars.

The estimation of the temperature of the Sun's surface was carried out in collaboration with Mr. P.L. Gray of Mason College in Birmingham and later with Dr Arthur A. Rambaut who was Director of Dunsink Observatory from 1892 to 1897. Wilson benefitted also from discussions with Fitzgerald with whom he corresponded regularly. The detector was a differential radiomicrometer, invented by C.V. Boys, which combined the functions of a bolometer and a galvanometer; it was constructed by Yeates of Dublin. A null method was used to compare the solar radiation with that from an electrically heated strip of platinum. The first estimate of 7073K was made in 1894 but after a revised determination of the absorption of the Earth's atmosphere the value was corrected in 1901 to 6863K which compares favourably with modern estimates. A large siderostat was loaned by the Royal Society for this investigation.

Apart from the electrical measurement of starlight, the other major achievement of Wilson was in celestial photography. From 1893 onwards the 24-inch telescope was used for photographing nebulae and clusters and his results equal the best obtained elsewhere at that time. Eleven stellar photographs appear in his collected papers and although Wilson claimed that the collotype process used was unable to reproduce the delicate nebulosities in the original negatives, the results are impressive. Photographs of the solar disk were taken regularly with a 4-inch Grubb refractor for the purpose of studying sunspots. In August 1898 a large sunspot was photographed on cine film for four hours at the rate of about 100 exposures an hour. Although Wilson reported that the spot did not show much change, this may have been one of the earliest applications of time-lapse photography to the study of solar phenomena.

A number of other investigations were carried out and are mentioned in the annual reports which appeared in the *Monthly Notices* of the Royal Astronomical Society in 1883 and from 1892 to 1908. These include observation of the transit of Venus in 1882 and a photographic search for a planet beyond Neptune in 1901/1902. In May

1900 Wilson, Rambaut, H. Grubb, C.J. Joly and Bergin took part in the successful joint Royal Irish Academy — Royal Dublin Society eclipse expedition to Plasencia in Spain.

In 1886 Wilson married Carolina Ada, the third daughter of Capt. R.C. Granville of Biarritz and the family visited there from time to time. He was elected a Fellow of the Royal Society in 1896 and was awarded an honorary D.Sc. by Trinity College Dublin in 1901. He died on 8 March 1908 at the age of 57 and was survived by his widow, one son and two daughters.

Monck's Telescope

The $7\frac{1}{2}$-inch lens of Monck's telescope was made by Alvan Clark and had an eventful history which is worth recounting.

Alvan Clark lived in Boston and was by profession an inventor and portrait painter. He had two sons — George and Alvan Graham — and two daughters. George, while a student at Andover College, attempted to build a small reflecting telescope. This led to his father becoming interested in optics and eventually to the founding of the firm of Alvan Clark and Sons in 1850. Five times the firm made the world's largest telescope with apertures ranging from 18.5 inches for Dearborn Observatory in 1862 to 40 inches for Yerkes Observatory in 1897. The Yerkes instrument is still the largest refractor in the world.

One of Clark's earliest customers was the Rev. William Rutter Dawes (1799-1868), the experienced English observer of double stars. After completing the $7\frac{1}{2}$-inch objective in 1853, Clark used it to discover the faint companion of 95 Ceti. When Dawes learnt about this discovery he asked Clark to carry out further tests on other close binaries and then asked if he could buy the telescope. Clark reluctantly agreed and sold it to Dawes in March 1854 for $950. Dawes was very pleased with its quality and used it for two years, mostly for observing Saturn. Between 1855 and 1859 Dawes bought four more Clark telescopes and thus helped to establish Clark's reputation as a telescope maker.

In 1856 Dawes sold the $7\frac{1}{2}$-inch lens only to Frederick Brodie who remounted it at Eastbourne. By 1869 the lens had been bought and remounted again by Dr Wentworth Erck of Sherrington House at Shankill, 16km south of Dublin. Erck made regular observations of sunspots from 1869 to 1888 and carried out more than 1900 measurements of 40 selected binaries between 1873 and 1880. In 1872 Charles E. Burton compared the optical performance of the $7\frac{1}{2}$-inch objective with a 9-inch mirror of silvered glass. At the opposition of Mars in 1877 Erck was one of the first in the British Isles to observe the moons of Mars which had been discovered a few weeks previously by Asaph Hall with the 26-inch Clark refractor of the U.S. Naval Observatory in Washington. He also observed Jupiter and Saturn and, according to A.A. Rambaut, he was one of the first, if not the very first, to notice the proper motion of Jupiter's Red Spot. Burton, shortly before his death at the age of 35, used Erck's telescope to take photographs of the Moon.

There is ample evidence that the $7\frac{1}{2}$-inch Clark lens had a remarkable optical quality. The noted amateur, William Lassell, was "astonished" by its performance. Rambaut wrote about the lens as follows:

> "It is in some respects a remarkable glass, being so full of bubbles that to one versed in such matters it might have appeared almost worthless for the delicate purpose for which it was intended; whereas it is remarkable for the exquisite definition it is capable of affording, as was proved on more than one occasion by the work done with it in the hands of Mr. Dawes and Dr. Erck".

According to Erck, Dawes considered the $7\frac{1}{2}$-inch telescope "more perfect than any of its successors".

After Erck's death in 1890, the telescope was acquired by Sir Howard Grubb who sold it to Monck for use from his residence at 16 Earlsfort Terrace where it was used for the pioneering measurements in August 1892. In 1912 Monck presented his complete telescope to the Queen's College Belfast (now Queen's University). According to Prof. K.G. Emeleus the lens was incorporated into a Littrow spectrograph by R.C. Johnson about 1921. As the spectrograph did not perform well, Emeleus sent the lens to Hilger's for testing; the two components were found to be off-centre but the main source of imperfection was eventually traced to a cheap prism. Although the spectrograph was still in existence in the mid-1960s, the subsequent fate of the lens is not known.

Wilson's Telescope

In 1871 Wilson built a small observatory in the garden of Daramona House for a 12-inch reflector made by Sir Howard Grubb in Dublin. There was attached a small room for photography and it also contained a small transit instrument made by Wilson himself and a sidereal clock. It was used for lunar photography on wet plates and for experiments on solar radiation with thermopiles.

In 1881 Wilson bought a 24-inch mirror and tube from Grubb and installed it on the old equatorial mount in a new dome which adjoined Daramona House. However, the original mounting proved unsatisfactory and the tube was remounted in 1892 and fitted with an electrically-controlled clock drive from Grubb. A laboratory and darkroom were added to the observatory in 1889. There was also a well-equipped workshop with a 6-inch Whitworth screw-cutting lathe, a shaping machine and a drilling machine. The tools were driven by a $1\frac{1}{2}$H.P. Crossley gas engine.

After Wilson's death in 1908, the telescope was scarcely used and in 1925 it was offered to the University of London by Wilson's son, J.G. Wilson, on condition that a dome was built to house it. A site was obtained at Mill Hill and the Observatory was opened in 1929 and used for spectroscopic research. From 1951 to 1974 it was used for practical instruction of astronomy students at University College. The telescope is now in the custody of Merseyside County Museum in Liverpool.

Daramona House, Streete, County Westmeath

Willaim E. Wilson, FRS (1851–1908)

Wilson's 24-inch reflector

Acknowledgements

Many people have helped in the search for information but we are particularly grateful to Prof. Denis Weaire of Trinity College for drawing our attention to the astronomical work mentioned in the Fitzgerald correspondence in the archives of the Royal Dublin Society. We thank the following for supplying photographs: Richard Morris of Brunel University, the Mary Lee Shane Archives of Lick Observatory, the University of New Brunswick at Fredericton, Trinity College Dublin, John McConnell and William Dumpleton.

References:

Anon. 1909, Mon. Not. Roy. Ast. Soc., **69**, 254.

Batt, E., 1979, *The Moncks and Charleville House*, Blackwater Press, Dublin.

Brodie, F. 1856, Mon. Not. Roy. Ast. Soc. **17**, 33: Notes on the Manufacture of Tubes for Refracting Telescopes.

Burton C.E. , Mon. Not. Roy. Ast. Soc. **52**, 423: Note on Lunar Photographs taken after Enlargement of the Primary Image by an Eyepiece.

Burton C.E., 1872, Astronomical Register **10**, 289: Refractors and Reflectors compared.

Butler, C.J. 1986, Irish Astron. J., **17**, 373.

Dawes, W.R. 1854, Mon. Not. Roy. Ast. **15**, 79: On the Telescopic appearance of Saturn with a $7\frac{1}{2}$-inch object glass.

Dixon, S.M., 1892, Astronomy & Astrophysics **11**, 844: The Photo-Electric Effect of Starlight.

Edgeworth, K. 1965, *Jack of All Trades – The Story of My Life*, Allen Figgis, Dublin.

Elliott, I. 1987, Irish Astron. J., **18**, 122.

Erck, W. 1869, Astronomical Register **7**, 176: On the Construction of Observatories.

Erck, W. 1877, Observatory **1**,135: Description of an Observatory erected at Sherrington

Erck, W. Mon. Not. Roy. Ast. Soc. **38**, 362 Observations of the outer satellite of Mars.

Larmor, J.(ed.) 1902, *The Scientific Writings of the Late G.F. Fitzgerald*, Dublin.

Loomis, E. *Recent Progress of Astronomy*, p 391.

McKenna, S.M.P. 1968, Vistas in Astronomy, **9**, 283.

McNally, D. and Hoskin, M. 1988, J. Hist. Astron., **19**, 146.

Mollan, R.C. *et al.*, 1985, *Some people and places in Irish Science and Technology*, Royal Irish Academy.

Mollan, R.C. *et al.*, 1990, *More people and places in Irish Science and Technology*, Royal Irish Academy.

Monck, W.H.S. 1892, Astronomy & Astrophysics **11**, 843: The Photo-Electric Effect of Starlight.

Rambaut, A.A. 1890, Mon. Not. Roy. Ast. Soc. **51**, 194: Obituary of W. Erck.

Warner, B. 1977, Sky and Telescope, **53**, 108.

Wilson, W.E. *Astronomical and Physical Researches*, published privately.

Photoelectric Photometry — The First Fifty Years

J.B. Hearnshaw

Department of Physics and Astronomy, University of Canterbury, Christchurch, New Zealand

Abstract

This historical review covers the first fifty years of the electrical measurement of starlight, which saw its beginnings in August 1892 when William Monck used a Minchin photovoltaic cell on his refractor in Dublin. The work of Stebbins using photoconductive cells from 1907 and of Guthnick, Stebbins and others using photoelectric cells from 1912 is also discussed. The advances brought about by thermionic amplification and the red-sensitive Cs-O-Ag photocathode in the early 1930s by respectively Whitford and Hall enabled new astrophysical problems to be tackled. Although many observers attempted photoelectric photometry, relatively few were successful during the 1920s and '30s. It was not until the introduction of the 1P21 photomultiplier tube that astronomers had a reliable and sensitive detector for photometric observations.

An Irish beginning

This colloquium is to celebrate the centenary of the first electrical measurement of starlight [1] on 28 August 1892 by William Monck, at his private observatory in Earlsfort Terrace, Dublin. The detector was a primitive photovoltaic cell comprising a selenium photocathode on an aluminium substrate immersed in acetone. A quartz window admitted light from Monck's $7\frac{1}{2}$-inch refractor and a quadrant electrometer recorded the emf produced. Ironically the real hero of the occasion, George Minchin, had returned to London from his native Ireland before the weather had cleared; he was not therefore a participant in the historic first observations which were made by Monck (a notable Dublin amateur astronomer and former philosophy professor) with the help of Professors Stephen Dixon and George Fitzgerald. Monck reported that although they failed to obtain any certain results from fixed stars, nevertheless the effects from the moon were 'of a striking character' (Monck 1892), while Dixon noted measurable deflections from Jupiter and Venus (Dixon 1892).

[1] Thermocouple observations had been made in astronomy as early as 1868 but are not considered in this article

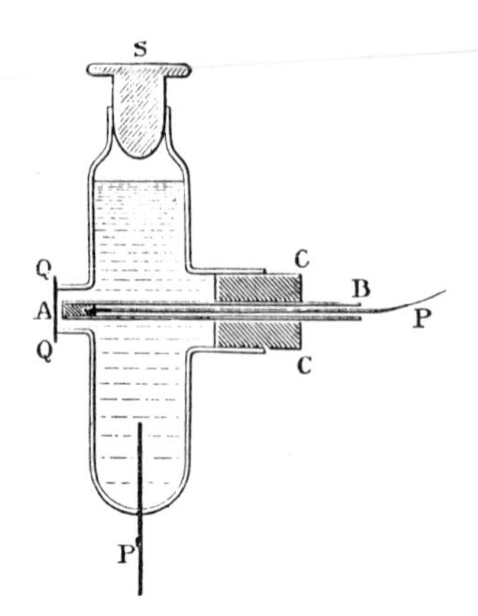

Figure 1 Minchin photovoltaic selenium cell.

Minchin was a genial professor of mathematics at the Royal Indian Engineering College at Cooper's Hill, who evidently enjoyed experimenting in his physical laboratory at the college. Here he developed photovoltaic cells from 1890 for astronomical use (Minchin 1895). From 1894–96 he made three visits to the Daramona House Observatory, Co. Westmeath, where more successful observations were made on the 24-inch reflector (Minchin 1896). Observations of at least ten stars and three planets were made, and Minchin concluded that '. . . there is little difficulty in obtaining fairly accurate measurements of the light of the stars of the first and second magnitudes' (Minchin 1895).

Stebbins' photoconducting cell at Illinois

The Irish photovoltaic observations had created practically no interest in electrical photometry in the astronomical community. Indeed the next electrical observations in 1907 involved a completely different type of detector, a photoconductive selenium cell made by F.C. Brown at the University of Illinois for Joel Stebbins, using the property of a decreased resistance of this element when under illumination. The earliest photoconductive cells were of low sensitivity and demanded great patience from the observer with a continuous stable current flow and short intermittent exposures to light. Stebbins found that the sensitivity improved at least twenty-fold on cooling, and by 1909 he was able to achieve $0^{m}.02$ precision on the 12-inch telescope. This was the first truly quantitative electrical photometry, and the device was used to produce the classical paper on the light curve of Algol, which showed the shallow secondary eclipse for the first time (Stebbins 1910). Other results for eclipsing binaries were published in 1910–11 as well as photometric observations of Halley's comet.

From 1907 we can say that a new era in photometry had been launched by Joel Stebbins. Yet the astronomical usefulness of this detector was as short-lived as Minchin's photocells. For in turn they were overtaken by the photoelectric cell, to which Stebbins himself, from 1912, quickly turned his attention. However Stebbins was not the only observer to employ the photoconductive cell. Stebbins referred to unpublished attempts by Pickering at Harvard to use such a detector, possibly as early as 1877. In Germany Ernst Ruhmer observed solar and lunar eclipses with a selenium photoconductive cell in 1902–03, and Fournier d'Albe used one to record star transits in his meridian telescope in Birmingham in 1913.

Early photoelectric photometry in Germany

The true photoelectric era in astronomical photometry began in 1912. Julius Elster and Hans Geitel had experimented with the photoelectric effect from 1889 using alkali metals. By 1910 they showed that the hydrides of potassium and sodium had significantly higher sensitivities than the metals themselves, and it was a KH photocell that was used in the earliest observations by Paul Guthnick at the Berlin Observatory in the summer of 1912. He can be regarded as the father of photoelectric photometry in astronomy. Guthnick's photocell was one made commercially by the firm of Günther and Tegetmeyer on the 31-cm Berlin refractor. He was able to demonstrate that the ratio of the cell's responses for α Lyrae to α Cygni was in accord with the ratio from visual photometry.

The photoelectric observations were continued on the 30-cm telescope of the new Berlin-Babelsberg Observatory from 1913. From 1914 Richard Prager collaborated in this work. The light curve of β Cephei was the first variable star observed, and in fact the first variable to be discovered by photoelectric means (Guthnick and Prager 1914). The Ap star α CVn was observed soon afterwards, as well as Nova Aquilae in 1918.

The German KH photocells were operated with 100 V across the cell and were filled with low pressure argon, which increased the sensitivity as a result of collisional ionization. A delicate string electrometer suspended from gimbals was used to collect the charge from the anode, the string being deflected sideways in an electric field. The rate of deflection was measured with a microscope and a stop-watch, and this was proportional to the brightness of the star.

By 1917 Guthnick and Prager had amassed 67 000 photoelectric measures on 50 stars and planets (Guthnick and Prager 1918), in the best cases with probable errors as small as $0\overset{m}{.}005$. A yellow filter was used to define a colour index as the difference between filtered and unfiltered magnitudes and the relationship between colour and spectral type was explored for 67 stars.

The fine Berlin tradition of photoelectric photometry was continued by Guthnick and his collaborators right up to the early 1940s. Most notable were Kurt Bottlinger, who made the first 2-filter (440 and 460 nm) colour-index observations in 1920–22 and who applied the results to astrophysical problems including interstellar reddening (at that time a controversial subject); Wilhelm Becker, who used filters at 425 and 475 nm for his colour indices in 1930–31 to obtain colour temperatures and Margarethe Güssow from 1931 to the early years of the Second War.

Hans Rosenberg with Edgar Meyer in Tübingen also attempted photoelectric photometry from 1912 (Meyer and Rosenberg 1913); although the cell was their design, their productivity was much less — in fact no useful photometry was reported by the time Rosenberg left Germany in 1933 for Chicago. However in 1920 Rosenberg pioneered the technique of thermionic valve d.c. amplification. His triode valve gave an amplification of at least 6×10^5, enabling the very low anode currents ($\sim 10^{-14}$ A) to be detected with a galvanometer off the telescope, instead of the delicate suspended string electrometer. These experiments were not generally copied at the time, although Bengt Strömgren in Copenhagen was using an amplifier to record star transits photoelectrically in 1925. Widespread use of d.c. amplifiers in photoelectric photometry only came after their introduction by Whitford at the Washburn Observatory from 1932.

Joel Stebbins and photoelectric photometry

Meanwhile Joel Stebbins had taken sabbatical leave from Illinois in 1912–13 and was able to witness at first-hand the installation of the photoelectric apparatus by Guthnick in Berlin in 1912. This good fortune, together with the arrival in 1911 of Jakob Kunz in the Physics Department at Illinois were the events which set Stebbins off on his illustrious career as a photoelectric photometrist. Kunz came originally from Switzerland, but in Germany he had learnt the techniques of KH photocell production. In the absence of Stebbins, Kunz and Schulz at Illinois constructed a photoelectric photometer and made successful stellar observations on Capella from December 1912 (Schulz 1913).

The photocells by Kunz were helium-filled and operated at about 300 V. They were far more sensitive than the selenium photocells used earlier by Stebbins (the peak quantum efficiency for the KH photocells was about 10 per cent), but suffered from a small bandwidth, the useful response being only 380 to 500 nm. A series of papers on the photometry of eclipsing binary stars appeared in the *Astrophysical Journal* by Stebbins and his associates (*e.g.* Stebbins (1920) on λ Tau). Perhaps the outstanding feature which distinguishes the Illinois publications from those of Stebbins' photometric contemporaries was his concentration on astrophysical problems of the stars themselves rather than the technical difficulties of operating a photoelectric photometer.

When Stebbins moved to the University of Wisconsin in 1922 he had access to the larger 15.6-inch telescope of the Washburn Observatory (see Stebbins (1928)). However the close collaboration with Kunz in Urbana (Illinois) continued. In fact Kunz supplied Stebbins with his best photocells throughout the 1920s and '30s, and provided cells for several other American observatories as well (including Edith Cummings and later Kron at Lick, Elvey at Yerkes, Smith at Mt Wilson, Calder at Harvard and Baker in Urbana). Two significant developments at Wisconsin were the change to a sensitive Lindemann electrometer in 1927 (this instrument had a needle on a torsion wire which rotated when charged, and was much more robust than the string electrometer used earlier), and the introduction of thermionic amplification in 1932 by Whitford, which therefore dispensed with the electrometer altogether. The amplifier was placed in an evacuated chamber to avoid problems with cosmic rays. It had a gain of over two million and enabled ninth magnitude stars to be reached.

Using two filters (426, 477 nm) Stebbins defined a colour index and tackled problems in interstellar reddening, intrinsic star colours and the colours of the globular clusters during the 1930s. During these years Morse Huffer, Albert Whitford and Gerry Kron were the outstanding photometrists that graduated from Wisconsin under Stebbins' tutelage. Kron obtained his doctorate at Berkeley after leaving Wisconsin. He worked at Lick where Stebbins had fostered close relations through his annual summer observing expeditions, to Lick from 1915 and later to Mt Wilson.

Stebbins lent his technical expertise to other institutions to enable them to set up photoelectric apparatus. Kron at Lick was one example of an outstanding instrumental innovator in his own right, but Christian Elvey at Yerkes also built a photometer in 1929 based closely on the Washburn design.

Other photoelectric photometrists before 1940

Stebbins was sometimes cited as the leader of the only successful photometric observers before the Second War. He was certainly the most productive, but many other observers attempted photoelectric work. Guthnick and his collaborators were nearly as successful as Stebbins and if it were not for political intervention they may well have become equally as productive.

In Britain the photometer of Adolph Lindemann and his son Frederick (professor of physics at Oxford and later Lord Cherwell) was especially notable, at least for its independent design and the fact that even the photocells were home-made, although extensive published observations were not forthcoming. Adolph Lindemann was a German civil engineer who settled in Devon in 1884, where he established his private observatory. The first Lindemann photometer was built in 1918 (Lindemanns 1919). The paper is remarkable for its foresight in design and in identifying the potential for photometry to solve astrophysical problems.

The robust Lindemann quadrant electrometer was described in 1924. This became a feature of the new Lindemann photometer in 1926, and the electrometer design was copied widely by many other observers, including Stebbins, Kron and John Hall.

By 1940 as many as 38 observers had attempted stellar photoelectric photometry at 22 observatories in 7 countries, and they had published either photometric results or instrumental descriptions in over 100 papers in the literature. The order in which these countries entered the arena was Germany (1912), USA (1912), United Kingdom (1918), France (1925), Canada (1929), Italy (1930) and the Soviet Union (1934), where the years are those of first involvement in photoelectric observing (see Table 1). In North America Kunz cells were used almost exclusively until the mid-1930s, while in Europe commercial cells from Günther and Tegetmeyer predominated. For example Smart in Cambridge, Maggini in Italy at the Collurania-Teramo Observatory and Nikonov with Kulikovsky from Leningrad all used the German cells for their early experiments. In addition Maggini acquired two Kunz cells. Only the last two of these observers produced successful results; at first they used a German built photometer on the 13-inch Abastumani Observatory in Georgia, but by 1936–37 a new Russian-built photometer had been installed. It featured a d.c. amplifier and a Fabry lens, the first photoelectric photometer to employ a Fabry lens to focus the primary onto the photocathode.

The Cs-O-Ag red-sensitive photocathode

A new red-sensitive photocathode was developed by Lewis Koller at the General Electric Company in 1929. It consisted of a caesium layer deposited on an oxygenated silver substrate and gave a broad spectral sensitivity to beyond 1μm in the near infrared, but very low quantum efficiency at all wavelengths. The new cathode (now designated S1) was incorporated into photocells by Charles Prescott at Bell Labs, and some of these were acquired by John Hall, then a graduate student at Yale, in 1930. Hall became the pioneer of near infrared photoelectric stellar photometry. By 1931 he had introduced cooling with dry ice to reduce the high dark current, enabling sixth magnitude stars to be observed on the 15-inch Loomis telescope (Hall 1934). Between 1934 and '42 a classical series of *Astrophysical Journal* papers were written from successively Yale, Columbia, Sproul and Amherst. Wratten filters enabled IR colour indices to be defined, and then coarse objective gratings and diffraction gratings were used for multicolour spectrophotometry. The power of the wide wavelength sensitivity allowed new astrophysical problems to be tackled, notably a new study of the interstellar extinction law, and studies of stellar continua and colour temperatures. By 1935 he had photoelectric photometry to support a $1/\lambda$ interstellar law, which was confirmed by Stebbins and Whitford using their 6-colour $UVBGRI$ photometry some eight years later, again using a Cs-O-Ag (S1) cell on a photometer at Mt Wilson (the U filter at 353 nm represented the first use of the ultraviolet in photoelectric photometry).

Arthur Bennett at Yale (from 1934), Kron at Lick (1939) and Nikonov at Abastumani were other early observers with the new red cell. In the U.S. the Western Electric cells were generally used, while Nikonov had one of Russian manufacture.

Undoubtedly the Cs-O-Ag photocathode added a new dimension to photoelectric photometry in the days before the photomultiplier, and in spite of its high dark current (when uncooled) and low efficiency, much valuable work was accomplished by a few diligent observers — mainly Hall and Whitford — on stellar colours and interstellar reddening.

Concluding remarks

The most successful early photoelectric photometrists were those who persevered with the intricacies of electronics at a time when electronic apparatus was generally absent from astronomical observatories. Guthnick and Rosenberg both commented on the reluctance of astronomers to acquire the necessary skills and know-how to operate electronic instruments.

The three most successful photometrists in terms of their output of photometric results were Stebbins, Guthnick and Hall (including their collaborators). It is note-

Table 1 Observatories engaged in photoelectric photometry before 1940

Observatory	Observers	Year
Berlin	Guthnick, Prager, Hügeler, Bottlinger, W.Becker, Güssow	1912
Tübingen	Rosenberg, Meyer	1912
Urbana, Illinois	Stebbins, Schulz, Kunz, Dershem, Wylie	1912
"	Baker	1928
Lick	Stebbins	1915
"	Cummings	1920
"	Fath	1931
"	Kron	1937
Sidholme, Devon	A.F. and F.A. Lindemann	1918
Washburn, Wisconsin	Stebbins, Whitford, Huffer	1922
Strasbourg	Rougier	*c.*1925
Yerkes	Elvey, Stebbins, Mehlin	1929
Dominion, Ottawa	Henroteau	*c.*1929
Cambridge, U.K.	Smart, Green	1929
Mt Wilson	Stebbins, Whitford	1930
"	Smith	*c.*1933
Collurania-Teramo	Maggini	1930
Goodsell, Minnesota	Fath	1931
Harvard	Calder	1931
Yale	Hall	1931
"	Bennett	1935
Rutherfurd, Columbia	Hall	1933
Pulkova	Nikonov	1933
Sproul, Swarthmore	Hall, Delaplaine	1934
Abastumani, Georgia	Nikonov, Kulikovsky	1934
Potsdam	Hassenstein	1938
Amherst	Hall	1938
Steward	Roach, Wood	1938

Notes:

1. observers separated by commas were collaborators
2. the year is the year of initial involvement
3. from 1913 the Berlin Observatory was resited at Berlin-Babelsberg

worthy that all three had the close collaboration of physicists to advise them on the operation of their photocells, namely of Kunz, of Elster and Geitel and of Charles Prescott respectively. Those who ventured into this new field without such help were often unproductive in their efforts. It should be remembered too that photometry in the 1890s and early twentieth century was principally photographic or even visual — techniques which required very different skills to those demanded by the photocell.

The principal technical developments in photoelectric instrumentation were hydrogenated photocathodes (1910), the Lindemann electrometer (1924), thermionic valve amplification (1920, 1932), the Cs-O-Ag photocathode (1929) and photocell cooling (1931). In 1936 P. Görlich in Dresden produced the Cs-Sb photocathode with a response to 580 nm and higher sensitivity than the KH cell. This was used by V. Zworykin at RCA in the earliest photomultiplier tube from 1936. The photomultiplier was used by Whitford and Kron as early as 1937 as the detector for an autoguider. The RCA 931-A photomultiplier was developed during the war years and was a precursor to the famous 1P21 that revolutionized photoelectric photometry and which allowed the first pulse-counting photometers to be constructed by Yates at Cambridge and Blitzstein at Pennsylvania from 1948. These developments belong to the second fifty years in the history of the electrical measurement of starlight.

References:

Dixon, S.M., 1892, Astron. and Astrophys., **11**, 844.

Guthnick, P. and Prager, R., 1914, Veröff. der königl. Sternw. Berlin-Babelsberg, **1**, 1.

Hall, J.S., 1934, Astrophys. J., **79**, 145.

Lindemann, A.F. and Lindemann, F.A., 1919, Mon. Not. R. astron. Soc., **79**, 343.

Meyer, E. and Rosenberg, H., 1913, Vierteljahrschr. der Astron. Ges., **48**, 210.

Minchin, G.M., 1895, Proc. R. Soc., **58**, 142.

Minchin, G.M., 1895, Proc. R. Soc., **59**, 231.

Monck, W.H.S., 1892, Astron. and Astrophys., **11**, 843.

Schulz, W.F., 1913, Astrophys. J.,**38**, 187.

Stebbins, J., 1910, Astrophys. J.,**32**, 185.

Stebbins, J., 1920, Astrophys. J., **51**, 193.

Stebbins, J., 1928, Publ. Washburn Observ., **15**, 61.

Session 1

Photometric Systems

Photometric Systems

Michael S. Bessell

Mount Stromlo and Siding Spring Observatories, Private Bag, Weston Creek P.O. Canberra, ACT 2611, Australia

Abstract

Multicolour photometry has historically been carried out within a variety of standard systems. With the advent of new detectors with different wavelength sensitivities to those of the original system and the use of subsets of secondary standards, many subtle and not so subtle changes have occurred to the original systems. However, by reverse engineering, the passbands of the modified standard systems can be determined which enables good realisation of theoretical colours and better passband matching for CCD-based photometry. In this paper, the various photometric systems will be discussed and compared and strategies will be outlined for ensuring that precise relative photometry is maintained in the future, in particular from area detectors.

1. Introduction

More than any other aspects of astronomy the subjects of magnitude scales and photometric systems are encumbered by history. Early astronomers compared star with star, a procedure that still retains great benefits. The temperatures of common stars range from 30000K to 3000K. Their brightnesses cover a range of almost a factor of 10^{10}, from the sky background upwards. (this range does not include the sun). The majority of stars are constant in total light output and in temperature. They must all be observed through the Earth's atmosphere. No laboratory sources of light have energy distributions similar to those observed in stars and it is natural that astronomers seek to use the standard candles in the sky rather than inferior and technically complex ones in the laboratory. Photometric systems represent attempts to define standard bandpasses and sets of standard sources, measured with these bandpasses, that are well distributed about the whole sky. Different photometric systems measure different wavelength bands. All photometric systems enable the measurement of relative fluxes, from which can be inferred particular properties of the emitting object, such as temperature and luminosity, but different systems claim to do so more precisely, more quickly or more easily compared with other systems. Some are suited for hot stars, others for cool stars. Most of the systems were developed and modified by different astronomers over many years and the literature contains confusing versions and calibrations. Some people have despaired that it is so confusing we should start again with a well defined ultimate system, but recent analysis has shown that modern versions of the existing photometric systems can be placed on a firm quantitive basis and that more care with passband matching will ensure that precise and astrophysical valid data can be derived from existing, though imperfect systems. This is not to say that better systems will not be devised but they face strong competition from more precisely defined and better calibrated old systems. The zero point of the V (visual) photometric system has undergone much refinement over time and although it was officially set by the specified visual magnitudes of stars in the *north polar sequence* the magnitude scale of all systems today is established by the contemporary whole-sky standard star catalogues.

2. New systems from new detectors

Technological advances over the last 50 years have provided a series of light detectors of ever increasing sensitivity and wavelength coverage. The advent of photography in the late 19th century revolutionised astronomy as did the introduction of photomultiplier tubes with their light sensitive photocathodes in the mid 20th century and detectors such as silicon charge-coupled-devices (CCDs) and infra-red detectors over the last 10 years. Light intensities or magnitudes measured with these new detectors naturally differ from the visual magnitudes and depend on the colour of the objects. Initially, there was only the difference between visual magnitudes and 'blue' photographic magnitudes to be considered, but three factors resulted in a proliferation of different passbands and photometric systems. These were the extension of photographic and photocathode sensitivities to a wider wavelength range, and the use of coloured glass filters and interference filters to sample the starlight in narrower bands within the total wavelength sensitivity range of detectors.

3. Rationale for multicolour photometry

Photometry of astronomical objects is carried out in order to measure the apparent total brightness of objects and their relative brightnesses at different wavelengths i.e. their energy distribution. It is possible to characterize the temperature of most objects from the overall shape of their energy distribution, and to infer the metal content of stars from depressions in the energy distribution at particular wavelengths due to the absorption of flux by lines principally of Fe and Ti, which have very rich line spectra, Ca, Mg and the molecules CN and CH, which also have very strong lines in the blue-violet region of cool stars. There are many other molecular absorption bands, such as TiO,CO and H20 which depress the continuum in very cool stars and such molecular features are also used to provide information on the temperature, abundance and luminosity. The energy distributions of galaxies and star clusters can be analysed to extract the relative numbers of different kinds of stars making up the composite object. Red shifts of very distant QSOs can also be measured from the positions of depressions or peaks in their energy distributions.

Multicolour photometry is best thought of as very-low-dispersion spectroscopy. The entire high resolution spectrum of a star or cosmic object contains a large amount of information, but when dealing with extremely faint objects or large numbers of objects it is a great advantage to measure a small number of wavelength bands in as short a time as possible. Such a minimal technique is invaluable if it enables the derivation of many of the same parameters derivable from a complete (and very redundant) description of the spectrum. A great deal of effort therefore has gone into accurately measuring the calibrating colours and depressions in terms of temperatures, metal abundances and other parameters, and investigating which of competing minimal descriptions of a star's spectrum is the most accurate or most practical.

4. Photometric systems - natural systems, standard systems

A light detector, a telescope, a set of coloured filters, and method of correction for atmospheric extinction makes up a *natural photometric system*. Each observer therefore has their own natural system. The *standard* system is indirectly defined by a list of standard magnitudes and colours measured for a set of typical stars with the natural system of the originator. These are often called the *primary standards*. Later lists comprising more stars and fainter stars but based on the primary standards are called *secondary standards*. However, in the case of all photometric systems, recently published secondary standards effectively redefine the standard system because they tend to be more accurately measured than the primary lists and to represent contemporary detectors, filters and practice. The term 'colour' is an abbreviation for 'colour index' which is the difference between the apparent magnitudes in two different spectral regions. Photometry is generally published as a series of colours and single magnitude. The zero points of many colour systems are set so that Alpha Lyrae (Vega) has zero colours. In the southern hemisphere (where Vega is inaccessible, and often also in the north) the zero point is set by requiring that an ensemble of unreddened A0 stars have zero colours.

The most influential of the early works of photoelectric photometry were the so-called broadband Johnson UBVRI and Kron RI systems, which covered the wavelength region between 310nm and 900nm. The natural systems of Johnson, Kron and coworkers served as *standard* systems for many other users who attempted with varying success (due to differences in detectors, filters, telescopes and techniques) to duplicate the originators natural systems. That is, using their own detectors and filters, astronomers measured stars from the Johnson and Kron lists and linearly transformed their natural magnitudes and colours to be the same as the Johnson and Kron colours and magnitudes. They then applied those same linear coefficients to transform the colours and magnitude of unknown stars onto the Johnson or Dron system.

The original blue and yellow filters were chosen by Johnson from readily available glasses so that when used with the IP21 photomultiplier tube they approximated the ordinary blue photographic response (~436nm) and the visual response (~545nm). A more violet magnitude U(~367nm) useful for very hot stars, was obtained by using a common violet glass. In retrospect, these choices should have been based more on astrophysics and less on glass availability, but so much work has been done in UBV that the weight of history assures its continuation. Intercomparison of much of the published broad-band photometry, in particular photometry taken more than 15 years ago, often shows scatter of more than 0.03 magnitudes, but more recent photometry obtained using better equipment, better matched natural systems and better secondary standard stars agree to better than 0.01 magnitude or 1 percent. Plans for automatic telescope based photometric systems aim for precisions of 0.001 magnitudes or 1/10 of a percent.

The 1P21 phototube was a remarkable invention and its high blue sensitivity dominated the development of photometric systems for over 30 years. There were red-sensitive devices available but observations were only made for bright stars because for many years the red detectors were much less sensitive, noisier and less reliable than the 1P21. In the mid-70s new detector materials became available, in particular the gallium-arsenide and multi=alkali phototubes which provided high ($>$15%QE) sensitivity between 300nm and 860nm, and the infra-red sensitive InSb photodiodes together with low-noise preamplifiers which revolutionised photometry between 1000nm and 4000nm., Both developments enabled red photometry to be done on faint objects, hitherto the province of blue detectors.

Photometry done with the new red sensitive tubes was placed on either the Kron or the Johnson standard systems again with mixed success and it has only been in the last few years that the Cousins RI 'near-natural' standard system (based on the Kron System) has gained wide spread acceptance. It has also been very useful that the Cousins system R($\sim$638nm) and I($\sim$797nm) bands are similar to the contemporary photographic R and I bands.

Johnson also introduced the infrared alphabetic JKLMN (approximately 1.22mm, 2.19mm, 3.45mm, 4.75mm, and 10.4mm) system in the mid-60s using PbS detectors and bolometers. The water vapour in the earths atmosphere defines a series of wavelength bands (windows) through which observations from the ground can be made. Johnson used interference filters (and unfortunately the atmospheric H20 absorption bands) to define what he called JKLM and N bands. Glass (1974) in his early observations with an InSb detector used the additional band H(1.63mm) between J and K, and in his choice of filters attempted to match the other Johnson bands. All IR observers have proceeded in a similar fashion and have concentrated mainly on copying the Johnson K magnitude scale. Identical detectors have been used but slightly different filters, dewar windows, focal plane shutters and observing altitudes have produced subtly different systems. The publication of sufficient numbers of stars in common from the different natural systems has helped delineate the differences and transformations between the systems are now quite reliable (eg Bessell & Brett 1988; Leggett et al 1992; McGregor 1992).

5. Passbands or sensitivity functions of standard systems

The most important specifications of a photometric system are the passbands or sensitivity functions of its magnitudes. For a variety of reasons, technical and historical, the passbands of the broad-band photometric systems have not been known with certainty and this has inhibited close matching of natural systems and prevented

Figure 1. The passbands of the UBVRIJHKL system. The spectrum is of an AO star

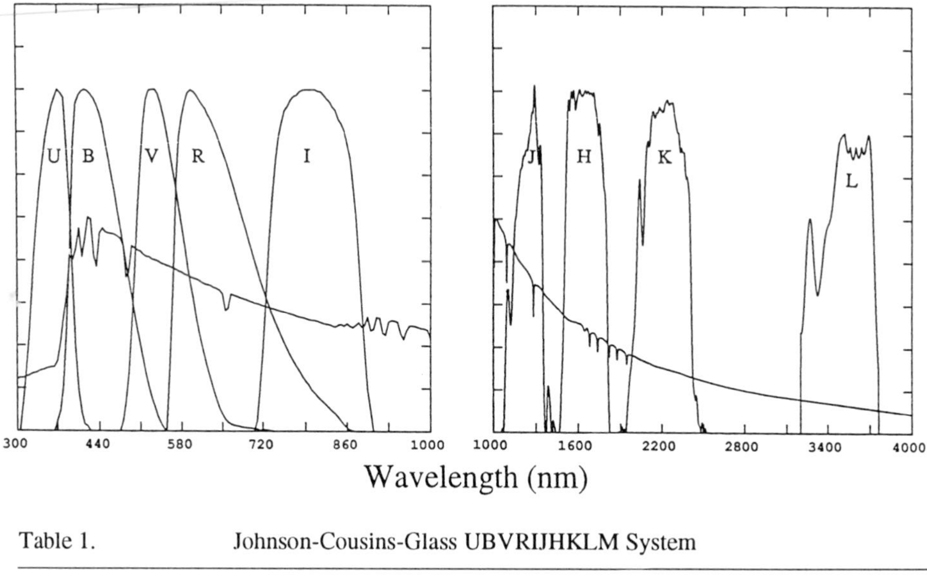

Table 1. Johnson-Cousins-Glass UBVRIJHKLM System

	U	B	V	R	I	J	H	K	L	M
λeff(nm)	367	436	545	638	797	1220	1630	2190	3450	4750
Δλ(nm)	66	94	88	138	149	213	307	390	472	460
$F_V(V=0)$ (10^{-30}W$cm^{-2}hz^{-1}$)	1780	4000	3600	3060	2420	1570	1020	636	281	154

Table 2. Effective Wavelengths (nm) and FWHM Bandpasses (nm) for Selected Systems

		λeff	Δλ			λeff	Δλ			λeff	Δλ
Geneva	U	343.8	17.0	Walraven	W	325.5	14.3	Wash'ton	C	391	110
	B	424.8	28.3		U	363.3	23.9		M	509	105
	B_1	402.2	17.1		L	383.8	22.7		T_1	633	80
	B_2	448.0	16.4		B	432.5	44.9		T_2	805	150
	V	550.8	29.8		V	546.7	71.9				
	V_1	540.8	20.2								
	G	581.4	20.6								
Strömgren	u	349	30	DDO	35	349.0	38.3	Thuan-Gunn	u	353	40
	v	411	19(12)		38	381.5	33.0		v	398	40
	b	467	18		41	416.6	8.3		g	493	70
	y	547	23		42	425.7	7.3		r	655	90
	β_w	489	15		45	451.7	7.6				
	β_n	486	3		48	488.6	18.6				

computation of accurate synthetic colours from theoretical spectra. The recent availability of spectrophotometry for many stars combined with the increased precision of second generation photometric catalogues has, however, enabled the passbands to be indirectly derived by computing synthetic colours from spectrophotometry of stars with well defined standard colours and adjusting the passbands until the computed and standard catalogue colours agree. This technique has enabled the passbands of the major systems to be well defined, which in turn has permitted filters to be designed which will still result in good passband matches with a variety of detectors. In addition, when it is not possible to exactly match passbands with detectors such as photographic plates it is possible to accurately predict the differences between photographic and p[hotoelectric magnitudes by computing the synthetic magnitudes.

The Vilnius optical spectrophotometry (Straizys & Sviderskiene 1972) has been used by Taylor (1986) and Bessell (1990) for passband analyses. Cousins (1987) uses spectra from Willstrop (1965). A more recent extensive catalogue of spectrophotometry is the Gunn- Stryker spectra (Gunn & Stryker 1983); corrections to these Gunn-Stryker absolute fluxes have been discussed by Rufener & Nicolet (1988) and Taylor & Joner (1990). The theoretical fluxes by Kurucz (1991) will also be extremely useful for passband analysis as well as calibration of colours. Excellent summaries of photometric systems are given by Lamla (1982) and Davis Philip (1979). More recent discussions of particular systems will be referenced individually below.

5.1 The UBVRIJHKL System

In Figure 1 the linear normalised passbands of the Johnson-Cousins-Glass UBVRI-JHKL system are shown. The F_ν spectrum of an A0 star is shown for orientation. Table 1 lists the effective wavelength λ_{eff}, the approximate bandwidth $\Delta\lambda$, which is the full width at half maximum (FWHM) of the band, and the absolute calibration of this system, based on the flux of Vega, for a zero magnitude A0 star. Note that the effective wavelengths of the broad bands change with the colour of the objects. The λ_{eff} listed are for an A0 star.

There are two large sets of photometric standards which represent the contemporary UBVRI system. One is in the southern hemisphere E and F regions (Cousins 1973, Menzies et al 1989) the other in the equatorial regions (Menzies et al 1991). The equatorial stars were previously established independently by Landolt (1983, 1992). Menzies et al (1991) has outlined some systematic differences between these representations and these are discussed by Menzies (1992). These differences have almost certainly arisen through the slight differences in the passbands used (in particular the B band and the method of U-B standardisation) and a lack of very red standards. Differences in passbands and transformations between various UBVRI systems have been discussed by Bessell (1979, 83, 86ab), Taylor (1986), Taylor et al (1989), Joner & Taylor (1990) and Bessell & Weis (1987). The variety of IR JHKL systems alluded to above has been discussed by Bessell & Brett (1988) who also give transformations between the various systems. New transformations have been derived by Leggett et al (1992) and McGregor (1992). Better calibration will now be possible through the absolute IR spectrophotometry of Cohen et al (1992).

5.2 Other Photometric Systems

Real or perceived drawbacks in existing photometric systems (the UBV system in particular) stimulated the design of other photometric systems better suited for measuring temperatures, metal-line blanketing, effective gravity and interstellar reddening. Some of these systems used broad bands comparable to the UBVRI system, others used narrower bands defined by different mixes of glass filters or interference filters. Effective wavelengths and other details of some of the better known systems are given in Table 2 and discussed below.

The Stromgren 4-colour system: The uvby system was devised by Stromgren to better measure the Balmer discontinuity, the metallicity and the temperature of A, B and F stars. The bands are essentially separate unlike the UBV bands which overlap. Fig. 2 shows the passbands realised by Olson (1974).

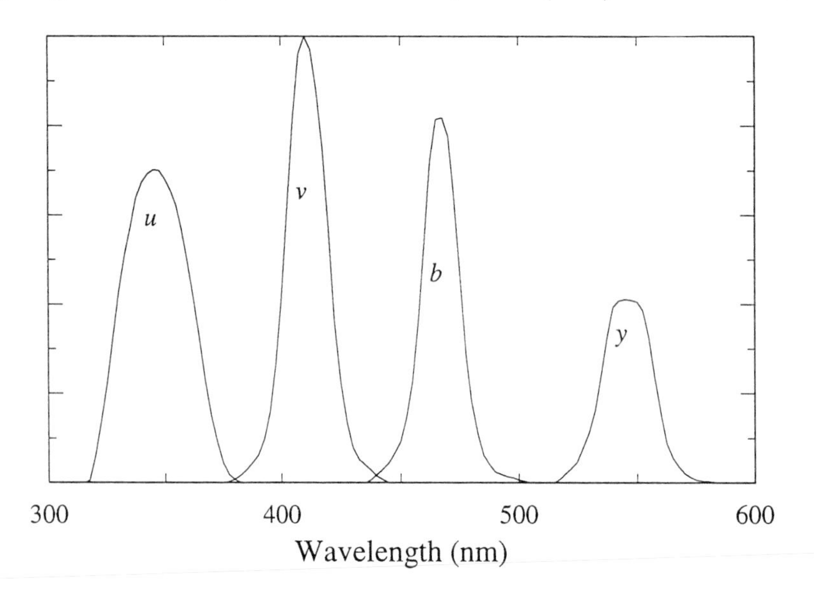

Figure 2. The uvby passbands of the Stromgren 4-colour system

The u band is completely below the Balmer jump; v measures the flux near 400nm, a region with much absorption due to metal-lines; b is centred near 460nm and is affected much less than B by metal-line blanketing; y is essentially a narrower V band. The u filter is coloured glass, the others are interference filters. Two special indices are derived m_1=(v-b)-(b-y), which measures metallicity and c_1=(u-v)-(v-y), which measures the Balmer discontinuity; (b-y) like B-V is used primarily as a temperature indicator. The system is capable of very high precision but unfortunately errors in the width of v filters manufactured some years ago resulted in non-standard filters being supplied to many users. Since then, published photometry has exhibited some systematic differences in c_1 and m_1 and there are difficulties in synthesizing c_1 and m_1 from theoretical spectra, particularly for cool stars. (For such late-standard catalogues (Olsen 1983, Cousins 1987) of new and more homogeneous observations are of high precision and internal consistency and it should now be possible to better define the v band; Cousins (1987) has investigated the theoretical realisation of the standard passbands for earlier stars than spectral type K. Systematic colour transformation effects are discussed by Manfroid and Sterken (1991, 1992). Anthony-Twarog et al (1991, 1992) discuss CCD photometry on the uvby system. Calibration of the Stromgren system for A, F and G supergiant stars is discussed by Gray & Olsen (1991) and Gray (1991).

Two additional interference filters (15nm wide and 3nm wide) centred on the $H\beta$ line are often used together with the 4 colours. The $H\beta$ index is used to derive luminosities in B stars and reddening in F and G stars (Crawford 1975). Some of the more recent $H\beta$ catalogues are Perry et al (1987), Cousins (1989, 90). The Stromgren system was one of the first photometric system devised to measure specific features in stellar spectra (eg Crawford 91). Because of the short wavelength range of its 4 colour filters, 1 percent photometry at least is required to realize the benefits the system has over the UBVRI system.

The DDO (35, 38, 41, 42, 45, 48) system: This system (also built around the sensitivity of the 1P21) was designed for the analysis of G and K dwarfs and giants. Figure 3 shows the passbands. The 35 filter is the u filter of the 4-colour system; the 38 filter is also a glass filter and better measures metal blanketing than the v filter, being further to the violet and wider; 41 measures the CN band; 42, 45 and 48 are continuum filters. The colour 35-38 (the 3538 index) measures the Balmer jump, 3842 the metallicity, 4245 and 4548 are used for gravity and temperature measurements. By restricting the measurements to the blue spectral region complicated blanketing corrections are necessary to derive temperatures and gravities. good results, especially for faint K stars, can be obtained by using V-I or R-I as the temperature indicator. Because of the narrow band width of the filters the DDO system has been mainly restricted to relatively bright stars. Standards are given by McClure (1976), Dean (1981) and Cousins (1992). The system is discussed by McClure (1976, 79) and Osborn (1979).

The uvgr system: This system was devised by Thuan & Gunn (1976) in the mid-70s from the UBVR system for use with an S20 detector and in order to avoid the strong Hg lines from city lights and [OI] lines in the night sky. The g and r bands are of similar width to the V and R bands while the u and v bands are about half

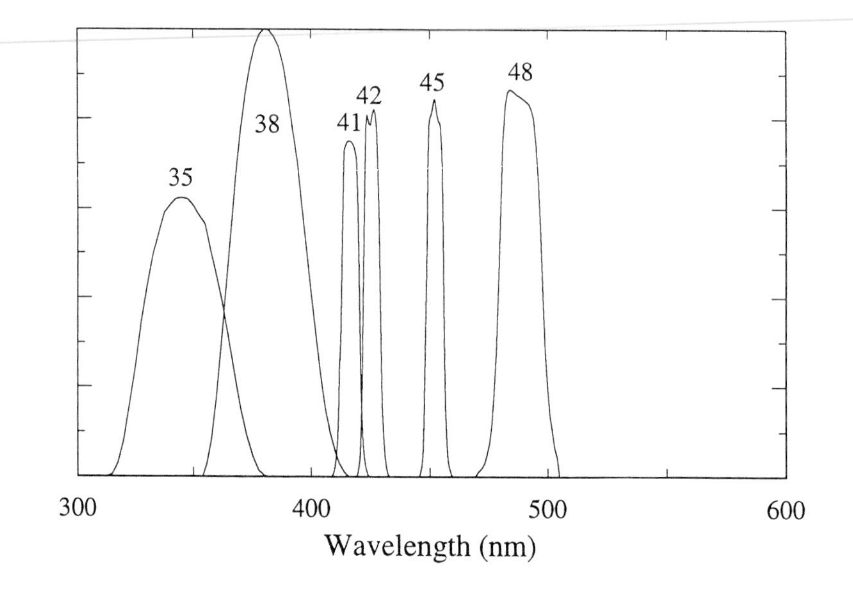

Figure 3. The passbands of the DDO system.

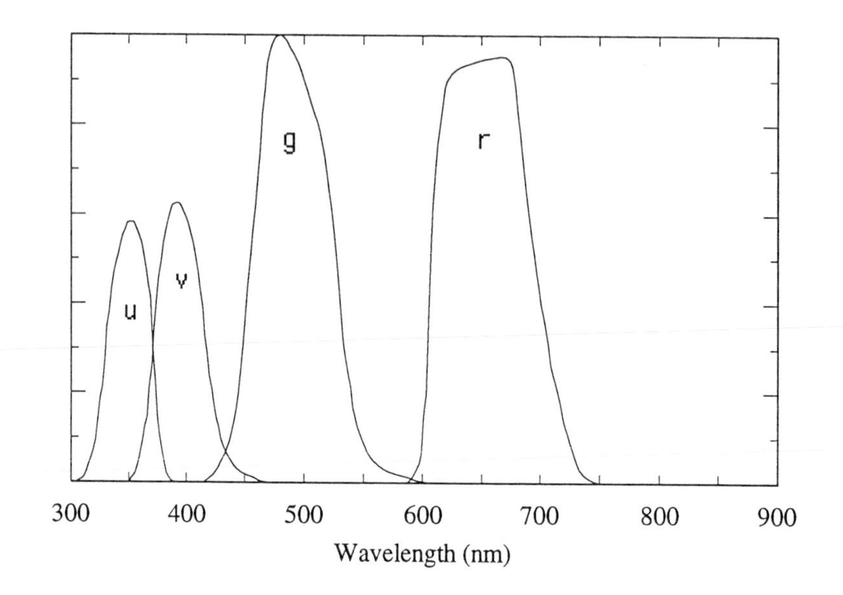

Figure 4. The passbands of the Thuan-Gunn uvgr system.

the width of the U and B bands. Unlike other systems the zero points have been set by the FG subdwarf CD+174708, which by definition has g = 9.50; g-r = u-v = v-g = 0. The g-r colour has a longer baseline than V-R but they transform well. The r band has also been used for photographic and CCD photometry. The relation between r and R is r = R + 0.35 - 0.148(R-I) + $0.122(\text{R-I})^2$ + $0.0118(\text{R-I})^3$ for (R-I) < 1.8. The relation between g-r and V- R is V-R = 0.29 + 0.585(g-r) + $0.060(\text{g-r})^2$.

The Geneva ($UBB_1B_2VV_1G$) System: Difficulties with matching natural systems have been eliminated by the strategy employed by proponents of the closed Geneva ($UBB_1B_2VV_1G$) system. This multiband photometric system is supervised by a small group who control the instrumentation and supervise the data reduction and calibration. The 4th edition of the catalogue (Rufener 1988) containing data for 29400 stars and is also available from the Stellar Data Centre in Strasbourg. Cramer (1991) discusses more recent data. Rufener & Nicolet (1988) discuss the passbands and their absolute calibration and the theoretical calibration of some colours in terms of temperature and gravity. In practice, three colours $U\text{-}B_2$,$B_2\text{-}V_1$ and $V_1\text{-}G$ are used together with linear combinations which are reddening free for a standard extinction law and E(B-V)<0.4 mag. The combination indices are d = $(U\text{-}B_1)\text{-}1.430(B_1\text{-}B_2)$, F = $U\text{-}B_2)$ - $0.832\ (B_2\text{-}G)$, g = $(B_1\text{-}B_2)$ - $1.357(V_1\text{-}G)$ and m_2 = $U\text{-}B_1)$ - $1.457(B_2\text{-}V_1)$. The various Geneva indices have been well calibrated in terms of gravity, temperature and abundance (eg Golay 1980). The indices d and m_2 are clearly closely related to the $[c_1]$ and $[m_1]$ indices of the Stromgren system (eg Eggen 1977).

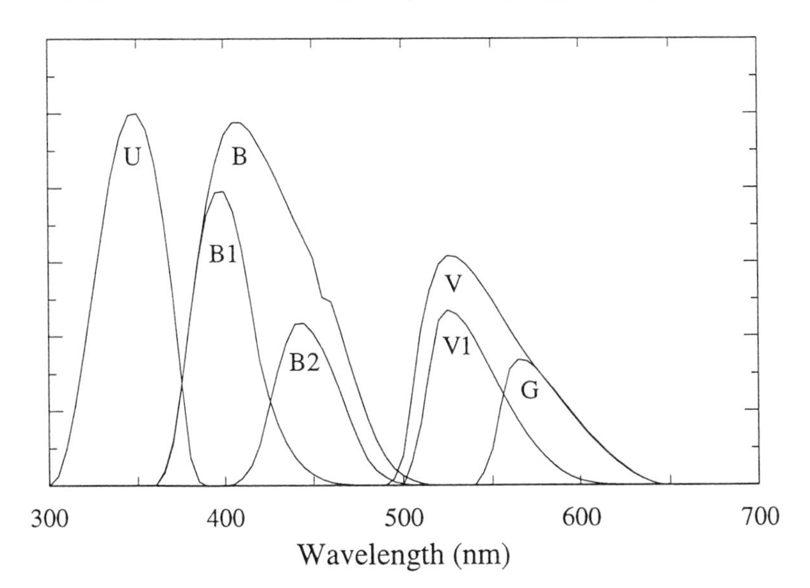

Figure 5. The isophotonic passbands of the Geneva system.

The Vilnius (U P X Y Z V S) system: The intermediate band Vilnius system was developed independently from the Geneva system but for similar reasons, namely to derive temperatures, luminosities and peculiarities in reddening and composition from photometry alone (eg Straizys 1979). The colours are normalised by the condition U-P = P-X = X-Y = Y-Z = Z-V = V-S for unreddened O-type stars. Therefore all colours for all normal stars are positive. Reddening free indices are constructed as with the Geneva system and these are Vilnius system has now been used with CCDs (Straizys 1991,92) and has also been extended to the southern hemisphere (Dodd et al 1992).

Straizys (1979) has also utilized a 7 colour VILGEN system comprising passbands from the Vilnius (P Z S) and Geneva (U B_1 B_2 V) system. Its advantage is that it uses wider bands than the original Vilnius system yet scarcely compromises the selectivity of the indices.

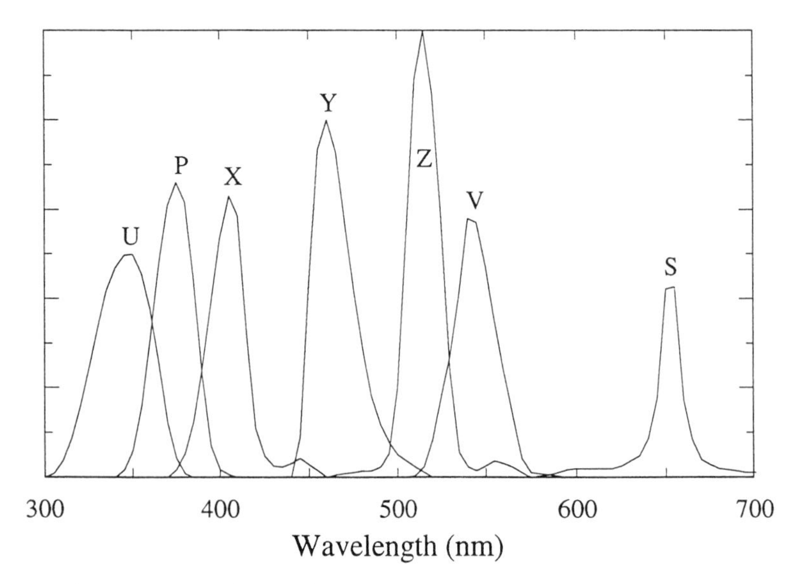

Figure 6. The passbands of the Vilnius system.

The Walraven (VBLUW) system: The properties of the system are described by Lub and Pel (1977). Unlike the other systems discussed here, Walraven photometry is obtained using a specially constructed spectrograph-photometer. Originally used in South Africa from 1960, the photometer was moved to La Silla in 1979 (Pel et al 1988). The VBUW bands are defined by a special filter of crystal quartz and calcite polarization optics and separated geometrically by a quartz prism spectrograph. The L band is taken from a beam that does not pass through the spectrograph. This technique provides great stability of passbands which has helped Pel (1991,92) to discern small systematic errors with right ascension and declination in the V photometry of Landolt (1992) and Rufener (1988). Much of the work in the VLBUW system has been on cepheids and RR Lyrae stars (eg Lub 1979, Pel 1985). Comparison of the VLBUW colours of stars with synthetic colours from theoretical fluxes has proved to be a very precise test of the model atmospheres and fluxes (Pel 1992).

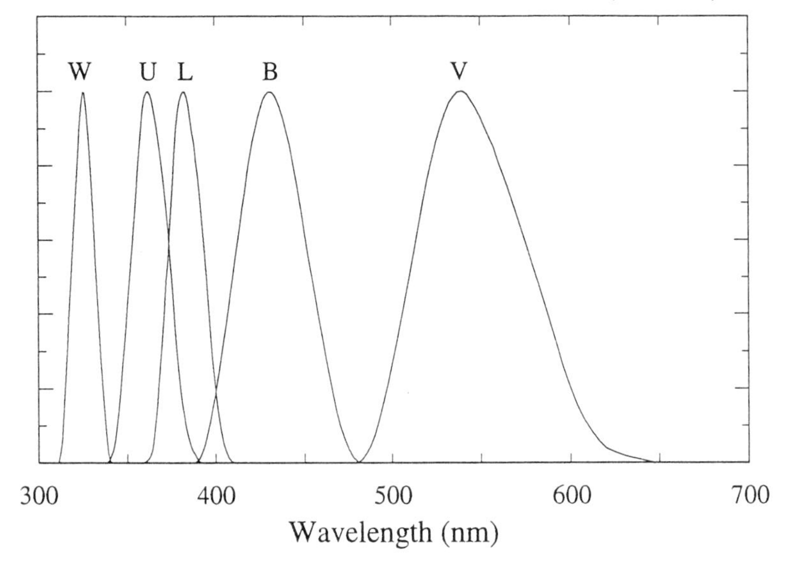

Figure 7. The passbands of the Walraven VLBUW system

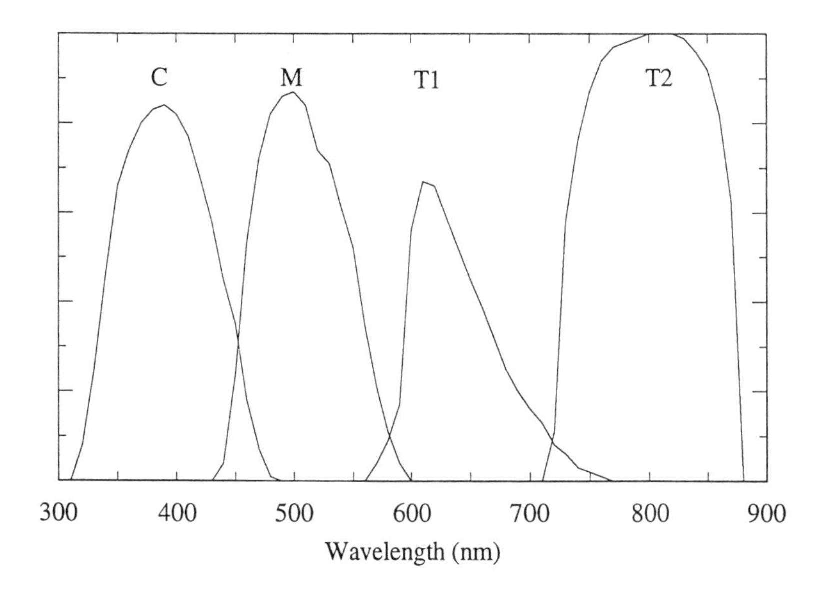

Figure 8. The passbands of the Washington CMT_1T_2 system

The Washington CMT_1T_2 System: The Washington system is a very broad-band system which was devised to use the wideband sensitivity of the extended-red detectors, to improve the sensitivity of blue-violet colours to metallicity and gather more violet light in cool stars, and to try and separate the effects of CN from other metal lines. New standards are given by Geisler (1990) and revised calibrations are given by Geisler et al (1991) together with many references to the successful work that has been done on the abundances of faint K giants in globular clusters and in external galaxies. Tyson (1991) discusses CCD photometry. We have found that the violet C band is a very useful metallicity indicator for faint K giants but that the M band contains little more information for giants than does V; T_1 and T_2 have no advantages over R and I. We find that a minimal CVI system is very useful for metal-weak K stars. It takes advantage of C being better than U and B for a metallicity sensitive band and uses the larger baseline of V-I compared to T_1 - T_2 as a temperature index. Most importantly, it reduces the proliferation of additional bands.

Good transformations can be derived between the Washington indices and VRI indices.

$$R-T_1 = -0.012 + 0.046(T_1 - T_2) - 0.082(T_1 - T_2)^2$$
$$V - I = 0.003 + 1.794(T_1 - T_2) + 0.517(T_1 - T_2)^2 - 0.402(T_1 - T_2)^3$$
$$V - R = 0.008 + 1.016(V-T_1)$$
$$M - V = 0.006 + 0.240(V-I)$$
$$(M-T_2) - (V-I) = 0.242(V-I) + 0.018(V-I)^2$$

The Photographic systems: Originally photographic emulsions were only sensitive to light blueward of 490nm. These were to O emulsions. Different chemical sensitising shifted the red sensitivity cutoff to longer wavelength, G 580nm, D 650nm, F 700nm and N 880nm, approximately. By using blue cutoff glass filters and the red cutoff of the emulsions various photographic passbands were made. Photographic U used a violet filter for both blue and red cutoffs. The photographic colours were normally converted onto the photoelectric UBVR system but the accuracy of the conversions were restricted by limitations in iris photometry and poor matches of the bandpasses. The Basel RGU system (Buser 1979) was used quite successfully in studies of galactic structure. In recent years astronomical photography has undergone a renaissance caused firstly by the development of new fine grain emulsions (Kodak IIIaJ and IIIaF) and the utilization of methods of greatly increasing the sensitivities of the J and F emulsions using hydrogen gas and the IVN emulsions using silver nitrate solution, and secondly by the use of new scanning microdensitometers and better methods of intensity calibration. Averaging of several wide field Schmidt plates of higher scale prime-focus plates can now produce photometry to a few percent to very faint limits. Recent experiments to digitally combine 64 limiting exposure Schmidt plates (Hawkins 1992) has resulted in an increase in the limiting magnitude by about 2 magnitudes, a very impressive result. Theoretical investigation of bandpasses enables better filter design for bandpass matching or predicts the relevant transformations and systematic differences between photoelectric and photographic photometry. Photographic photometry these days is usually restricted to attempted matches to the Johnson U and B or the Thuan and Gunn g using IIIaJ plates; Cousins R or Thuan- Gunn r using IIIaF plates; and Cousins I using IVN plates. Direct photographic calibration from step-wedges is usually supplemented by direct magnitude measurements of stars in each field using a CCD array. A discussion of some of the new photographic passbands is given in Bessell (1986b).

CCD photometric systems: The high QE of CCDs and their inherent linearity has made them the detectors of choice for most area photometry, especially for colour- magnitude diagrams of clusters. Unfortunately the advantages of the CCDs were not initially fully attained because many users paid insufficient care in defining their passbands and in standardizing their photometry. This resulted in internally precise results but an inability to relate these results with much confidence to the photoelectric system data or to the theoretically derived magnitudes and colours. Most astronomers now realize the importance of matching the CCD passbands to the photoelectric passbands and measuring their instrumental passbands.

Standardization, at the time, was not made easy by a paucity of suitable standards for CCD photometry, namely stars that were faint enough, had a good range in colour and many of which could be observed on a single CCD frame. That problem has now been addressed by RCCD-sized fieldsS of UBVRI standards of Landolt (1992) and Jones (1992) and CMT_1T_2 standards of Geisler (1990). It would be very useful were these CCD fields to be standardized for JHKL, uvby, DDO and other systems as well.

There can be additional problems associated with interference filter photometry from imaging systems. Firstly, the filters need to have uniform transmission across the filters and to be specified as image quality from the manufacturer. Secondly, there are shifts in passbands associated with off-axis imaging and these shifts are more significant in narrow- band work. Photoelectric photometry is done on-axis and at long (f/18) focal ratios. Imaging is done at shorter focal ratios, f/8 at 1m class telescopes, f/3.5 at 4m class telescopes or with focal reducing cameras. In the collimated sections of cameras the passband shifts to the blue as one moves away from the field centre, whilst in short focal ratio telecentric sections the bandpass does not vary with field position but is distorted compared to that on-axis in a near-parallel beam. These complications can be accounted for, but users should be aware of such problems.

Broad band glass filter defined systems are advantageous for imaging photometry as they provide robust passbands together with a large range of possible aperture sizes and can be used to the faintest magnitude limits. Care must be taken however, to ensure that band passes of the broad-band CCD systems are good matches to the photometric system within which the standards were established so that astrophysically sound transformations can be made and so the same colour calibrations can be used. Bessell (1990) discusses these problems and suggests some glass mixes for UBVRI which are appropriate for some CCDs. The U and the B bandpasses are the most critical for most stellar work and the significant differences in the blue-violet sensitivities of coated, uncoated, thinned and coated, thinned and uncoated CCDs necessitate different glass mixes being designed for different CCDs. The stellar flux atlases and theoretical fluxes discussed at the beginning of section 5 are very useful for synthesizing colours from trial glass mixes with various CCDs when designing new filters which will enable best matches to the standard system colours.

One possible remaining problem with CCD bandpass matching concerns the I band. The photoelectric I band has a steep red cutoff due to the GaAs photocathode; however, the cutoff of the CCD is much less steep with the CCD sensitivity extending beyond 1m. This does not cause transformation problems for most stars, but cool, strongly banded stars such as late M and C stars will be affected. Walker (1991) also notes that the telluric emission is brightest beyond the cutoff of the photoelectric I band which will result in brighter sky backgrounds and possible problems with sky flat-fielding. He therefore recommends removing the extended red sensitivity tail of the CCD using an interference filter. One such filter (an infrared mirror No 60.5050) is readily available from the Rolyn Optics Company.

References:

Anthony-Twarog, B.J. & Twarog, B.A., 1991 in Precision Photometry: Astrophysics of the Galaxy (Eds: A.G. Davis Philip, A.R. Upgren & K.A. Janes, L. Davis Press, Schenectady 1991), p. 127.

Anthony-Twarog, B.J., Twarog, B.A. & Suntzeff, N.B. 1992 AJ 103, 1264.

Bessell, M.S. 1979 PASP 91, 589

Bessell, M.S. 1983 PASP 95, 480

Bessell, M.S. 1986a PASP 98, 354

Bessell, M.S. 1986b PASP 98, 1303

Bessell, M.S. 1990 PASP 102, 1181

Bessell, M.S. & Brett, J.M., 1988 PASP 100, 1134-1151

Bessell, M.S. & Weis, E.W. 1987 PASP 99, 642

Buser, R. 1979 Dudley Observatory Report No. 14, (Ed: A.G. Davis Philip, Dudley Observatory, Schenectady 1979).

Cohen, M., Walker, R.G., Barlow, M.J., Deacon, J.R., Witteborn, F.R., Carbon, D. & Augason 1992 (these proceedings)

Cousins, A.W.J. 1973 Mem RAS 77, 223

Cousins, A.W.J. 1987 Mem SAAO Circ 11, 93

Cousins, A.W.J. 1989 Mem SAAO Circ 13, 15

Cousins, A.W.J. 1990 Mem SAAO Circ 14, 55

Cousins, A.W.J. 1992 (Private communication)

Cramer, N. 1991 in Precision Photometry: Astrophysics of the Galaxy (loc cit), p 139.

Crawford, D.J., 1975 AJ 80, 955

Crawford, D.L. 1991 in Precision Photometry: Astrophysics of the Galaxy (loc cit), p 121

Davis Philip, A.G. 1979 Dudley Observatory Report No. 14, (Ed: A.G. Davis Philip, Dudley Observatory Schenectady 1979), Appendix, p 409

Dean, J.F. 1981 SAAO Circ 6, 10

Dodd, R.J., Forbes, M.F. & Sullivan, D.J., 1992 (These proceedings)

Eggen, O.J., 1976 PASP 88, 732

Eggen, O.J., 1977 PASP 89, 706

Geisler, D., 1990, PASP 102, 344

Geisler, D., Claria, J.J. & Minniti, D. 1991 AJ 102, 1836

Glass, I.S., 1974 MNASSA 33, 53

Golay, M., 1980 Vistas in Astronomy 24, 141

Gray, R.O. 1991 A & A 252, 237

Gray, R.O. & Olsen, E.H. 1991 A & AS 87, 541

Gunn, J.E. & Stryker, L.L. 1983 ApJS 52, 121

Hawkins, M.R.S. 1992 (Private communication)

Joner, M.D. & Taylor, B.J. 1990 PASP 102, 1004

Jones, D.H.P. 1992 (These proceedings)

Kurucz, R.L. 1991 in Precision Photometry: Astrophysics of the Galaxy (loc cit), p 27

Leggett, S.K., Oswalt, T.D. & Smith, J.A. 1992 (These proceedings)

Lamla, E., 1982 in Landolt-Bornstein, New Series Volume IV/2b (Eds: K. Schaifers & H.H. Voigt, Springer-Verlag, Berlin) p 35

Landolt, A.U. 1983 AJ 88, 439

Landolt, A.U. 1992 AJ 104, 340

Lub, J. & Pel, J.W., 1977 A & A 54, 137

Lub, J 1979 in Dudley Observatory Report No. 14, (Ed: A.G. Davis Philip, Dudley Observatory, Schenectady, 1979), p 193

McClure, R.D. 1976 AJ 81,182

McClure, R.D., 1979 in Dudley Observatory Report No. 14, (Ed: A.G. Davis Philip, Dudley Observatory, Schenectady 1979), p. 83

McGregor, P.J. 1992 (Private communication)

Manfroid, J., & Sterken, C., 1991 in Precision Photometry: Astrophysics of the Galaxy (loc cit), p 139

Manfroid , J. & Sterken, C. 1992 A & A 258, 600

Menzies, J.W., Cousins, A.W.J., Banfield, R.M. & Laing, J.D., 1989 SAAO Circ 13, 1

Menzies, J.W., Marang, F., Laing, J.D., Coulson, I.M. & Engelbrecht, C.A. 1991 MNRAS 248, 642

Menzies, J.W., 1992 (These proceedings)

Olsen, E.H., 1983 A & AS 54, 55

Olson E.C., 1974 PASP 86, 80

Osborn, W., 1979 in Dudley Observatory Report No. 14, (Ed: A.G. Davis Philip, Dudley Observatory, Schenectady 1979), p 115

Pel, J.W., 1985 in RCepheids: Theory and ObservationsS, IAU Colloq. 82, (Ed. B.F Madore, Camb Univ Press), p1

Pel, J.W., Trefzger, C.F., & Blaauw, A., 1988 A & AS 75, 29

Pel, J.W., 1991 in Precision Photometry: Astrophysics of the Galaxy (loc cit), p 165

Pel, J.W., 1992 Private communication

Perry, C.L., Olsen, E.H. & Crawford, D.L. 1987 PASP 99, 1184

Rufener, F. 1988 A & AS 78, 469

Rufener, F., Nicolet, B., 1988 A & A 206, 357

Straizys, V. 1991 in Precision Photometry: Astrophysics of the Galaxy (loc cit), p 341

Straizys, V. 1979 Dudley Observatory Report No. 14, (Ed: A.G. Davis Philip, Dudley Observatory, Schenectady 1979), p 217, Appendix, p 485

Straizys, V., & Sviderskiene, Z., 1972 Bull.Vil.Astron.Obs No. 35

Taylor, B.J., 1986 APJS 60, 577

Taylor, B.J., Joner, M.D., & Johnson, S.B., 1989 A.J. 97, 1798

Taylor, B.J., & Joner, M.D., 1990 AJ 100, 830

Thuan, T.X., & Gunn, J.E., 1976 PASP 88, 543

Tyson, N.D., d1991 in Precision Photometry: Astrophysics of the Galaxy (loc cit), p 193

Willstrop, R.V., 1965 MemRAS 69, 83

Discussion

D. Crawford: *Strömgren and I talked about red extensions to uvby, but decided that the existing R and I were perfectly adequate for any who needed a red extension, (eg. for B,A and F stars). It was a conscious decision not to add two new filters; R,I are, in fact, very useful supplements to uvby for many applications.*

Bessell: That is interesting. But I feel it is likely that, because you and your collaborators have never combined the uvby and R,I systems in your published work, (unlike Eggen), it has inadvertently led others to neglect the possibility of using redder colours.

W.Z. Wisniewski: *H. Johnson was well aware how the UBV filters should match the solar spectrum. But after the Second World War the number of available filters was limited. One would have had to pay for development and there were severe financial limitations. We could afford to spend just a few hundred dollars.*

Bessell: As you have pointed out to me in the past, we should not judge the originators of photometry by modern criteria. But one should ensure that their inadequacies are not still evident in the modern secondary standards.

G. Śzécsényi-Nagy: *As CCD's are mushrooming in astronomical observatories too and manufacturers offer a wide choice of ultraviolet and violet-blue sensitive coatings, do you see any possibility of standardizing the short wavelength sensitivity of thinned, back illuminated CCD's?*

Bessell: It is unlikely that the ultraviolet sensitivity of CCD's will be standardized in the near future; however as most of the flux in the U systems, (the flux which is astrophysically significant), lies redward of 350 nm, it is possible to get good transformations for unreddened stars by concentrating on matching the red side of the passband, and using standards with a good range of spectral types.

S.B. Howell: *What differences have you found between the manufacturers, supplied, 'generic' QE curves for CCD's and the actual curves? Also, can you comment on any problems when using a filter set designed for a specific CCD (type and QE curve) with another CCD of different type?*

Bessell: Alistair Walker can give you better information but I believe that the main differences between the manufacturers' curves and the actual curves occur shortward of 350 nm and longward of about 800 nm. The long wavelength response is the most temperature sensitive and this is usually set by observatory engineers to minimize the dark current and will be different from the temperature relevant to the manufacturers' curves.

One can reliably calculate the differences by convolving the different responses with the spectrophotometric atlases and plotting these differences against spectral type, luminosity etc.

R.F. Garrison: *The small differences in passband definition can be disastrous for unusual objects like supernovae, for which the emission lines develop and recede in addition to moving in radial velocity, so as to move in and out of the boundaries of the passbands. These differences can lead to differences of 0.4 magnitude or more which is not small for theoreticians.*

D.H.P. Jones: *About ten years ago the KPNO supervised a contract to buy in a uniform set of BVRI filters to be used at several observatories throughout the world. On La Palma we have three of these sets still in use. Does anyone else still use them?*

Bessell: It is now recognised that the rectangular bandpasses of the KPNO BVRI interference filter sets was very unfortunate. Although these filters have high throughputs, they provide very poor transformations to the BVRI system and with the paucity of suitable standards at the time they resulted in very poorly standardized photometry.

The Homogeneity of the UBV(RI)$_C$ System

J W Menzies

SAAO, PO Box 9, Observatory 7935, South Africa

Abstract

There are two large sets of high precision standard star photometry which claim to represent the UBV(RI)c system, one in the E regions in the south established by Cousins and one at the equator due to Landolt. There appear to be systematic differences between them. Astrophysical interpretations which depend on photometric data may in consequence depend on which set of standards was used.

Introduction

Many photometric systems have been established in the visible wavelength range, often designed to address specific problems. The UBV system is undoubtedly the most widely used general-purpose system in the blue-visual part of the spectrum. In the red, Johnson's RI system is commonly used, but since it is based on the low-quantum-efficiency S1 photocathode, and observers normally use more sensitve tubes with different response functions, (nonlinear) transformation equations with relatively large coefficients are required. The Cousins (RI)$_C$ system is based on a GaAs photomultiplier and is somewhat easier to reproduce so it has become increasingly popular since its introduction.

Why reduce photometric data to a standard system? First, it makes intercomparison with other observations more straightforward, but more importantly it permits the use of standard relations to derive astrophysically interesting parameters like spectral type, reddening and temperature for faint stars, and ages of globular clusters from main sequence fitting. It also facilitates the comparison of population synthesis models with observation. The standard relations have only to be derived once, which consequently saves considerable effort and valuable observing time. To use them, one needs to have one's data on the standard system. It is usually necessary to transform the raw data via non-linear relations. To derive the transformations accurately, many stars covering the whole colour range of interest must be observed along with the programme stars.

Standard stars should not be concentrated in a small region but spread widely over the sky so that a representative sample of them is available to observers at all latitudes and at all times of the year; unreddened stars of some spectral types may be difficult to find in a group of standards confined to a small region. The system needs to be homogeneous over the sky for consistent interpretation of observations between different locations and at different times. As Johnson(1963) pointed out there is a risk of inhomogeneity if different sets of standard star data are combined. However, the standard system is normally an instrumental one, and if the original photometer no longer exists, observations obtained with other photometers must be used when fainter standards, or a better sky- or colour coverage is needed. The original UBV standards were relatively few and bright and their precision low compared with what is currently attainable. Two large sets of stars widely distributed in right ascension and extending to fainter magnitudes than the original, with an internal precision of a few millimagnitudes, are currently available as supplements to the primary UBV standards and also provide standards for photometry on the Cousins $(RI)_C$ system.

Systems

The E-region UBV system established by Cousins and Stoy (1962) (ROB49) is ultimately based on the Johnson UBV system as represented by the magnitudes and colours tabulated by Cousins (1971) (ROA7). The latter is a compilation of various series of photometry of bright stars in the equatorial regions and was produced in response to the adoption by IAU commission 25 in 1970 of the stars brighter than magnitude 5.0 in the band between declinations +10° and -10° as new primary standards for the UBV system to supplement those of Johnson & Harris (1954). In setting up the $V(RI)_C$ system, Cousins (1971) used the Johnson V magnitude and the natural R and I magnitudes of his photometer, with the zero point being set to give colours of 0.0 for a typical A0V star. Several series of observations made at the SAAO in Cape Town and in Sutherland have been used to augment the original E-region compilations (UBV and $V(RI)_C$) and the current working list of standard magnitudes and colours in use at SAAO incorporates the new data (Menzies *et al.* 1989.) The data in the current list conform to the original Cousins $UBV(RI)_C$ system (Cousins 1990) and can be considered now to define the E-region system.

Cousins has made continual checks on the degree of conformity of SAAO UBV photometry to the Johnson system. Most recently, in discussing some new photometry of stars in the equatorial region, he concludes that for V and B-V, the ROA7 system, and hence that of the E-regions, is essentially the same as that of Johnson & Harris (1954) and that of Johnson *et al.* (1966). However, there is evidently a systematic difference between the SAAO representation of (U-B) and that of Johnson such that $(U-B)_J = (U-B)_{E-regions} - 0.003 + 0.015(B-V)$.

Intending to establish a set of standards on the UBV system accessible to both northern and southern observers and with a consistent zero point around the sky, Landolt (1973) observed stars in the equatorial Selected Areas. His results were tied to the Johnson & Harris (1954) standards and he presented evidence that he had succeeded in reproducing the UBV system. In his second series, Landolt (1983) measured $UBV(RI)_C$ for a set of stars in the magnitude range $7 \leq V \leq 12.5$, and referred them the E-region standards. On finding non-linear differences between the new results and those from the 1973 series for B-V and U-B he applied corrections to the new data. Thus, Landolt's 1983 results for UBV are on his 1973 system and the $(V\text{-}R)_C$ and $(V\text{-}I)_C$ colours are tied to the E-regions.

Comparisons

With a view to checking that the Landolt (1983) and E-region (Menzies *et al.*1989) systems were the same, an observing program was carried out at Sutherland in which the Landolt stars were treated as programme stars and the measurements were referred directly to the E-regions. The analysis of the results revealed marked systematic differences between the SAAO and Landolt sets (Menzies *et al.* (1990). In an independent programme of equatorial star photometry, Cousins (1984) also included some of Landolt's (1983) stars.

Landolt's transfer of the $(RI)_C$ system to the equator worked well, with only small zero-point differences and very small colour terms between his and our results. Cousins (1984) found similar small differences. There is a hint of a change of slope in the $(V\text{-}R)_C$ differences for the reddest stars. Landolt evidently used a linear colour transformation, and the non-linearity is probably the result of a small bandwidth mismatch between his R filter and the standard one – a common problem resulting from the relative difficulty of matching the extended redward wing of the Cousins R passband.

The situation is not nearly so good for the differences in (B-V) and (U-B), which both show marked non-linear trends. That Cousins (1984) found similar trends from observations made at a different site and with a different photometer suggests that the origin of the differences probably lies in Landolt's use of linear transformation equations in his 1973 series, and in selection effects coupled with the relatively low precision of the original list of UBV standards.

Polynomials have been derived to allow Landolt's data to be transformed to the E-region $UBV(RI)_C$ system and these are listed in Table 1.

Table1. Conversion of Landolt System to Cousins System

$$\Delta(U-B) = 0.002 - 0.257(B-V), (B-V) \leq 0.08$$
$$\Delta(U-B) = -0.029 + 0.136(B-V) - 0.093(B-V)^2, (B-V) \geq 0.08$$

$$\Delta(U-B) = 0.013 - 0.014(U-B) + 0.026(U-B)^2 - 0.019(U-B)^3$$

$$\Delta(B-V) = -0.004 + 0.002(B-V) + 0.048(B-V)^2 + 0.032(B-V)^3$$
$$\Delta(V-R)_C = -0.002 + 0.005(V-R)_C + -0.018(V-R)_C{}^2$$
$$\Delta(V-I)_C = -0.001 - 0.005(V-I)_C$$

where $(B-V)_{Cousins} = (B-V)_{Landolt} + \Delta(B-V), etc.$

Two different regressions are given for Δ(U-B). The fits for Δ(U-B) as a function of both (B-V) and (U-B) are illustrated in Figure 1. These also illustrate some of the pitfalls in making this kind of transformation. In both plots, the open symbols represent reddened B stars, for which the wrong correction would be given by the fitted curve using (B-V) as the independent variable. The colours must first be approximately dereddened before the correction is derived from the polynomial. Also, most of the stars with (B-V) $\geq$ 1.0 are giants and in this case the correction to (U-B) should be derived from the (U-B) curve. Better still, both curves should be used and an inconsistent pair of corrections should lead to further investigation as to the reason.

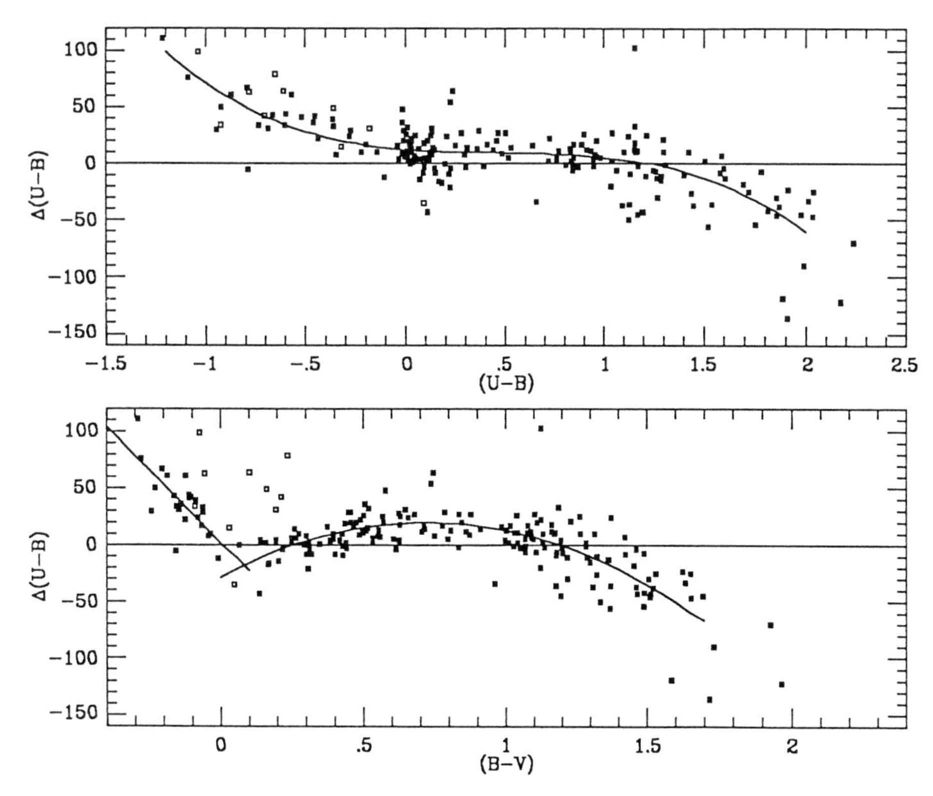

Figure 1. (U-B) differences, SAAO - Landolt, as a function of B-V and U-B. The open symbols represent reddened B stars.

A zero-point difference was found in V, which was due to some unspecified right-ascension-dependent effect. Cousins (1984) did not find any such difference in his measurements of the Landolt stars. Subsequently, Cousins & Menzies (in preparation) have compared the magnitudes from a compilation of Walraven photometry over the whole sky provided by Pel (private communication) with those for stars in both the E regions and the equatorial sample (ROA7) and find no evidence for any right-ascension-dependent differences. Thus it is most likely that the error is in the Sutherland photometry, though the origin is still not clear.

Consequences

The Landolt and E-region systems are clearly not the same. Does this really matter? The differences are not very large. Landolt's data can be transformed to the E-region system via the equation in Table 1. The changes for V-I, V-R and B-V are small, except for the reddest stars in V-R and B-V. Even for U-B over most of the colour range, changes of less than 0.02 mag are required. The red stars will usually give trouble - in V-R because the Cousins system is not well defined in this colour range, although Bessel (1990a) has attempted to improve the situation in this range; and in B-V and U-B because of severe bandwidth effects, low signal at U and probable small scale variability amongst the majority of red stars.

What are the consequences of using the different systems? The standard two-colour diagram should be corrected by means of the equations in Table 1 before comparison with colours obtained on Landolt's system. The effects are small except for the bluest stars where errors of no more than about 0.02 mag would be incurred in determining $E_{(B-V)}$ if the standard curve were to be used instead of the transformed one. On the other hand, for the E-region system significant differences appear only for the very reddest stars where photometry becomes more uncertain anyway. Ages of globular clusters are determined by fitting unreddened cluster colour-magnitude diagrams to theoretical isochrones in the M_V,(B-V) plane. The turn-off region appears near $(B\text{-}V)_0 \sim 0.3$, where there is no significant difference between Landolt and the E-regions. Small differences appear on the lower main sequence and on the sub-giant branch which might lead to slightly poorer fits if the isochrones are not transformed before use with data on the Landolt system. Much larger errors due to uncertainties inherent in the reddening determination would swamp any problems due to mismatching photometric systems.

Conclusions

The E-region system and that of Landolt are clearly different though the two sets of standards probably have about the same internal precision. To preserve the high internal precision of one's own observations, care should be taken in matching the passbands used to those of the standard system to which the data are reduced. Nonlinear transformations should always be expected. If Bessel's (1990b) proposed glass filter combinations in practice give linear transformations with small coefficients to the E-region system then there would be some advantage in adopting them as the standard set for future use. The temptation to merge the Cousins and Landolt systems should be resisted considering the non-linearities involved in the transformation between them as the inherent internal precision of each would be compromised.

References:

Bessel, M.S., 1990a. *Astron. Astrophys. Suppl.*, **83**, 357.

Bessel, M.S., 1990b. *Pub. astron,. Soc. Pacific*, **102**, 1181.

Cousins, A.W.J., 1971. *Roy. Obs. Ann.*, No. 7.

Cousins, A.W.J., 1984. *SAAO Circulars*, No. 8, 69.

Cousins, A.W.J., 1990. *Mon. Not. astr. Soc. S. Africa*, **49**, 2.

Cousins, A.W.J & Stoy, R.H., 1962. *Roy. Obs. Bull.*, No. 49.

Johnson, H.L., 1963. in *Basic Astronomical Data*, ed. K. Aa. Strand, Univ. Chicago Press, p204.

Johnson, H.L. & Harris, D.L., 1954. *Astrophys. J.*, **120**, 196.

Johnson, H.L., Mitchell, R.I., Iriarte, B. & Wisniewski, W.Z., 1966. *Comm. Lunar & Planet. Lab.*, **4**, 99.

Landolt, A.U., 1973. *Astron. J.*, **78**, 959.

Landolt, A.U., 1983. *Astron. J.*, **88**, 439.

Menzies, J.W., Cousins, A.W.J., Banfield, R.M., & Laing, J.D., 1990. *SAAO Circulars*, No. 13, 1.

Menzies, J.W., Marang, F., Laing, J.D., Coulson, I.M. & Engelbrecht, C.A., 1990. *Mon. Not. R. astr. Soc.*, **248**, 642.

Discussion

E. Budding: *If you fold theoretical flux distribution (e.g. Kurucz) with available transmission functions you ought to be able to check how the measured discrepancies compare with theoretical ones. Has this been done — and, for example, can your empirically derived transformation equation be theoretically substantiated?*

Menzies: We used several different photomultipliers and filter sets in the course of our work. For each combination we derived linear transformation equations and non-linear corrections, but did not measure tube response functions or filter pass bands. What you ask ought to be possible in principle but we don't have the information needed to do it.

J.D. Fernie: *If one wants to do Cousins RI photometry from the north, what is the best source of (secondary) standards?*

Menzies: You should see the paper by Cousins, 1984, SAAO Circulars No. 8, p59, for a discussion on this point. You could use the data in the paper I have been talking about, although you should be careful with the reddest stars.

K. Oláh: *It seems everybody is happy with the equatorial standards you have as an extension of the system to the northern hemisphere. However, as my observatory is situated at about 48° N, the standard star measurements are quite complicated and, because they are so low in the sky, an additional error is introduced into the standardization. Do you plan to extend your otherwise really well defined, excellent system further north? It would be very important for many observatories, I think.*

Menzies: For us to set up standards north of the equator would be almost as difficult as for you to observe standards on the equator. It would be an ideal project for a small telescope at a good site in N. America, and would take only about a year to do. It would certainly be worth the effort.

T. Oja: *Are your observations of the equatorial stars available?*

Menzies: Yes, the reference is: J W Menzies et al, 1991. MNRAS, **248**, 642.

M.J. Bessell: *It would be very helpful for attempts to reproduce theoretically the $UBV(RI)_c$ system, and attempts to match your natural system, were you to publish your transformation techniques and transformations equations as well as the filter transmissions.*

Menzies: I agree that it is worth having such information in print. I shall investigate the feasibility of publishing it when I return to Cape Town.

A.J. Penny: *There is a distinct lack of metal-poor standards. I would urge that the list of $UBV(RI)_c$, standards is extended to include a large number of such stars.*

Menzies: We will consider this.

Multicolour Photometric System for Investigation of the Galaxy Population

V. Straižys

Institute of Theoretical Physics and Astronomy, Goštauto 12, Vilnius 2600, Lithuania

Abstract

Many photometric systems for classification of stars have been proposed. However, not all of them are equally suitable for investigation of distant and reddened stars without additional information from their spectra. Earlier it was shown that the Vilnius photometric system is optimum for purely photometric classification of stars in spectral types, luminosities, metallicities and peculiarity types, when interstellar reddening is present. When realized with a CCD detector, the system permits to classify stars down to 17 mag with a 1.5 m class telescope or down to 20 mag with a 4 m telescope. A possibility of realizing the system with the WF camera of the Hubble Space Telescope is investigated. The problem of unification of broad-band and medium-band photometric systems is discussed.

For population study of the Galaxy in its different subsystems, very faint and reddened stars have to be classified in spectral types, luminosities and metallicities. For this one must use a sufficiently sensitive area detector and the optimum photometric system. CCD detectors are what we need. Another question: which photometric system is to be used. To be effective, such system must have bandpasses as broad as possible, the number of bandpasses must be minimum and the system must be able to classify stars in a galactic field, where objects of different temperatures, luminosity classes, populations, peculiarities and interstellar reddenings are mixed.

The optimum bandpasses of the system for two- and three-dimensional classification of stars were selected at the Vilnius Observatory in 1962–1965. This process was based on the energy distributions in the spectra of stars of different spectral types and on the interstellar extinction law. It was found that for photometric classification of stars of all spectral types, the system must include seven bandpasses at 345, 374, 405, 466, 516, 544 and 656 nm with half-widths of the order of 20 nm. The eighth bandpass at 625 nm was also found to be useful (but not essential) for identification of M-type stars. Detailed justification of each bandpass is described in our earlier papers and in the Russian and English versions of my book "Multicolor Stellar Photometry" (Straižys, 1977, 1992). Here the functions of every bandpass will be explained briefly.

The bandpass at 345 nm measures the continuum intensity beyond the Balmer jump. The bandpass at 374 nm measures the common absorption by higher members of the Balmer series and is very sensitive to luminosity of early-type stars. The bandpass at 405 nm measures the continuum intensity before the Balmer jump for early-type stars and the metallic-line blanketing for later-type stars. The bandpass at 466 nm is placed near the break-point of the interstellar reddening law and is very important in distinguishing the temperature and the interstellar reddening. The bandpass at 516 nm measures the absorption intensity of Mg II triplet lines together with the MgH band. This bandpass is very important as a discriminator of luminosities of G5–K–M stars. The bandpass at 544 nm measures the continuum at the mean wavelength of the V bandpass of the BV system. It helps to transfer directly the magnitude scale from BV to the Vilnius system. The bandpass at 656 nm is placed on the Hα line measuring its absorption or emission intensity for early-type stars or the pseudocontinuum for late-type stars. The bandpass at 625 nm measures the absorption intensity of the TiO band and is useful for recognition of M-type stars. Not all bandpasses are always necessary. For two-dimensional classification of early-type stars the bandpasses at 345, 374, 405, 466 and 516 (or 544) nm are sufficient. For two-dimensional classification of late-type stars only the bandpasses at 405, 466, 516, 625 and 656 may be used. If we know which stars are of spectral class M, the bandpass at 625 nm may be omitted.

The system makes it possible not only to classify normal stars with the accuracy of ± 1 spectral subclass and to determine M_V with the accuracy of ± 0.5 mag but also to recognize Be, Am, Ap stars, subdwarfs, metal-deficient giants, carbon-rich stars, white dwarfs, horizontal-branch stars, T Tauri and Herbig Ae/Be stars, as well as many combinations of unresolved binaries.

The Vilnius photometric system has been so far successfully applied to investigate fields of the Galaxy, open and globular clusters, star forming regions, different types of peculiar stars, etc. More than 7000 stars have been observed photoelectrically down to $V = 14$ mag (Straižys et al., 1989; Straižys and Kazlauskas, 1992). The system also was realized with a CCD camera on the 90 cm telescope of the Kitt Peak Observatory (Boyle et al., 1990a,b, 1991; Smriglio et al., 1991). Using the exposure times of the order of 5 min for five filters in the visible and about 20–30 min in the ultraviolet we were able to classify stars down to 17 mag. Even fainter stars can be reached, when using more transparent interference filters and increasing exposure time. It seems that there is no difficulty to accomplish two- and three-dimensional classification of stars down to 18 mag with a 1.5 meter class telescope or down to 20 mag with a 4 meter telescope, with exposure times of the order of 1 hour in the ultraviolet. Exposures in all 7 colours for one field can be obtained during one night.

Very faint stars in crowded regions of the Milky Way can be observed from outside the atmosphere. The Wide-Field Camera of the Hubble Space Telescope contains filters which have their mean wavelengths at 336, 375, 413, 469, 517, 547 and 656 nm, i.e. very close to the mean wavelengths of the bandpasses of the Vilnius photometric system. However, four of them at 375, 469, 517 and 656 nm are much narrower. This HST system gives a possibility to classify stars in spectral and luminosity classes, as it is shown by Straižys and Valiauga (1992) using reddening-free Q, Q diagrams with different combinations of bandpasses. Of course, we do not recommend to use the

HST with the present Wide-Field Camera for stellar classification but we show that this is possible in principle and may be useful in the future.

Speaking about the HST, we must remember that in the first WFC all existing photometric systems have been neglected. This prevents using the ground-based standards of the systems and their calibration. Surely, there is no necessity to place into the HST the filters of all ground-based photometric systems. However, it would be important having there the filters at least of two systems: a "*UBVRI*-like" system and a medium-band system suitable for photometric classification of stars. For this I suggest to form a working group with the representatives of all most widely used photometric systems and to decide which filters are necessary for optimum photometric classification both from the ground and from the space. Introduction of such a system would help to make economy of the observing time and to increase our knowledge about the Galaxy.

The first step in this direction has been done by the VilGen system (Straižys et al., 1982, North et al., 1982) which combines the best properties of the Vilnius and Geneva systems. A revision of the *UBV* system proposed by Straižys (1973, 1983) was also realized, and 13600 stars down to $V = 7.2$ have been measured in the *WBVR* system (Khaliullin et al., 1985; Kornilov, 1992). The magnitude W, replacing the ill-defined Johnson's U magnitude, has an exactly known response function without red leak, and the $W - B$ indices are correctly transformed outside the atmosphere with the extinction coefficient dependent on energy distribution functions of stars (Moshkalev and Khaliullin, 1985). Realization of the *WBVR* system on the HST would be one of the first steps towards the unification of photometric systems. Such systems (both broad- and medium-band ones) must be defined by exact response functions, not by a list of standard stars as it was the case with the *UBV*, *RI* and *JHKLMN* systems.

References:

Boyle, R.P., Smriglio, F., Nandy, K. and Straižys, V. 1990a, A&AS, 84, 1.

Boyle, R.P., Smriglio, F., Nandy, K. and Straižys, V. 1990b, A&AS, 86, 395.

Boyle, R.P., Dasgupta, A.K., Smriglio, F., Straižys, V. and Nandy, K. 1991, A&AS, in press.

Khaliullin, Kh., Mironov, A.V. and Moshkalev, V.G. 1985, Ap&SS, 111, 291.

Kornilov, V.G. 1992, in Stellar Photometry, IAU Coll. No. 136, Book of Abstracts, Dublin.

Moshkalev, V.G. and Khaliullin, Kh.F. 1985, AZh, 62, 393 = Soviet Astron., 29, 227.

North, P., Hauck, B. and Straižys, V. 1982. A&A, 108, 373.

Smriglio, F., Nandy, K., Boyle, R.P., Dasgupta, A.K., Straižys, V. and Janulis, R. 1991, A&AS, 88, 87.

Straižys, V. 1973, A&A, 28, 349.

Straižys, V. 1977. Multicolor Stellar Photometry, Mokslas Publishers, Vilnius, Lithuania.

Straižys, V. 1983, Bull. Inform. CDS, Strasbourg, No. 25, 41.

Straižys, V. 1992. Multicolor Stellar Photometry, Pachart Publ. House, Tucson, Arizona.

Straižys, V., Jodinskienė, E. and Hauck, B. 1982. Bull. Vilnius Obs., No. 60, 50.

Straižys, V., Kazlauskas, A., Jodinskienė, E. and Bartkevičius, A. 1989. Results of Photoelectric Photometry of Stars in the Vilnius Photometric System (magnetic tape). Bull. Inform. CDS, Strasbourg, No. 37, 179.

Straižys, V. and Kazlauskas, A. 1992. Results of Photoelectric Photometry of Stars in the Vilnius Photometric System (magnetic tape).

Straižys, V. and Valiauga, G. 1992, Baltic Astron., 1, No. 4 (in press).

Discussion

A.J. Penny: *A limitation on using the HST for n-dimensional classification may be the low accuracy.*

Straizys: Of course the accuracy of photometry obtained with the WF camera of the HST must be of the same order as for the ground-based photometry if we want to classify very faint stars in two or three dimensions. Let us hope that the present problems with low accuracy of the WF camera will be overcome.

Vilnius Photometry in the Southern Hemisphere

R.J. Dodd[1], M.C. Forbes[2], D.J. Sullivan[2]

[1] *Carter Observatory, P.O. Box 2909, Wellington, New Zealand.*
[2] *Victoria University of Wellington, New Zealand.*

Abstract

The seven intermediate passband filters of the Vilnius photometric system enable a more comprehensive and discriminating classification of stellar objects than is possible in broadband systems. To date, a lack of suitable standard stars has inhibited its use in the Southern Hemisphere. We have an established programme for extending the network of northern Vilnius standards into the southern skies. This paper briefly summarizes the characteristics of the Vilnius system, provides a synopsis of the data obtained on over 200 standard stars so far and presents preliminary CM and CC diagrams for the open cluster Omicron Velorum (IC2391).

1. Introduction

A programme of Vilnius intermediate band stellar photometry for the southern hemisphere was first mooted at the Royal Observatory Edinburgh in 1985. A suitable set of filters was kindly provided by Professor V. Straižys (Institute of Theoretical Physics and Astronomy, Vilnius, Lithuania) and observations begun by Sullivan and Dodd in 1988. The initial goal of the observing programme was to establish standards both near the South Celestial Pole and for selected bright stars ($V < 7$) across the entire sky south of $-20°$. The rate of data gathering increased markedly when Forbes (Ph.D. research student) joined the group and to date (August 1992) some 109 Vilnius primary standards, 225 southern hemisphere secondary standards and 90 cluster stars have been observed at least once.

2. Instrumentation and Observations

The passbands of the New Zealand set of Vilnius filters are shown in figure 1 superimposed on the spectra for an early (A) and a late (M) type star, obtained from the spectrophotometric atlas of Gunn & Stryker (1983). The rationale for using this particular set of filters is given in detail by Straižys (1992). The Vilnius photometry allows the determination of spectral type, luminosity class, chemical composition parameter and interstellar reddening.

Observations have been made using the 61cm telescopes at the Mt. John Observatory of the University of Canterbury near Lake Tekapo in New Zealand. The thermoelectrically cooled photometer uses an EMI 9558B tube operated at 1700v. The most commonly used photometer aperture was 32 arcsec diameter. The form

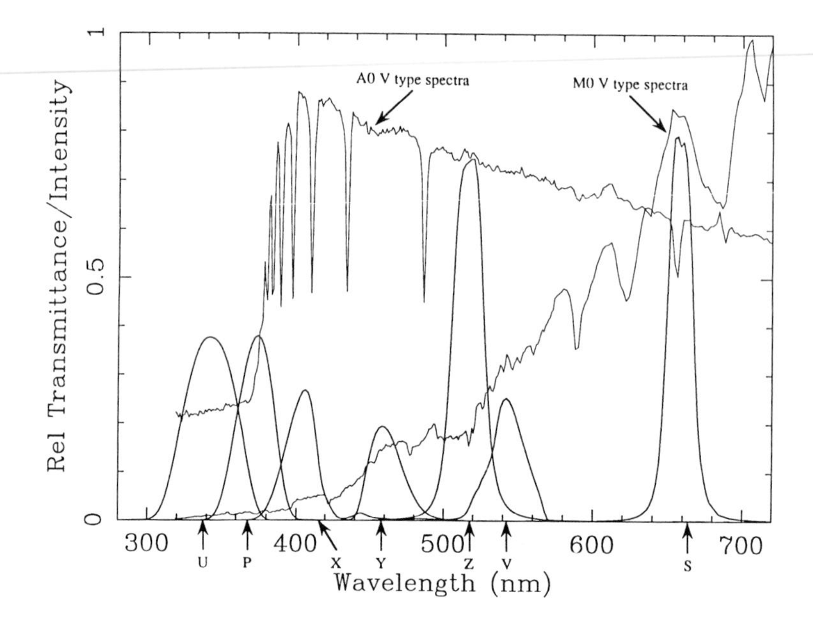

Figure 1: Southern Hemisphere Vilnius filter set passbands superimposed on an A and M type stellar spectra.

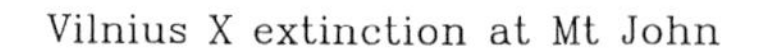

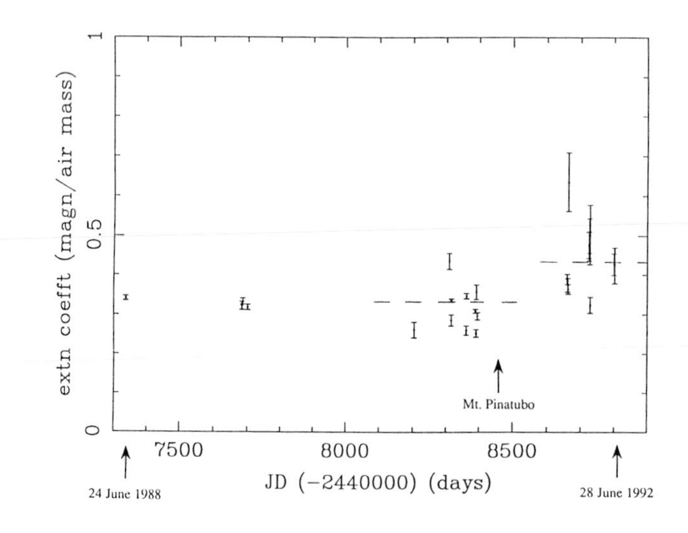

Figure 2: Vilnius X-band extinction at the Mt. John Observatory 1988-92.

of observing followed that set out by Straižys (1992) so extinction and control stars were observed in addition to standard and programme stars. The filter observing sequence being :

(UPXYZVS)$_*$ check star centering (UPXYZVS)$_*$ (UPXYZVS)$_{sky}$.

Integration times varied from 5 to 100 seconds, with three integrations per colour for integrations up to 20s and two integrations per colour for longer integrations. For the longer integrations, image centering was checked more frequently than every filter set run. The preliminary reduction of the observations presented in this paper used the method described by Hardie (1962). A more comprehensive reduction of the data is in progress, including removal of colour and time–dependent extinction effects.

To obtain a measure of the reliability of the southern standards and to tie the southern to the northern hemisphere standards, a series of observations of selected northern standards, 109 stars listed in the various Vilnius catalogues in all, was carried out at Mt. John. The relationships between southern (S) and northern (N) measurements of magnitude and colour for two typical nights (pre and post Mt. Pinatubo eruption) were as follows :

1989 June 8	1992 June 27
$V_S = (0.971 \pm 0.009)V_N - (1.075 \pm 0.062)$	$V_S = (0.993 \pm 0.012)V_N - (1.114 \pm 0.084)$
$(U{-}P)_S = (1.037 \pm 0.027)(U{-}P)_N + (0.409 \pm 0.007)$	$(U{-}P)_S = (0.840 \pm 0.141)(U{-}P)_N + (0.518 \pm 0.024)$
$(P{-}X)_S = (0.963 \pm 0.020)(P{-}X)_N + (0.266 \pm 0.012)$	$(P{-}X)_S = (0.936 \pm 0.050)(P{-}X)_N + (0.544 \pm 0.029)$
$(X{-}Y)_S = (1.089 \pm 0.007)(X{-}Y)_N + (0.520 \pm 0.004)$	$(X{-}Y)_S = (1.043 \pm 0.019)(X{-}Y)_N + (0.748 \pm 0.015)$
$(Y{-}Z)_S = (0.847 \pm 0.012)(Y{-}Z)_N - (0.357 \pm 0.010)$	$(Y{-}Z)_S = (0.769 \pm 0.025)(Y{-}Z)_N - (0.369 \pm 0.026)$
$(Z{-}V)_S = (1.108 \pm 0.017)(Z{-}V)_N + (1.516 \pm 0.021)$	$(Z{-}V)_S = (1.152 \pm 0.041)(Z{-}V)_N + (1.640 \pm 0.048)$
$(V{-}S)_S = (0.886 \pm 0.011)(V{-}S)_N + (0.816 \pm 0.006)$	$(V{-}S)_S = (0.868 \pm 0.015)(V{-}S)_N + (0.678 \pm 0.005)$

Using these equations, which were determined from 12 standard stars, the standard deviations about the mean values of magnitude ($V_N - V_S$) and colours ($C_N - C_S$) were as follows :

	$\sigma(\Delta V)$	$\sigma(\Delta(U-P))$	$\sigma(\Delta(P-X))$	$\sigma(\Delta(X-Y))$	$\sigma(\Delta(Y-Z))$	$\sigma(\Delta(Z-V))$	$\sigma(\Delta(V-S))$
1989 June 8	±0.034	±0.028	±0.035	±0.019	±0.014	±0.009	±0.019
1992 June 27	±0.039	±0.047	±0.046	±0.042	±0.019	±0.018	±0.016

The standard deviations about the mean values of the magnitudes and colours in the Mt. John system only (i.e. internal comparison between measures of the same star made on the same night corrected for first order extinction) are typically smaller than $\pm 0\overset{m}{.}01$. The eruption of Mt. Pinatubo in the Philipines in June 1991 increased the average extinction at Mt. John by some $0\overset{m}{.}1$ in the X passband (see figure 2). Also evident is the increased scatter in the measurements of atmospheric extinction after the eruption.

The presence of a colour term is well illustrated in figure 3 which shows the residuals in the V magnitude plotted against the composite (Y–V) colour. The pairs of crosses are for the same standard star observed at different times.

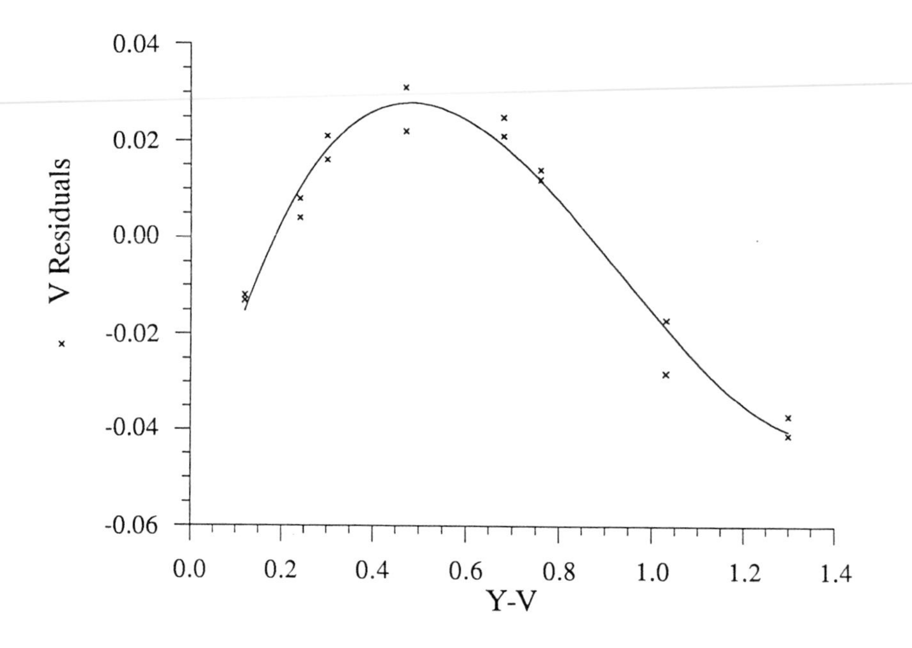

Figure 3: Colour dependence of standard star residuals 8 June 1989.

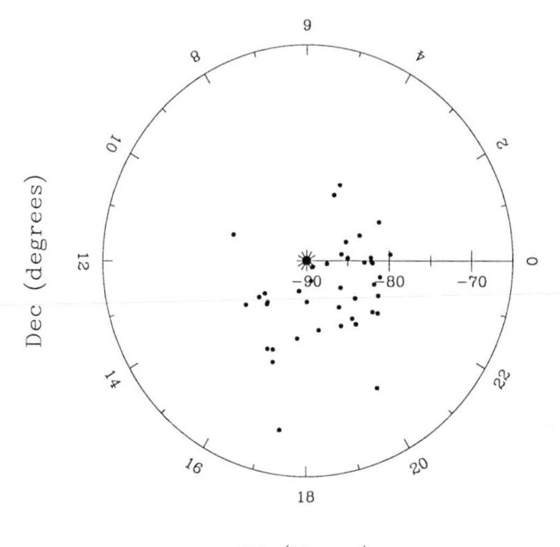

Figure 4: South celestial pole Vilnius secondary standards.

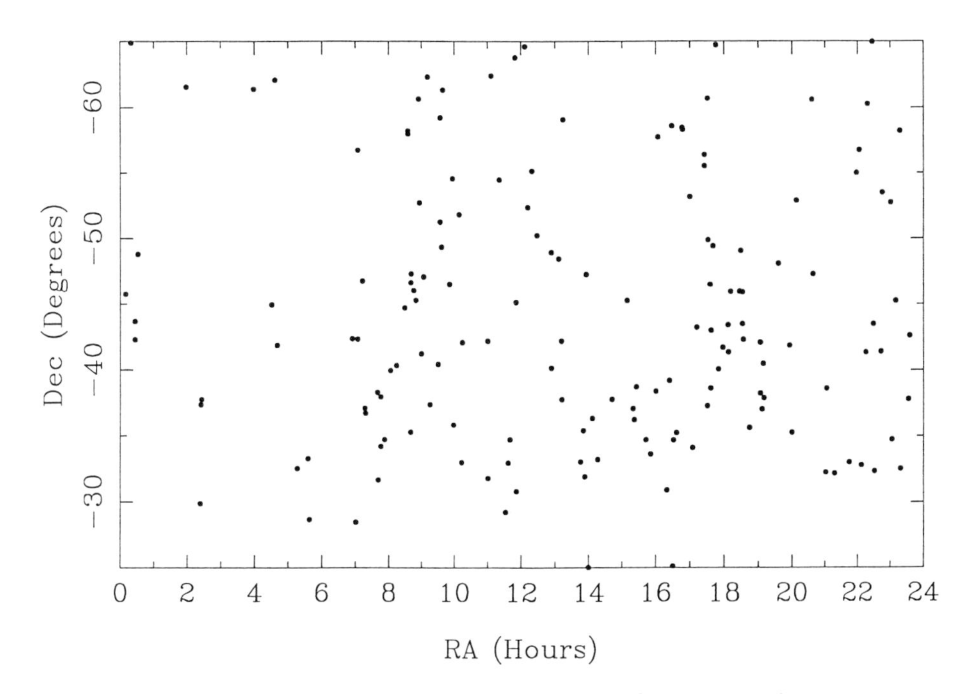

Figure 5: Vilnius secondary standards between $\delta = -20°$ and $-70°$.

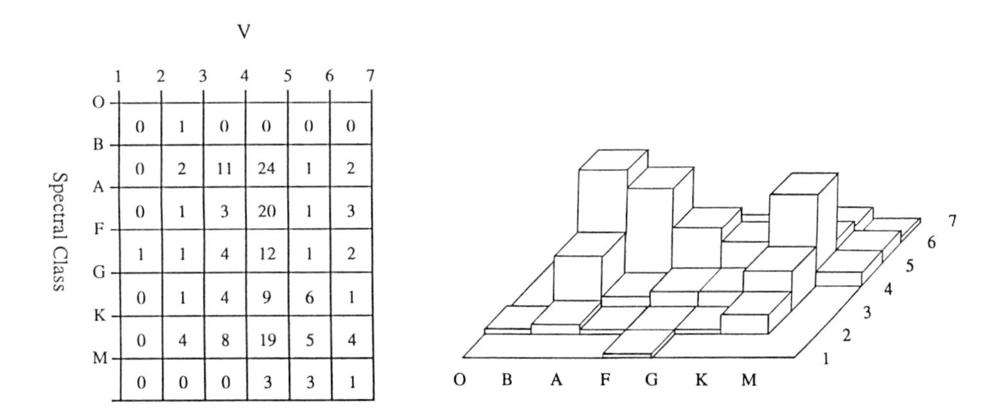

Figure 6: Magnitude - Spectral type histogram for Vilnius southern hemisphere secondary standards.

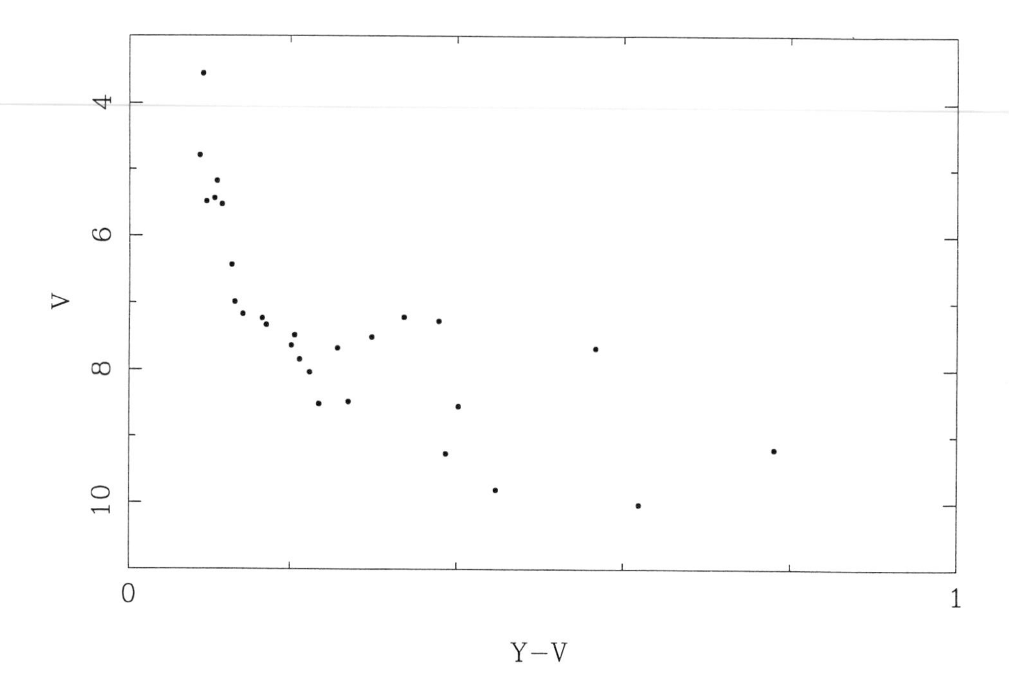

Figure 7: CMD for O Vel galactic cluster.

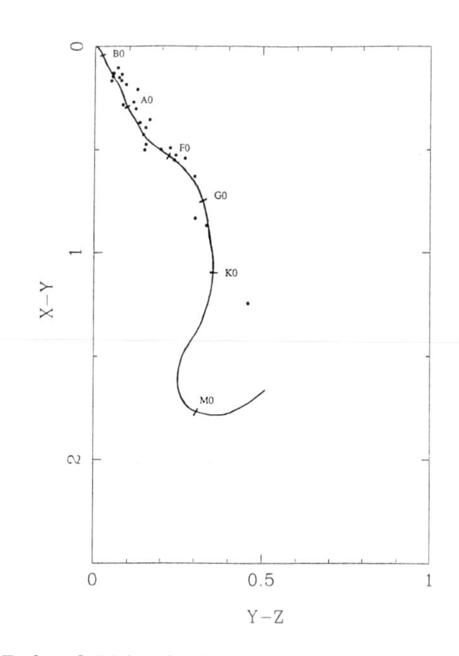

Figure 8: CCD for O Vel galactic cluster with superimposed ZAMS.

3. Southern Hemisphere Secondary Standards

In all some 225 stars have been observed so far in setting up a southern hemisphere network of secondary standards. Figure 4 shows the distribution of stars observed near the south celestial pole and figure 5 those between declinations −20° and −70°. The stars range in magnitude from V = 1.90 to 6.90 with spectral types from O5 to M3 and luminosity classes Ia to V. Figure 6 shows a two dimensional plot of numbers of stars in particular magnitude and spectral ranges. The next phase of the southern hemisphere standards programme will be to measure the magnitudes and colours of selected E–region stars.

4. Star Cluster Programme

To date observations have been made of 8 stars in M41, 39 in Kappa Crucis and 43 in IC2391. Figure 7 shows the CM diagram (V vs (Y–V)) for the nearby, young cluster *O* Velorum (IC2391) so far obtained and figure 8 shows a (X–Y),(Y–Z) CC plot for the same cluster. A ZAMS is superimposed on the CC diagram. More observations using both photoelectric and CCD photometry have yet to be made of these and other galactic clusters.

5. Acknowledgements

We wish to thank the Foundation for Research, Science and Technology and the V.U.W. Internal Research Committee in New Zealand for partial funding of the project; the Physics and Astronomy Department of the University of Canterbury for generous allocation of telescope time at Mt. John Observatory; and the Institute of Theoretical Physics and Astronomy, Vilnius, Lithuania for encouragement and help.

References:

Hardie R.H., 1962, Photoelectric reductions. In *Astronomical Techniques*, W. A. Hiltner, editor, volume 2 of *Stars and Stellar Systems*. The University of Chicago Press, Chicago, Illinois, U.S.A.

Straižys V., 1992, *Multicolor Stellar Photometry*, volume 15 of *Astronomy and Astrophysics Series*. Pachart Publishing House, 1130 San Lucas Circle, P.O. Box 35549, Tucson, Arizona 85740, 1992.

Gunn, J.E. and Stryker, L.L., 1983, *Astrophys. J. Suppl.* **52**, 121.

Discussion

W. Wisniewski: *How do you handle extinction? Vilnius filters are narrow and there are seven of them. With longer exposures between the first and last filter, there will be a large airmass difference.*

Sullivan: UPXYZVS observations are made of extinction stars approximately hourly. For the longest integrations (100s), the time required for a full set of observations is approximately 12 minutes. Unless the star is at a large zenith distance, the change in airmass through a single observation is small but in any case, it is allowed for by correcting each colour integration separately for extinction.

W. Tobin: *Are the stars you're observing standards in other photometric systems, so that you know that they aren't variable?*

Sullivan: Generally the standards so far observed are taken from the Bright Star Catalogue and have been observed in other photometric systems. We deliberately avoided those stars in the Catalogue denoted as variables. In addition, we cross-checked the selected stars using the SIMBAD data base.

R.F. Garrison: *It would make transformations between systems a lot easier if you could use the E region standards as your standards. If the Vilnius system can be used to observe 17th magnitude stars with a 1.5m telescope, it should be no problem to reach 10th magnitude in the E regions! Why set up a whole new and different system of stars?*

Sullivan: We agree and will begin measuring E region standards later this year. We felt that in order to get the programme underway we should start with a selection of bright stars, given the vagaries of the New Zealand weather and our likely access to telescopes of a given aperture.

V. Straižys: *What is the magnitude and spectral type range of the Vilnius system standards in the Southern Hemisphere?*

Sullivan: The magnitude range is $1.90 < V < 6.90$ with spectral types from O5 to M3.

Absolute Spectrally Continuous Stellar Irradiance Calibration in the Infrared

Martin Cohen[1,2], R.G. Walker[2], M.J. Barlow[3], J.R. Deacon[3], F.C. Witteborn[4], D.F. Carbon[5], G.C. Augason[6]

[1]*Radio Astronomy Laboratory, University of California, Berkeley* [2]*Jamieson Science and Engineering, Inc.* [3]*Department of Physics and Astronomy, University College London* [4]*Astrophysics Branch, NASA-Ames Research Center* [5]*NAS Systems Development Branch, NASA-Ames Research Center* [6]*Space Instrumentation and Studies Branch, NASA-Ames Research Center*

Abstract

We present first efforts to establish a network of absolutely calibrated continuous infrared spectra of standard stars across the 1–35μm range in order to calibrate arbitrary broad and narrow passbands and low-resolution spectrometers from ground-based, airborne, balloon, and satellite-borne sensors. The value to photometry of such calibrated continuous spectra is that one can integrate arbitrary filters over the spectra and derive the stellar in-band flux, monochromatic flux density, and hence the magnitude, for any site. This work is based on new models of Sirius and Vega by Kurucz which were calculated by him, for the first time, with realistic stellar metallicities and a customized finely-gridded infrared wavelength scale. We have absolutely calibrated these two spectra and have calculated monochromatic flux densities for both stars, and isophotal wavelengths, for a number of infrared filters. Preliminarily, the current IRAS point source flux calibration is too high by 2, 6, 3, and 12% at 12, 25, 60, and 100μm, respectively.

1. Introduction

In his critical review of the optical absolute calibration of Vega, Hayes (1985) states of the corresponding situation in the infrared: "The calibration of the IR, and the availability of secondary standard stars in the IR, is yet immature, and I recommend more effort...". Unfortunately, infrared astronomical calibration has been developed from the completely erroneous assumption that normal stars can be represented by Planck functions at their effective temperatures (although local fits to some blackbody in a restricted region may be an adequate approximation for some purposes). Recently, Cohen *et al.* (1992a) have demonstrated from ratios of cool stellar spectra to that of Sirius that even early K-type stars such as α Boo are far from featureless blackbodies. In order to develop spectrally continuous absolute standards in the infrared, Cohen, Walker, and Witteborn (1992) have devised a technique for splicing together absolutely calibrated versions of existing spectral fragments and have demonstrated the method by producing a complete 1.2–35μm absolutely calibrated spectrum of α Tau. Their method depends in part upon correct normalization of spectral fragments in accordance with infrared stellar photometry. We summarize the independent effort on broadband infrared calibration that supports this spectral calibration scheme.

2. New spectra of Sirius and Vega by Kurucz

Both these A dwarf stars are sufficiently hot that molecules could not survive in their atmospheres and both have been modeled in the past (Kurucz 1979; Dreiling and Bell 1980; Bell and Dreiling 1981). What distinguishes our latest Kurucz (1991) models from all previous efforts are the metallicities inherent in Kurucz's new work. After critical examination of detailed high resolution ultraviolet and visible spectra of Vega, Kurucz finds definite support for the idea that Vega has less than solar metallicity. Sirius, because of mass transfer from its companion, is metal-rich compared with the sun (Latham 1970). It is the presence of dust around Vega and the greater brightness of Sirius that renders the latter a more desirable standard for infrared work. Consequently, we have chosen to work with both Vega – the canonical standard at UV-optical wavelengths – and Sirius.

3. Calibration of the new spectra

We normalize the Vega model to Hayes' 5556A weighted average monochromatic measurement, and integrate it through a variety of infrared filters using transmission profiles taken at their operating temperatures, and detailed model calculations for terrestrial atmospheric transmission. We take Vega as zero magnitude at all wavelengths up to 20μm and use existing infrared photometry differentially to scale the new Sirius model absolutely, to provide the calibration past 20μm because of Vega's dust shell. This effectively determines the angular diameter of Sirius to be 6.04 mas, well within the 1σ uncertainty (0.16 mas) of the Hanbury Brown, Davis, and Allen (1974) measurement (5.89 mas). We hope that the stellar interferometry programme, described in Session 2 of this meeting by Shobbrook, will result in a more precise measurement of Sirius' diameter than the 1974 one. Figure 1 presents the calibrated Vega spectrum in the form of a $\lambda^4 F_\lambda$ plot and compares the model with absolute measurements attempted from mountaintops. Note that nowhere do these deviate by more than 2σ from our calibrated model spectrum.

Using detailed terrestrial atmospheric codes supported by the 1991 release of the HITRAN database (Rothman *et al.* 1987), we calculate isophotal wavelengths and monochromatic flux densities for both stars from different ground-based sites. Details of these flux density calibrations and the calibrated hot stellar spectra appear in Cohen *et al.* (1992b).

Another application of these calibrated spectra is to the IRAS point source flux density calibration. This calibration is clearly of interest given the wealth of data provided by that satellite. Until we have constructed a detailed spectrum for every star in Table VI.C.3 of the IRAS Explanatory Supplement (1988: pg.VI-19*ff*) we cannot definitively address the IRAS point source calibration. However, direct comparisons are possible with Vega and Sirius from which Cohen *et al.* (1992b) conclude that the current IRAS absolute calibration is too high by 2.4, 6.5, 2.9, and 11.6% in the four wavebands, respectively. These estimates could be refined by a more rigorous explanation of the procedures actually carried out in the calibration of IRAS.

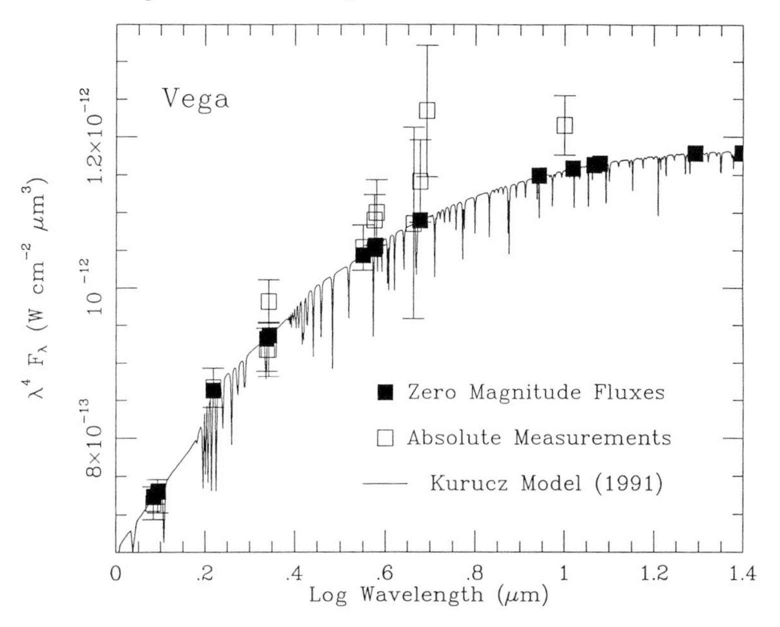

Figure 1 The calibrated spectrum of Vega based on the new Kurucz model. Filled squares indicate our "zero magnitude flux calibrations" for the UKIRT filter set. Open squares signify absolute measurements of Vega that were attempted from mountaintops.

4. Construction of complete continuous calibrated spectra

Below we define the several salient procedures undergone in order to construct an absolutely calibrated infrared spectrum of a given star.

a) Define flux density calibrations only for filters with known cold-scanned transmission profiles using terrestrial atmospheric models based on the updated 1991 release of the "HITRAN" database.

b) Cull photometry with errors from the literature only if magnitudes are given for standard stars (initially only Vega and Sirius were acceptable; after constructing our own spectra of other standard stars, however, we were able to extend this acceptance to literature that designated system magnitudes for α Tau, β Peg, and α Boo).

c) Locate and examine all relevant spectral fragments with known calibration pedigree (clearly, we still prefer the ratio of a star to Sirius's spectrum, when available).

d) Normalize fragments with respect to photometric bands that lie wholly within the wavelength range covered by the fragments.

e) Single-sidedly splice one spectrum to an overlapping spectrum.

f) Double-sidedly splice to two flanking spectra, each overlapping the fragment.

All such splicing operations are implemented by χ^2 minimization techniques over the wavelength regions common to overlapping spectral fragments. These techniques result in the best-fit rescaling of one fragment with respect to another, and generate a bias (a wavelength-independent uncertainty) from the process that is incorporated (in a root-sum-square sense) into the existing wavelength-dependent uncertainties.

Figure 2 compares the complete spectrum of α Tau with a simple continuum approximation with the dominant opacity source as H^- free-free absorption (Engelke 1990).

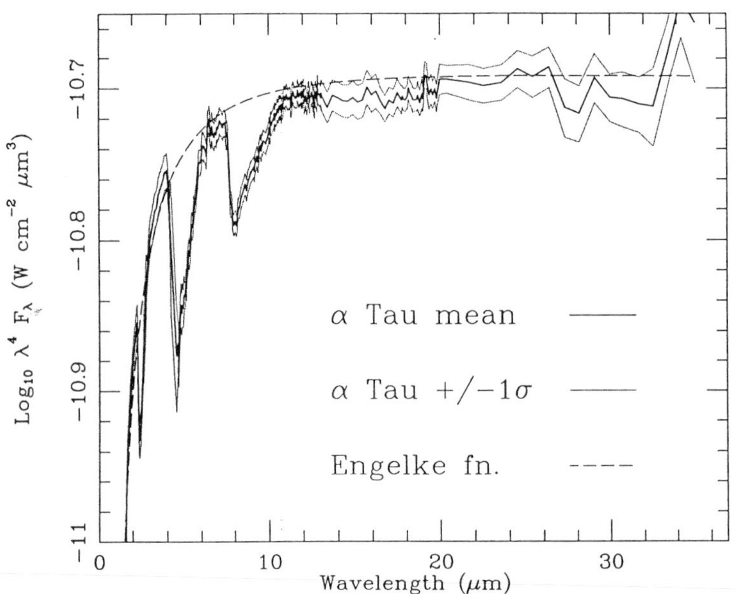

Figure 2 The complete spectrum of α Tau built according to the procedures just described. Note the absorptions of CO (1st overtone), CO (fundamental), and SiO (fundamental), from left to right across the spectrum. The 1σ error bounds are plotted. These include the 1.45% uncertainty inherent in the Hayes' determination of Vega's visual flux: this component underpins all other absolute uncertainties in this method of spectral assembly.

5. Tables of magnitudes based on Vega and Sirius

Infrared astronomy suffers from a plethora of filters, none of them truly standard, and an apparent reluctance on the part of observatories to publish the filter transmission profiles *taken at their actual cryogenic operating temperature.* Attempts have been made to relate magnitudes in these different filters to one another (e.g., see the paper by Leggett, Oswalt, and Smith in Session 1 of this meeting; Bessell

Table 1. Near-infrared magnitudes of calibration stars derived from our continuous spectra. Upper stars have calibrated Kurucz models; lower stars represent our complete, calibrated, observed spectra.

Star	Jn	Kn	Ln	J	H	K	L	L'	M
α Lyr	0.00	0.00	0.00	0.00	0.00	0.00	0.00	0.00	0.00
α CMa	-1.39	-1.37	-1.36	-1.39	-1.40	-1.37	-1.36	-1.36	-1.36
β And	-0.89	-1.94	-2.06	-0.79	-1.74	-1.90	-2.03	-2.04	-1.78
α Tau	-1.97	-2.93	-3.06	-1.84	-2.74	-2.90	-3.04	-3.05	-2.77
β Peg	-1.18	-2.33	-2.49	-1.09	-2.09	-2.29	-2.45	-2.47	-2.20
α Boo	-2.27	-3.08	-3.17	-2.14	-2.96	-3.05	-3.16	-3.16	-2.91

Table 2. Mid-infrared magnitudes of calibration stars derived from our continuous spectra. Upper stars have calibrated Kurucz models; lower stars represent our complete, calibrated, observed spectra.

Star	8.7	N	11.7	Q	IRAS12	IRAS25
α Lyr	0.00	0.00	0.00	0.00	0.00	...
α CMa	-1.35	-1.35	-1.35	-1.34	-1.35	-1.34
β And	-1.96	-2.04	-2.11	-2.11	-2.09	-2.11
α Tau	-2.95	-3.02	-3.07	-3.08	-3.08	-3.10
β Peg	-2.37	-2.44	-2.49	-2.51	-2.48	-2.51
α Boo	-3.09	-3.14	-3.17	-3.17	-3.17	-3.16

and Brett 1988). Young, Milone, and Stagg (Session 4, this meeting) will present an improved set of infrared passbands which it is hoped the worldwide infrared community will adopt. These new passbands have several merits, such as the ability to extrapolate more readily to zero airmass, and reduced sensitivity of isophotal wavelength on altitude of observing site. Having constructed absolutely calibrated stellar spectra based upon best-fitting pre-existing photometry, we can now invert the process and derive refined photometry for these stars by integrating the products of filter-plus-atmospheric profiles over the calibrated spectra. We present the infrared magnitudes of stars for which we have either calibrated Kurucz models, or have already constructed complete spectra. In Tables 1 and 2, we present magnitudes for the three narrowbands defined by Selby *et al.* (1988:Jn,Kn,Ln), the UKIRT filter set (JHKLL'M,[8.7],N,[11.7],Q), and IRAS (12,25). Of course, there is always still more work to be done to extend these bright standards, suitable for spectroscopy, to much fainter standards appropriate to two-dimensional infrared arrays (see the papers by Leggett *et al.* in Session 1, and by Glass in Session 3). We would be happy to provide isophotal wavelengths and corresponding magnitudes for stars using the new passbands proposed by the Working Group on IR Filters of IAU Commission 25 (see the paper by Young *et al.* in Session 4), as well as the fundamental monochromatic or in-band quantities from our calibrated Vega spectrum.

References:

Bell, R.A., and Dreiling, L.A. 1981, Astrophys. J., **248**, 1031.

Bessell, M.S., and Brett, J.M. 1988, PASP, **100**, 1134.

Cohen, M., Walker, R.G., and Witteborn, F.C. 1992, Astron.J., in press [CWW]

Cohen, M., Walker, R.G., Barlow, M.J., and Deacon, J.R. 1992b, Astron.J., in press

Cohen, M., Witteborn, F.C., Carbon, D.F., Augason, G.C., Wooden, D.H., Bregman, J., and Goorvitch, D. 1992a, submitted to Astron. J.

Dreiling, L.A., and Bell, R.A. 1980, Astrophys. J., **241**, 737.

Engelke, C.W. 1990, in: *Long Wavelength Infrared Calibration: Infrared Spectral Curves for 30 Standard Stars* (Report of Group 51, Lincoln Labs., MIT)

Hanbury Brown, R., Davis, J., and Allen, L.R. 1974, Mon. Not. R. Astron. Soc., **167**, 121.

Hayes, D. S. 1985, in Proc. IAU Symposium 111: *Calibration of Fundamental Stellar Quantities*, eds. D.S. Hayes, L.E. Pasinetti and A.G. Davis Philip, D. Reidel, Dordrecht, Holland, p225.

IRAS Explanatory Supplement 1988, *Catalogs and Atlases. Volume 1*, NASA RP-1190.

Kurucz, R.L. 1979, Astrophys. J. Suppl., **40**, 1.

Kurucz, R.L. 1991, private communication.

Latham, D.W. 1970, Ph.D.dissertation, Harvard University.

Rothman, L.S. *et al.* 1987, Appl. Optics, **26**, 4058.

Selby, M.J., Hepburn, I., Blackwell, D.E., Booth, A.J., Haddock, D.J,, Arribas, S., Leggett, S.K., and Mountain, C.M. 1988, Astron. Astrophys. Suppl., **74**, 127.

Discussion

A.T. Young: *How do you extend this calibration from α Tau to the whole sky?*

Cohen: If we assume that every K5III star looks like our absolutely calibrated α Tau spectrum then we can proceed as follows. Take the completely observed α Tau spectrum, deredden it for the known interstellar extinction and normalize the result. Now, to transfer this normalized 'template' shape to another K5III star not spectrally observed, we apply the known interstellar extinction of this other K5III star, to the intrinsic template and then fit the resulting shape to independently calibrated IR photometry of the star in question. This provides us with a first order calibrated spectrum for each K5III star across the sky with known A_V and IR photometry. We have to observe, completely, one star of each type from K0-M0III then apply the template technique to tens, indeed hundreds, of stars across the sky in this spectral type range.

E.F. Milone: *You've discussed absolute calibration but do you want to say something further — on say the stability of the cavity?*

Cohen: We hope that some national laboratory will build for us and accredit, a black body furnace that we can put in a vault and compare with our current airborne cavity, both before and after a calibration flight. It's apparently quite difficult to find a lab that can accredit a Black Body across the entire 2-35 mm range - NIST(the old US 'NBS') cannot yet do this. We'll be talking to the NPL in the UK next.

S.K. Leggett: *First, NPL did calibrate the blackbody used by Selby et al in their calibration. Second, how does the new Kurucz model compare to the old Dreiling and Bell model for Vega that was previously used for calibration?*

Cohen: That old Oxford furnace has been located recently but is no longer working. We'd like NPL to design and/or build us a newer, even better one as our accredited absolute reference blackbody.

Kurucz ran about 28 different Vega models, sampling phase space (log T, log g, [He/H], microturbulent velocity) before he settled on the 9400K model for our calibration work. Roger Bell did not search phase space in the same way, nor did he examine very high resolution UV-optical spectra to validate his model. Consequently, he didn't really *determine* T_{eff} in the way Kurucz did; at least that's what Roger told me last year. The Dreiling and Bell Vega model had T_eff = 9650 K, so it's hotter than our calibrated Kurucz model; consequently it falls below ours in $\lambda^4 F_\lambda$ space, and is even further below the absolute mountain top measurements.

I.S. Glass: *Can you comment on the alleged variablity of Vega at the 0.02 mag level?*

Cohen: In an infant science like IR astronomy I don't think we can comment about variations at the 2% level. Of course, we have now transferred our attention to Sirius as our primary spectral standard and it's that star that we hope to hear is non-variable.

Infrared Photometric Systems, Standards and Variability

S.K. Leggett[1,2,4], J.A. Smith[3,4], T.D. Oswalt[3,4]

[1]*U.S. Naval Observatory Flagstaff Station, USA,* [2]*Universities Space Research Association, USA,* [3]*Florida Institute of Technology, USA,* [4]*Guest observers at the CTIO, IRTF and KPNO*

Abstract

The accuracy with which infrared photometry can be carried out is currently limited by poor definition of the instrumental system and by the accuracy of the available standard stars. We present new colour transformations between the CIT J, H, K photometric system and systems currently in use at Cerro Tololo, Mauna Kea and Kitt Peak. The precision of the J, H, K data for some of the stars observed by Elias *et al.* is improved and the system extended to fainter stars suitable for use with larger telescopes and/or infrared arrays. Evidence of infrared variability has been detected for one M dwarf star in our programme.

1. Introduction

In this paper we will describe the current status of J, H, K stellar photometry. The effective wavelengths of these passbands are $\sim 1.2\mu m, 1.6\mu m$ and $2.2\mu m$. At longer wavelengths the problems encountered are similar to, but worse than, those that we will be describing here. A useful review of JHKLM photometry is given by Bessell & Brett (1988).

For the last 10 years the pioneering work by Elias *et al.* (1982) has defined the most commonly used J, H, K photometric system (the CIT system), and provided the standard stars for infrared photometry. However, over several years of trying to obtain absolute photometry of isolated stars, we have found that the limiting accuracy of JHK measurements is only about 3%. To achieve even 3% takes much effort. It is not the detector or the terrestial atmosphere that limits the accuracy of these measurements, but instead the limits are due to a lack of definition of the natural system being used, and also due to the poor quality of some of the standard stars.

In the early days of infrared astronomy, 5% or even 10% photometry allowed one to do useful science, but it is important now that we achieve the potential 1% accuracy of modern day instruments. For example, in our work on cool dwarf stars this factor of three improvement would allow us to determine the metallicity of the atmospheres of such stars, and improve the determination of fundamental parameters such as temperature and luminosity by a similar factor.

We have used observations of Elias *et al.* stars, made during several observing runs, to derive new colour transformations between the CIT J, H, K photometric system and systems currently in use at Cerro Tololo, Mauna Kea and Kitt Peak. The precision of the J, H, K data for some of the stars observed by Elias *et al.* is improved and the system extended to fainter stars suitable for use with larger telescopes and/or infrared arrays.

2. The Sample

The observing runs whose results are used here were: two runs at NASA's Infrared Telescope Facility on Mauna Kea in 1990 and 1991 using a single aperture detector; a run in February of 1992 on the 50-inch telescope at Kitt Peak using a single aperture detector; and a run on the 4-m telescope of the Cerro-Tololo Inter-American Observatory in April 1992 using an infrared array.

TABLE 1. NUMBER OF JHK OBSERVATIONS OF ELIAS *ET AL.* STARS

Name	J−K	IRTF 1990 PRIMO 1 3/30−4/2	IRTF 1991 PRIMO 1 2/15−2/17	CTIO 1992 IR IMAGER 4/24	KPNO 1992 OTTO 2/28−2/29	Total
HD 161743	0.005	0	0	1	0	1
HD 44612	0.020	0	0	0	2	2
HD 130163	0.020	2	0	5	0	7
HD 75223	0.045	2	0	1	0	3
HD 106965	0.060	4	13	0	6	23
HD 129653	0.060	0	0	0	1	1
HD 77281	0.075	4	1	1	2	8
HD 129655	0.125	11	3	0	0	14
HD 161903	0.150	5	0	1	2	8
Gl 105.5	0.715	0	2	0	2	4
Gl 299	0.740	9	8	1	1	19
Gl 748AB	0.770	0	0	0	1	1
Gl 347A	0.780	0	1	1	1	3
Gl 390	0.825	12	0	0	1	13
Gl 811.1	0.825	0	0	1	0	1
G 77-31	0.900	0	1	0	1	2
Gl 406	0.980	4	1	0	1	6
BD+0°1694	1.055	0	0	0	4	4

Table 1 lists the 18 Elias *et al.* stars used in this work, which range from A-type to M-type stars. We are preaching to the converted here, but of course to carry out accurate photometric measurements several standard stars must be observed each night, where the stars cover a range of colour and airmass. Such measurements take a good fraction of the available telescope time (at least 10%), and furthermore the major part of the data reduction involves using the standards to properly calibrate the data, and to determine the extinction and transformation coefficients. On average we made 12 standard star observations every night.

3. Colour Transformations

Table 2 gives the colour coefficients required to convert to the CIT system, determined by us using the observing runs described above. The coefficients listed are defined as follows:

$$J-H_{CIT} = A(J-H) + B$$
$$H-K_{CIT} = A(H-K) + B$$
$$J-K_{CIT} = A(J-K) + B$$
$$K_{CIT} - K = A(J-K)_{CIT} + B$$

In cases where no second coefficient value is given B=0. Transformations published by other authors for various systems are also given in Table 2.

TABLE 2. COEFFICIENTS FOR TRANSFORMATION TO CIT SYSTEM

System	Coefficients A, B for:			
	J−H	H−K	J−K	K
IRTF90[1]	0.923	1.047	0.960	−0.05
IRTF91[2]	0.88	1.13		
IRTF[3]	0.847	1.050		0
UKIRT[4]	0.920	0.960	0.936	−0.018
UKIRT[5]	0.929	0.893	0.908	0.05
CTIO[6]		0.94	0.88	0
CTIO[7]	1.00	0.94	1.00	0
ESO[8]			0.874	−0.01, 0.019
KPNO[9]	0.995	0.959	0.985	−0.023
HCO[10]	0.92	1.00		0
AAO[11]	0.876, 0.013	0.954	0.897	0, −0.014
Johnson/Glass/Carter/SAO[12]	0.89[±.03]	0.94[±.03]	0.91[±.01]	0, −0.01

1) This work, using PRIMO 1.
2) This work, using PRIMO 1. Less well defined than the 1990 transformation.
3) RC1 photometer, Humphreys *et al.* 1984, A.J. **89**, 1155.
4) UKT9 photometer, Casali & Krisciunas, 1991, private communication.
5) UKT9 photometer, Leggett & Hawkins 1988, M.N.R.A.S. **234**, 1065.
6) This work, using IR IMAGER.
7) Elias *et al.* 1982, A.J. **87**, 1029.
8) Bouchet *et al.* 1991, A & Ap. Suppl. Ser. **91**, 409.
9) This work, using OTTO.
10) Persson *et al.* 1977, A.J. **82**, 729.
11) Elias *et al.* 1983, A.J. **88**, 1027.
12) Bessell & Brett 1988, PASP **100**, 1134; Leggett 1992, Ap.J. Suppl.

The main points to notice in Table 2 are that the various 'natural' systems differ significantly from each other, and that there is now no commonly used system that looks like the old CIT system. Not correcting properly from the standard star system to that of the instrument used can lead to errors in the determined colours of 5–10% for red stars.

Differences between systems can in most cases be easily understood, as observatories use different filters, especially for the J band. Moreover the edges of the JHK filters are often determined by the terrestrial atmosphere, and so the effective bandpass can differ from site to site. In some cases however the difference is not easily understood, for example the UKIRT and IRTF observatories on Mauna Kea use the same filter set at equivalent sites. Presumably in this case the colour term is due to some optical path difference, such as perhaps the gold-coated dichroic employed by UKIRT.

We are currently investigating the cause of time variations in the colour transformations, such as those demonstrated for the IRTF and UKIRT in Table 2. We are investigating how the transformation depends on when the mirror was resurfaced or cleaned. Some time dependence could perhaps also be caused by slowly changing atmospheric conditions which alter the effective filter bandpasses.

As the CIT system does not represent any system used today another system should be adopted as the 'standard'. However there is no overlap or agreement between the various systems commonly in use. Clearly a better defined filter set is required that avoids the edges of the terrestial windows, and this system would then be the obvious one to adopt. Milone *et al.* will be suggesting new infrared filter profiles later in this meeting.

4. New Standards

The 6th magnitude Elias *et al.* standard stars are too bright for use with the new infrared arrays on larger telescopes. Table 3 lists 14 stars that we suggest could be adopted as primary or secondary JHK standards. These stars are taken from compilations of published photometry for white dwarf and red dwarf stars by Leggett (1989, 1992) and also from a study of the South Galactic Pole area by Leggett & Hawkins (1988).

TABLE 3. FAINT PRIMARY OR SECONDARY STANDARD STARS

Name	Source[a]	RA 1950	Dec 1950	Proper "/yr	Motion θ	Opt.[b] mag.	J−H	H−K	K	No. Obs.
SGP 174	LH	00 47 55.9	−26 55 43			12.0R	0.83	0.23	7.92	25
SGP 157	LH	00 50 28.6	−27 22 18			19.6R	0.63	0.43	12.47	4
SGP 69	LH	00 59 41.2	−25 47 16			18.4R	0.58	0.30	12.67	4
SGP 50	LH	01 01 29.9	−26 36 03	0.302	165.0	12.0R	0.60	0.22	8.54	26
Hyad 214	LH	04 16 49.8	+16 38 12			19.9R	0.61	0.34	13.43	6
LHS 211	L92	05 45 14	+08 22 00	1.218	135.4	14.11	0.53	0.20	10.41	2
LHS 212	L89	05 53 47	+05 22 12	1.056	207.0	14.10	0.20	0.11	12.63	2
LHS 216	L92	06 11 07	+15 11 48	1.399	152.8	14.66	0.47	0.24	10.59	2
LHS 39*	L92	11 03 02	+43 46 42	4.531	281.9	14.40	0.52	0.29	7.85	3
LHS 2978	L92	14 44 23	−12 31 42	0.545	246.7	12.09	0.61	0.20	8.01	2
LHS 399	L92	15 33 08	+17 52 48	1.219	263.0	12.37	0.53	0.22	7.97	2
LHS 421*	L92	16 33 27	+57 15 06	1.620	316.0	12.91	0.47	0.28	7.79	2
LHS 429*	L92	16 52 55	−08 18 12	1.190	222.5	16.80	0.58	0.37	8.82	5
LHS 474*	L92	19 14 31	+05 04 48	1.461	203.1	17.50	0.66	0.44	8.80	4

a) Sources are Leggett & Hawkins (1988), Leggett (1989), Leggett (1992)
b) Optical magnitude is V, or R from Leggett & Hawkins (1988) if so indicated
*LHS 39 is an emission line flare star
*LHS 421 is an emission line, optically variable, eclipsing binary
*LHS 429 is an emission line star and may be a flare star
*LHS 474 is an emission line flare star

All except two of the stars listed are red dwarfs. We believe that these stars are stable in the infrared, even if they vary in the optical. However we have found evidence of infrared variability in at least one red dwarf star, as discussed in the next section.

5. Infrared Variability and Problem Standards

We have found that one of the Elias *et al.* stars, Gl 105.5, varies from night to night by ~10% at J, H and K. The star appears hotter when it is fainter in the IR, and based on our photometry varies from a dK8 type to dK5. We have searched the literature (including a SIMBAD database search) and have not found any reference to this star flaring or being variable. The known flare star Gl 406 (Wolf 359) does not show any sign of infrared variability. Table 4 lists the JHK values observed by us for Gl 105.5 on four different nights, as well as the values given by Elias *et al*, which are based on 22 measurements.

TABLE 4. OBSERVATIONS OF GL 105.5

Date	J	J−H	H−K	Error
Elias *et al.* 1982	7.240	0.605	0.110	±.007
16 Feb 1991	7.237	0.539	0.075	±.025
17 Feb 1991	7.179	0.571	0.162	±.040
28 Feb 1992	7.398	0.641	0.049	±.020
29 Feb 1992	7.243	0.621	0.118	±.010

This variability poses a problem for those of us trying to establish infrared standards; however for those interested in variable stars this could be an exciting opportunity. The effect we have seen may be an example of a 'negative infrared flare' as described by Gurzadyan (1988). He suggests that such flares would occur simultaneously with a positive optical flare, in the case of flares due to fast electrons appearing in the outer regions of a star.

Besides Gl 105.5, we have found problems at the ≥3% level for two other Elias *et al.* stars. Our revised values for these two stars, Gl 390 and BD+0°1694, are given in Table 5. Our estimated uncertainties and those quoted by Elias *et al.* are given in the Table. Elias *et al.* made many more observations of BD+0°1694 than we did; this red giant may be another infrared variable.

TABLE 5. REVISION OF ELIAS *ET AL.* VALUES

Name	Elias *et al.*					This Work				
	K	J−K	H−K	No. Obs.	Error, mag.	K	J−K	H−K	No.	Error, mag.
Gl 390	6.045	0.825	0.205	11	0.007	6.085	0.769	0.196	13	0.020
BD +0° 1694	4.585	1.055	0.225	27	0.005	4.620	1.027	0.219	4	0.015

6. Conclusions

We have found that about 20% of the commonly used Elias *et al.* (1982) standard stars display errors of ≥0.03 magnitudes. New and fainter standards are needed for use with infrared arrays on the larger telescopes. Such work is in progress at the major observatories; for example Casali *et al.* at UKIRT and Elias *et al.* at CTIO have started observing Landolt standards to define equatorial infrared standard stars. In this paper we have presented 14 additional infrared-faint stars that could be adopted as standards.

A new standard system, with better defined JHK filter bandpasses, is required before the full potential of modern day detectors can be routinely achieved. Such a filter set will be described by Milone *et al.* later in these proceedings.

Finally, we make a plea to the major observatories that they monitor their users' data reductions, recording not only for example changes in instrumental zero point and atmospheric extinction, but also to make note of any standard stars that deviate from the adopted calibration. The problems we had with some of the 10-year-old Elias *et al.* stars should have been made public before now.

This work was supported in part by NSF grant AST-9016284.

References:

Bessell, M.S. & Brett, J.M., 1988, PASP **100**, 1134.

Bouchet, P., Manfroid, J. & Schmider, F.X., 1991 Astr. & Ap. Suppl. **91**, 409.

Casali, M.M. & Krisciunas, K., 1991, private communication.

Elias, J.H., Frogel, J.A., Matthews, K. & Neugebauer, G., 1982, A.J. **87**, 1029.

Gurzadyan, G.A., 1988, Ap.J.**332**, 183.

Humphreys, R.M., Jones, T.J. & Sitko, M.L., 1984, A.J.**89**, 1155.

Leggett, S.K., 1989, Astr. & Ap. **208**, 141.

Leggett, S.K., 1992, Ap.J. Suppl. **82**, in press.

Leggett, S.K. & Hawkins, M.R.S., 1988, M.N.R.A.S. **234**, 1065.

Milone, E.F. *et al.*, 1992, these proceedings.

Persson, S.E., Aaronson, M. & Frogel, J.A., 1977, A.J. **82**, 729.

Discussion

E.F. Milone: *You haven't talked about extinction corrections. Could the difference between your ITF and UKIRT results be due to different* H_2O *content, from night to night, for the two sets of observations?*

Leggett: No we are happy with our extinction measurements. The difference appears to be in the telescope or instrument optical path.

R.R. Shobbrook: *If the J, H and K band-passes are defined by the atmosphere, they must also change with air mass. Is this a main part of the problem?*

Leggett: We try to observe at high airmasses. Although the terrestrial atmospheric windows are narrow, and do affect the filter cut-offs, I don't think airmass effects are the main part of the transformation problem.

M. Cohen: *Now that we have these absolutely calibrated Sirius and Vega IR spectra, they are available as a resource for anyone wishing to calibrate their own IR filters, new or old. If they can send us (digitally) their filter transmission profile at its cryogenic operating point, we can provide the 'zero magnitude flux calibrations' by integrating over our calibrated Vega spectrum.*

I.S. Glass: *I think your estimate of errors is unduly pessimistic, especially amongst southern hemisphere observations. We have found rms differences with Elias at the 0.01 magnitude level, at H K at least, and a little worse at J.*

Leggett: The northern hemisphere systems are perhaps not as well defined as the southern ones. Also you do have to be careful which Elias standards are used.

CCD Standard Fields

D.H.P.Jones

Royal Greenwich Observatory, Cambridge

Abstract

There are several standard fields for CCDs available in the literature. Some of these have been observed many times with the CCD cameras of the Isaac Newton Group of Telescopes on La Palma. These observations were made in order to establish the zero-points and colour equations for the *UBVRI* filters. Now that many observations are available in the data archive it is possible to combine them and search for inconsistencies in the published magnitudes and colours. This discussion will lead to standard fields of a higher quality. Furthermore astrometric plates have been taken of several of these fields which will be used to provide accurate positions as well as magnitudes and colours. These will provide the scale, orientation and distortion of any CCD camera.

1. The Observatorio del Roque de los Muchachos

Our observatory on the Canary Island of La Palma comprises three telescopes operated by the Royal Greenwich Observatory with apertures of 1.0, 2.5 and 4.2 metres. They are operated on the 'common user' principle for an international community including British, Dutch, Spanish and Irish astronomers; numbering about a thousand active members.

2. The La Palma Data Archive

Since first light in 1984 the great majority of observations have been kept in the La Palma Data Archive (Zuiderwijk 1991, Zuiderwijk and Meikle 1992). Only the unreduced data, together with attendant flat fields, bias frames and the like are held in the archive. Descriptions of the 165 000 observations made to date are held in a disk file at Cambridge which can be examined from virtually any computer worldwide connected to a network.

3. Standard Fields

About half the observing on the 1-metre and 2.5-metre telescopes has been devoted to direct imaging and the photometric calibration fields of the six clusters observed by Christian *et al* (1985) and of M67 observed by Schild (1983) have been frequently observed. The archive reflects the popularity of these fields and contains the following numbers of observations:

NGC 2264	74
NGC 2419	76
NGC 4147	138
M92	422
NGC 7006	298
NGC 7790	221
M67	730

4. The M92 Field

As part of our support work for the La Palma telescopes I have embarked on an examination of these data. I discuss here only the R, I observations of M92 which are typical of the whole. A comparison between R magnitudes based on 10 exposures with a GEC chip on the Jacobus Kapteyn 1-metre telescope over a week in 1992 June and those published by Christian *et al.* is shown in Fig.1.

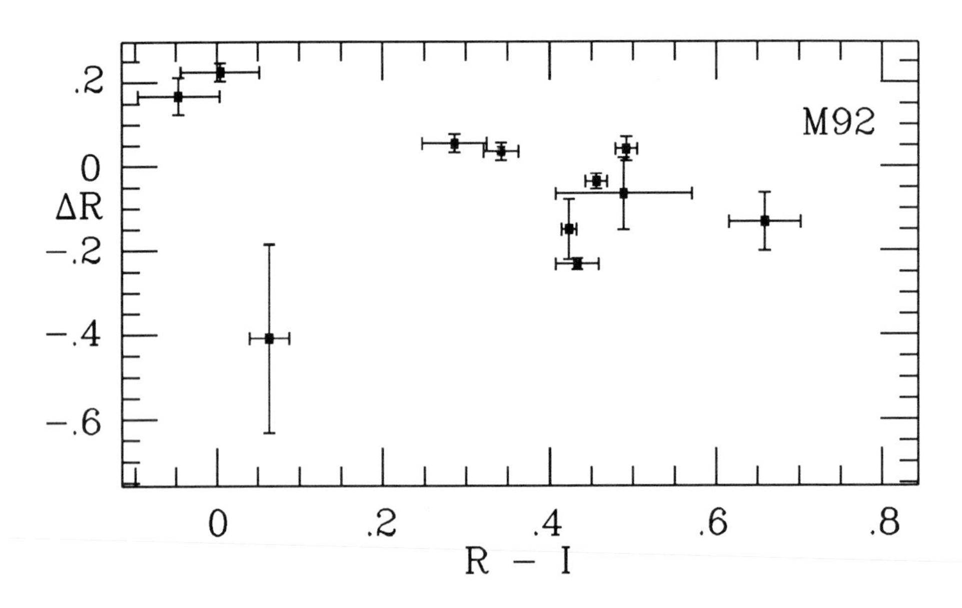

Figure 1 Difference in R (Jones - Christian) for stars in M92 plotted as a function of $(R-I)$ Christian.

The CCD observations were reduced with simple aperture photometry in a 10 arcsec diaphragm to make them directly comparable with the photomultiplier observations of Christian *et al.* The median error of the CCD observations which total 640 sec is 0.011 mag compared to 0.029 for Christian *et al.* Even making due allowance for the errors, the two series do not agree well; in fact they imply a colour equation $\sim 0.4(R-I)$. The origin of this colour equation remains a mystery because the colour equation for this chip-filter combination is given as $0.045(R - I)$ by Unger *et al.* (1988); based on a comparison with the work of Schild (1983) in M67 with a scatter of ± 0.009 mag.

5. The Kron-Cousins System

The La Palma R, I photometry is designed to be on the Kron-Cousins system. Christian *et al.* nowhere explicitly state what their system is. However they used the filters recommended by Bessell (1979) and the standard stars of Landolt (1983), both of which refer to the Kron-Cousins system. Fig.2 shows a comparison between the La Palma CCD $(R - I)$ colours and those of Christian *et al.*

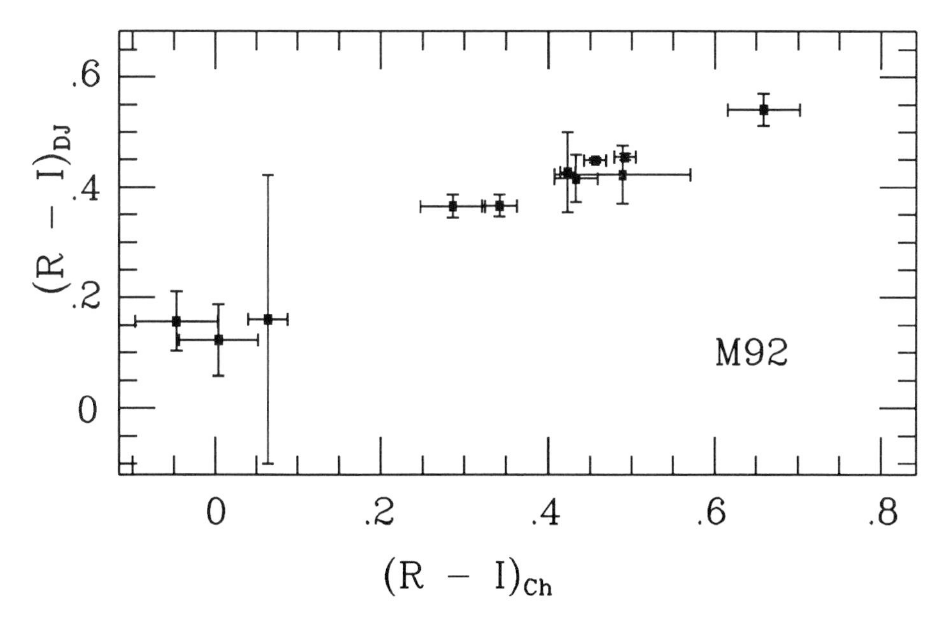

Figure 2 $(R - I)$ (Jones) plotted against $(R - I)$ (Christian) for stars in M92.

The observations agree within their errors but there is a clear systematic difference.

6. Flatfielding

There is always the possibilty that the scatter in Fig.1 and 2 is caused by some defect in the process of flatfielding. Our usual procedure on La Palma is to use the twilight sky on a field chosen to be devoid of bright stars. Faint stars are eliminated by offsetting the telescope by a few arcseconds between exposures and taking the median of series of exposures, sometimes continuing over several evenings and mornings. The marginal distributions in sensitivity of the R flatfield used in reducing the observations in Fig.1 is shown in Fig.3.

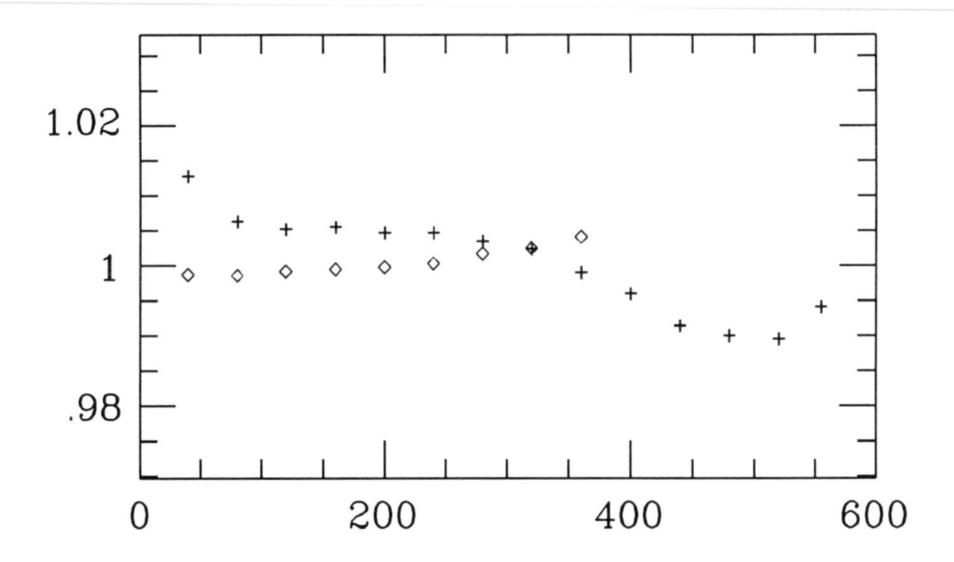

Figure 3 Marginal averages of a typical median twilight flatfield in *R*. The diamonds are along the rows 40-360 and the crosses along the columns 40-560; each symbol embraces 40 rows or columns.

It is difficult to believe that failure to flatfield properly can have introduced more than a scatter of a few thousandths of a magnitude into Fig.1.

7. Astrometric Calibration

The reason for studying these fields in detail is to improve the internal accuracy and to extend the photometry to fainter limits suitable for larger telescopes. Some visiting observers require astrometric calibration as well as photometric of their CCD observations. There are little suitable data in the modern literature so I have recently taken photographic plates with our 1-metre telescope of all these fields. However I have also found that the bifilar micrometer measures by Barnard (1931) and the photographic observations of Schlesinger (1934) give comparable results.

References:

Barnard, E.E., 1931, Pub.Yerkes Obs., **6**.

Bessell, M.S., 1979, Pub. Astr. Soc. Pacific, **91**, 589.

Christian, C.A., Adams, M., Barnes, J.V., Butcher, H., Hayes, D.S., Mould, J.R. and Siegel,M., 1985, Pub. Astr. Soc. Pacific, **97**, 363.

Landolt, A.U., 1983, Astron. J., **88**, 439.

Schild, R.E., 1983, Pub. Astr. Soc. Pacific, **95**, 1021.

Schlesinger, F., 1934, Astron. J., **44**, 21.
Unger, S.W., Brinks, E., Laing, R.A., Tritton, K.P. and Gray,P.M., 1988 *La Palma Observers' Guide*, p86.
Zuiderwijk, E.J., 1991, La Palma Technical Note No.69.
Zuiderwijk, E.J. and Meikle, W.P.S., 1992, Gemini, **36**, 12.

Discussion

W. Tobin: *Do you see any evidence for distortion in your CCD cameras?*

Jones: Not down to the one tenth of a pixel level but there may be smaller distortions which we haven't been able to detect.

T.J. Kreidl: *With very large telescopes (8-10m) coming on-line this decade, the need for faint standards is very important. Please comment on your efforts to improve the present evident discrepancy between the standard calibrations you are comparing your data with, and what precision you hope to achieve for, say, 16-18 magnitude standards.*

Jones: The immediate aim is to improve the internal consistency of the Christian *et al* sequences which are around magnitude 16. It is difficult to use archive data to check the external accuracy of the work of Christian *et al.* I agree that it is very important to provide standards for the new generation of large telescopes.

E. Budding: *My impression was that the chief contribution to the large discrepancies you show is related primarily to the filters. This was presumably the KPNO set which were commented on previously. This point gave rise to thoughts about the role of another cut-off effect which may have an important bearing on filter choice, ie the financial cut-off. Is this speculation relevant to your situation?*

Jones: We are in the process of replacing the KPNO interference filters with the Harris recipe for combinations of Schott glasses, published in the Kitt Peak newsletter about 3 years ago. these glasses are comparatively cheap and so we aren't constrained financially.

R. Florentine-Nielsen: *Was the ΔR a function of the R magnitude itself, i.e. did you test for any nonlinearity in your CCD photometry?*

Jones: At the beginning of each CCD run we routinely check the linearity of the CCD with a constant light source and exposures of increasing length. This does not verify the linearity of the CCD in the exact sense of your question but we find the number of electrons is linearly related to exposure time up to 50K for all our chips, and 64K for some of them. As a by-product we also derive the ratio of electrons to ADU and the read-out noise.

A group standing in front of the remains of *The Leviathan of Parsonstown*, the great six foot reflecting telescope built by the fourth Earl of Rosse in the grounds of Birr Castle, Co. Offaly, in 1845.

Session 2

High Precision Photometry

High-Precision Photometry

A.T. Young

European Southern Observatory, Garching bei München, Germany

and

San Diego State University, San Diego, California, USA

Abstract

The current precision of differential comparisons, within a fixed instrumental system, reaches about a millimagnitude. The current accuracy of transformations between systems is on the order of 0.01 mag, but is worse in some cases. The transformation errors are due to mis-matches between instrumental and standard systems, and cannot be transformed away without further information. A simple geometric model is proposed, which illuminates the transformation problem, and suggests means of minimizing it. Misplaced band edges are the main cause; tighter specifications on filters will help to reduce these errors. The ultimate solution is to use sampled systems in which some bands look like the derivatives of their neighbors. This automatically produces enough band overlap to comply reasonably well with the sampling theorem.

1. Introduction

By ordinary standards, the phrase "high-precision photometry" is an oxymoron. Basic physical quantities like mass, length, and time are routinely measured in *absolute* SI units with an accuracy of a few parts per million with off-the-shelf commercial equipment these days. Even cheap digital watches are accurate to one second per day, or about a part in 10^5. But photometrists are proud to make *relative* measurements with a single instrument agree to 1 part in a thousand, and are hard pressed to make relative measurements with different instruments agree to one per cent — as Sterken and Manfroid (1987), Menzies et al. (1991), and Pel (1991) have shown. In contrast, absolute measurements of length having such accuracy are regularly made by undergraduates using meter sticks.

So perhaps we should not be surprised to hear observational astronomers called second-rate scientists; or that "astrophysical accuracy" is a bad joke even among astrophysicists. Given the meager progress in photometric precision and accuracy in the last 20 years, it is hardly surprising that observatory directors close down photometric telescopes, and regard photometry as an obsolete art to be relegated to the amateurs these days. If we are to overcome this erosion in the status of photometry, we must do substantially better than we have done in the past. Just doing *more* photometry of modest precision is not enough, even if it is useful.

2. Precision and accuracy

As all the measurements we call "photometry" are relative comparisons between stars, the term "precision" strictly applies to all our work. We rarely refer our observations to SI units, so "accuracy" seems inappropriate. But, as it is useful to distinguish between differential and all-sky photometry, one can regard the discrepancies in differential comparisons of stars made with a single instrument as estimates of *precision*, and the discrepancies in comparisons between different instruments as estimates of *accuracy*. However, as the first Workshop on Improvements to Photometry (Borucki and Young, 1984) concluded, rather than talk about "absolute" or "relative" measurements, we should specify the typical differences in time, position, and spectral functions of the measurements being compared.

The current state of photometric precision and accuracy is well represented in the papers of the first (Borucki and Young, 1984) and second (Borucki, 1988) Workshops on Improvements to Photometry, and in "Precision Photometry" (Philip et al., 1991). The first Workshop dealt with fundamentals. The second concentrated more on hardware, primarily fiber optics, multichannel systems, and silicon detectors. Dave Philip's meeting concentrated on actual results. Together with the papers at our current meeting, these proceedings give an up-to-date picture of the state of our art.

The future of photometric precision is discussed in hopeful terms by Young et al. (1991); there appear to be techniques within reach that should allow some further improvement. Rapid chopping between stars; photometer designs that promote better uniformity of response across the field, together with automatic centering; and improved modelling of extinction and other instrumental variables, should all contribute to better results.

Briefly, one can say that differential comparisons of neighboring stars can be done to about a millimagnitude. This is possible with CCDs, provided that one is very careful about handling the nonuniformities of the chip (Gilliland et al., 1991). This may require keeping the same stars on the same diodes, in addition to very careful flat-fielding. Similar results in night-to-night comparisons of neighboring stars have been done with photomultipliers, and with silicon diodes; and, as laboratory work with PMTs has demonstrated short-term stability of a part in 10^4 or so, and rapid chopping between stars should remove most of the atmospheric transparency variations, there is reason to hope for still better results with larger telescopes.

In the laboratory, silicon diodes seem to be somewhat more stable than photomultipliers, and are the detector of choice if adequate light is available. At the telescope, our signals are weak; giving up photomultipliers means giving up 15 magnitudes or more of electrical sensitivity, and returning to the days of electrometers and DC amplifiers (though now in modern guise), with their long time constants, slow readouts, and linearity problems. I am not sure this tradeoff is wise, except for some special purposes.

Furthermore, commercial laboratory spectrophotometers using photomultipliers have been good to 0.1% — that is, a millimagnitude — for twenty years (Hawes, 1971); and the best laboratory instruments, using 10-second integrations with photomultipliers (very much like ordinary photometric observations) have achieved a level of accuracy of one part in 10^4 (Mielenz et al., 1973), and a precision of a few parts in 10^5. Therefore, even though photomultipliers *can* have problems at the 1% level, I do not think photomultiplier problems are the main obstacle to better photometric accuracy, which is typically 100 times worse than the best that PMTs are capable of.

CCDs are complex devices whose properties are not as well understood as either silicon diodes or PMTs. They have limited dynamic range, and undersample the focal plane. However, their specialized ecological niche is becoming better appreciated: they are an excellent replacement for the photographic plate in crowded fields. But their long readout times and complex data-handling problems make them less than optimal for general photometry.

Still, as millimagnitude or better comparisons have now been done with CCDs, PMTs, and silicon diodes alike, and all have performed better in the laboratory than any has yet done at the telescope, I believe we can say the limits of current detectors have not been reached, and that all of these types are suitable for high-precision photometry. In any case, I do not think that detectors are limiting our precision, nor that we will do substantially better in the immediate future with any new detector. If there are advances to be had on the instrumental front, they must come from areas other than detector technology.

One area that has received considerable attention is the question of nonuniform responsivity across the focal plane, and the concomitant centering errors. There is no doubt that this is a serious problem when inclined photocathodes are used, because the field lens in the photometer transforms positions in the focal plane to angles of arrival at the detector, whose angular response is non-uniform, particularly well off normal incidence. This explains the attention given to fiber optics at the Second Workshop, and is discussed at length by Young et al. (1991).

Charles KenKnight's contribution at the first Workshop, on the extended wings of the star image, is especially important. The wings are due to polishing errors in primary mirrors, and require large focal-plane apertures to secure insensitivity to seeing and guiding variations, quite apart from the perfection of the photometer itself. Large field stops permit very reproducible photometry; guiding and centering must be reproducible within a moderately small fraction of the stop size, but this is entirely possible (especially with automatic control).

As far as the photometer itself is concerned, conventional optics of generous diameters, especially if used to image the pupil on the filters as well as the detector, certainly can give millimagnitude uniformity over some fraction of a minute of arc. If the highest precision is required, moderately efficient scramblers have been designed that provide uniformity of tens of micromagnitudes, but at the cost of a magnitude or so in lost light. Again, I note that sub-millimagnitude precisions have actually been achieved at the telescope. In short, there are several ways to achieve good uniformity even without going to fiber optics, which present serious coupling problems, especially those arising from dust particles and small surface defects.

Thus, "precision" is in relatively good shape — maybe not by comparison with other physical measurements, but at least by comparison with photometric "accuracy". There, we still have unresolved systematic discrepancies of several hundredths of a magnitude. This dismal problem has remained practically unchanged for twenty years or more, and has gradually grown more acute, partly as experience has accumulated, and partly as more severe errors have appeared in work done with non-standard detectors and filters. Systematic discrepancies of several per cent remain between the results of photometric projects that have taken tens of man-years of effort, with good instrumentation, at excellent photometric sites — I remind you again of the studies by Sterken and Manfroid (1987), Menzies et al. (1991), and Pel (1991).

Yet each group had demonstrated *internal* precision of a few tenths of a per cent — roughly an order of magnitude smaller than the disagreements among them. And, as both Wes Lockwood and V. Grossmann report here, even sub-millimagnitude precision can be obtained with a single, conventional instrument. The contrast between internal precision and external accuracy is quite striking.

Recently, Russ Genet suggested to me that the difference between precision and accuracy is that precision is an instrumental problem, but accuracy is an astronomical problem. This is a useful distinction to make. Even so, our best precision is only about an order of magnitude better than our best accuracy, and falls far behind that of (for example) the people who compare masses with mechanical beam balances, where the best precision is now about a part in 10^{11} (Quinn, 1992). Our poor showing, compared to other physical measurements, suggests that, although the best ground-based photometric precision has been limited by scintillation, it may soon be limited by small variations of the effects now limiting our accuracy. The "precision" or instrumental side was recently reviewed in some detail by Young et al. (1991), and is treated further in the session on New Techniques at this Colloquium, so I shall emphasize the "accuracy" or astronomical side of the problem here.

While there is certainly a difference between instruments that are well designed and built and those that are not, my current impression is that the instrumental side of this problem is in much better shape than the astronomical one. The Workshops on Improvements to Photometry (Borucki and Young, 1984; Borucki, 1988) did not turn up any instrumental innovations likely to make order-of-magnitude reductions in photometric errors, though many interesting ideas were discussed.

Instead, I believe that it is the multidimensional property of photometric measurements that is the basic cause of our difficulties. All the physical quantities that can be measured so much more accurately than starlight, such as mass, length, and time, are simple and one-dimensional. But what we are trying to measure is really a (spectral) function, not a single point value.

3. Photometry as spectroscopy

At the Buenos Aires General Assembly, I argued that we do not understand our data well enough to reduce them correctly, and that the real problems are connected with spectral distributions: of response in our instruments, and of irradiance in the starlight we observe. Once we have converted the light to an electrical signal, the electrical measurement is straightforward, and can be made to much higher accuracy than we now reach. But it does little good to have excellent transducers, if we do not know what we are transducing — namely, these spectral distributions.

To understand photometric measurements in detail, we need a detailed and accurate model for measurements of stellar spectra. Unfortunately, the theory we have is relatively crude and undeveloped — and even this theory is often ignored by observers, who resort to purely empirical reduction methods. Yet, as a great scientist once said, "There is nothing more practical than a good theory" (Hampel et al., 1986).

Photometry is a sort of spectroscopy, done at a resolution so low that individual spectral lines are completely lost by blending. Worse, each instrument blends things together in a different way, so that measurements made with different instruments cannot be compared directly. Thus, the first thing we need is a good theoretical model for heterochromatic measurements. Until we have such a theory, we will remain unable to remove instrumental signatures completely from our measurements. I believe this problem is the most serious limitation of photometry today.

However, it seems to be a tractable problem. There already exists a mathematical description (Kolmogorov and Fomin, 1961; Oden, 1979) of the kind of operation we perform when we make a photometric measurement. Unfortunately, it seems to be taught only to mathematicians, not to astronomers and physicists.

4. Transformations

The problem is that while minutely examining the stability of *individual* instruments, we have neglected the reproducibility from one instrument to the next. That is, we have failed to investigate the differences between different instruments that are intended to reproduce the same system.

These differences are often swept under the rug of some empirical "transformation" from one instrument's measures to those of another, using a color index as the only independent variable. These transformations are usually represented as linear by textbook writers, who begin with the unrealistic assumption that stars are black bodies obeying Planck's law (or worse yet, Wien's); apply two or three more crude approximations; and end up "deriving" a linear equation. The worst part of this is that photometrists are led to believe there is something wrong with their data if their transformations are not linear and single-valued.

The careful photometrist, on the other hand, usually finds that a straight line won't quite fit, and is reduced to using piecewise-linear approximations of 2 or 3 parts (or cubics; or worse!). Different relations turn out to be necessary for stars of different luminosity, metallicity, or reddening. However, closer examination (King, 1952; Young, 1992) shows that these transformations are *inherently* non-linear. The latter investigation in fact shows that cubic or even quartic terms must be significant; furthermore, the relationships are necessarily multi-valued, if a single color-index is used as the only independent variable.

The large size of the high-order terms in the classical series expansion is evidence that we have chosen a model ill-suited to the job at hand. Furthermore, current photometric systems fail to capture essential astrophysical information, such as derivatives, that any model needs for accurate transformations. I wish to show that a different approach should lead to simpler — indeed, exactly linear — transformations. However, the price of linearity is to abandon our favorite nonlinear transformation: the logarithmic one from intensities to magnitudes.

5. Functional analysis of photometry

I must admit there is an additional price. One must delve into a fairly obscure corner of mathematics, to pick up the necessary tools. This area, known as functional analysis (Kolmogorov and Fomin, 1961; Oden, 1979), is standard stuff for mathematicians, but rarely encountered by astronomers. I have room here only to sketch the main results we will need.

Functions of the sort we encounter in photometry, such as instrumental spectral response functions and stellar spectra, are quite well-behaved. Their integrals over wavelength exist and are finite, as are the integrals of their squares and products. Such functions belong to a class that can be mapped onto the infinite-dimensional vector space known as Hilbert space (Kolmogorov and Fomin, 1961; Oden, 1979). This is a straightforward extension of ordinary Euclidean space to a countably infinite number of dimensions. One might think of the projection of a function onto the nth axis of Hilbert space as the value of the function at the nth of the denumerable infinity of rational values its argument can assume.

Let us represent an instrumental spectral response function $R(\lambda)$ by a vector in Hilbert space, from the origin to the point corresponding to the function. The squared length of this vector is just

$$\int_0^\infty R^2(\lambda)\, d\lambda.$$

This integral is the analog of summing the squares of the coordinates of a point in a finite-dimensional space to find its Euclidean distance from the origin. It will be convenient to assume the vector is normalized, so that this length is unity.

Similarly, a stellar spectral irradiance $I(\lambda)$ corresponds to a vector. But what we actually measure is not $I(\lambda)$, but the integral

$$L = \int_0^\infty I(\lambda) \cdot R(\lambda)\, d\lambda.$$

This integral is instantly recognised by the mathematicians as the inner product of I and R. Its geometrical representation in Hilbert space is the length of the *projection* of the vector $\mathbf{I}(\lambda)$ onto the normalized vector $\mathbf{R}(\lambda)$.

Now, suppose we have two response functions — say, those for the standard B and V bands. Their vectors span (i.e., determine) a two-dimensional subspace. The **B** and **V** vectors form a *basis* (though not an orthogonal basis) for this subspace. In fact, we can compute the angle, θ, between these vectors. In general, if two functions (i.e., vectors) are $f(\lambda)$ and $g(\lambda)$, the angle θ between them is given by

$$\cos\ \theta = \frac{\int f(\lambda)g(\lambda)\ d\lambda}{[\int f^2(\lambda)\ d\lambda \cdot \int g^2(\lambda)\ d\lambda]^{1/2}},$$

where the integrals all run from 0 to ∞. Because of the small overlap of the B and V response functions, the projection of one on the other is very small — that is, they are nearly orthogonal. The angle between the **B** and **V** vectors in fact turns out to be about 84°.

When we measure a stellar irradiance distribution with this system, we are projecting the vector $\mathbf{I}(\lambda)$ onto the subspace spanned by the **B** and **V** basis vectors. The lengths of the subsequent orthogonal projections of this projection **P** on the basis vectors are in fact the standard measurements. Because a fixed color index corresponds to a fixed ratio of B and V signals, all stars with the same B–V color index have projections that lie in the same direction in the (B, V) subspace.

Stars with different spectra have irradiance vectors with different directions in Hilbert space. If stars were black bodies, their irradiance vectors would define a warped surface in Hilbert space. Then transformations would be easy. But the line features of stellar spectra form rapidly-varying functions of wavelength that are nearly orthogonal to all the Planck functions. Thus, the spectral lines expand the regime of stellar irradiance vectors from a warped surface to a higher-dimensional manifold that encloses the black-body surface. Similarly, the effects of interstellar reddening add new dimensions to the subspace spanned by actual stellar irradiances. It is precisely these deviations of stellar spectra from black bodies that contain the interesting astrophysical information about stars.

If we have an instrumental system, with bands b and v, it will generally define a two-dimensional subspace that differs from the standard one. The projection $\mathbf{P}'$ of **I** onto it will differ somewhat from the projection **P** of **I** onto the (B, V) subspace. Because of the difference between the two subspaces, there is a component $\mathbf{P}_\perp$ of $\mathbf{P}'$ that is orthogonal to the (B, V) subspace; that is, orthogonal to **P**.

Then different stars whose irradiance vectors all lie in a plane through **P** and normal to the (B, V) subspace, all have projections on the (B, V) subspace in the direction of **P**; so these stars must all have the same color in the (B, V) system. But, as their plane is, in general, *not* orthogonal to the (b, v) subspace, their different components along $\mathbf{P}_\perp$ have non-zero projections on (b, v); so they have *different* colors in the instrumental (b, v) system, by amounts proportional to the length of $\mathbf{P}_\perp$. Clearly, the transformation from one system to the other cannot be one-to-one. Accurate transformation is impossible.

Such transformation errors represent the loss of the astrophysical information contained in the $\mathbf{P}_\perp$ component of the stellar irradiances. They necessarily appear in comparisons between instrumental systems whose passbands differ in their ability to capture the astrophysical details of stellar spectra.

The size of the transformation errors is proportional to the length of the perpendicular vector $\mathbf{P}_\perp$. The foot of this perpendicular marks the projection $\mathbf{P}$ of the instrumental vector onto the standard subspace. But that projection is just the least-squares approximation to the instrumental vector by a linear combination of the standard vectors; and the orthogonal component $\mathbf{P}_\perp$ is just the residual vector of the least-squares fit. The smaller the difference between the subspaces, the shorter will be $\mathbf{P}_\perp$, the least-squares residual vector. Ideally, the difference should be zero, so that the orthogonal component vanishes. So to make the transformation between photometric systems one-to-one, their basis vectors must span the *same* subspace.

What is the mathematical condition that ensures this? It is that the basis vectors or functions of one system must be a *linear combination* of the basis vectors or functions of the other. Then the measured *intensities* in one system will be that same linear combination of the intensities in the other! This unique, linear transformation of intensities between the two photometric systems is exact. It is valid for all stars, regardless of metallicity, reddening, and all those other astrophysical phenomena that complicate transformations in existing systems.

In fact, such linear combinations of intensities and response functions were already used by Harold Johnson (1952) in one of the most important papers he ever wrote. Johnson showed that the old International Photographic System and some of its successors could be closely approximated by a linear combination of the (still unnamed) ultraviolet and blue passbands of what later became the UBV system. Perhaps because the U and B bands were not familiar to photometrists at the time, or perhaps because of Johnson's emphasis on a single spectral feature (the Balmer continuum) rather than on general principles, the deeper lesson of this paper was not widely appreciated.

We also have natural additive linear combinations of intensities in the composite light of double stars, clusters, and galaxies; the approach advocated here obviously yields exact transformations for these objects. Furthermore, subtractive linear combinations of intensities have long been used to correct for red leaks. Gilliland et al. (1991) have used additive combinations most recently. So linear combinations of intensities have been around for a long time, but only in a supporting role. We should now move them to center stage.

6. Practical considerations

Given the limitations on real filters and detectors, we cannot guarantee to make one instrument's response functions exact linear combinations of another's. Nevertheless, we can work in that direction, knowing that the better we can approximate this condition, the better we can transform our data. At least, we now know what to aim for.

Geometrically, we must strive to keep the difference between the spaces spanned by different instruments as small as possible, for the transformation errors are proportional to the length of the perpendicular from a point in one subspace to the other subspace. In particular, we want to keep each response vector in an instrumental system as close to its projection on the standard system as possible.

Let an instrumental response function be $r(\lambda)$ and its least-squares approximation in the standard system be $R(\lambda)$. Because the square of the perpendicular distance is

$$P_{\perp}{}^2 = \int_0^\infty [r(\lambda) - R(\lambda)]^2 \, d\lambda,$$

we simply want each passband of one system to be accurately approximated, in the least-squares sense, by a linear combination of the passbands in the other system, to minimize this distance. This least-squares criterion is very simple to apply, and useful in guiding the design of photometric passbands.

Clearly, in regions where a response function has a steep edge, it is imperative to maintain the placement of that edge very exactly in the right place; for a small error in wavelength, multiplied by a large slope, means a big difference $r - R$ between the intended and realized response functions; a large contribution to the squared-difference integral; and, consequently, large transformation errors. Unfortunately, it turns out that the exact placement of sharp edges is technically difficult. But, now that we know this is a weak spot, we can concentrate attention on shoring it up.

Notice that it is more important to keep a steep edge in the right place than to maintain the effective wavelength of the passband as a whole in the right place. Therefore, one should worry more about steep-sided filters than about the gentler slopes imposed by variations in spectral response from one detector to another.

If we use absorbing glass filters, as in the UBV system, the sharp-cutoff glasses are colored by heat treatment. The thermal history of each piece of glass determines the edge placement, which can therefore vary substantially within a melt. If one looks at the tolerances in the Schott glass catalog, one finds that the stated wavelength tolerances (typically ± 6 nm) correspond to a variation in glass thickness by a factor of 2 on either side of the nominal thickness.

The obvious step to take here is to measure the spectral transmittance of the actual piece of glass to be used, and have it ground to exactly the thickness needed to put the cutoff where we want it. This is, in fact, a relatively inexpensive operation.

There are also technical problems in placing sharp edges exactly in interference filters. But one can negotiate with manufacturers to try to achieve the best possible placement. As most of the cost of making interference filters is setup and tooling charges, we can obtain somewhat better than "stock" accuracy in band placement for only a modest increase in price — perhaps 15 or 20 per cent. Ask your favorite filter maker what it would cost to cut the usual errors in half.

I have said nothing about extinction corrections. They are simply a transformation between an instrument that contains the yellow atmospheric filter and one that does not. At least in the visible, the atmosphere modifies the shape of our passbands only slightly; the typical angle by which atmospheric reddening rotates the B vector of the UBV system is only about four degrees. Notice that the atmosphere modifies the spectral distribution within the passband, but does not displace its edges.

Here again, we see the importance of edge placement. Thus, the Hilbert-space approach shows why atmospheric extinction is so much more tractable than the general transformation problem.

In the infrared, on the other hand, water-vapor absorption can produce much larger effects. This has led to a proposal for modified passbands for infrared photometry; see the paper by Milone and myself at this meeting for the details. Again the emphasis is on band edges: one should try to keep them gently sloping rather than steep, and far enough from strong telluric absorptions to avoid displacement by the atmosphere.

7. The future

In the design of future photometric systems, we need to keep transformability in mind. Because the astrophysically important spectral features make stellar irradiance vectors span a multidimensional manifold in Hilbert space, we must be able to measure significant parts of this whole manifold if we are to capture the astrophysical information in stellar spectra. As missing astrophysical information is what causes transformation errors, we must try to capture all the information available at our chosen spectral resolution.

Although correlations among spectral features reduce the number of necessary dimensions, photometry clearly requires several passbands to span a significant number of the dimensions in which the subspace of interest extends. The practical consequence of this requirement is that we must design systems so that linear combinations of their bands are easily realizable with actual filters and detectors, allowing for realistic manufacturing variations.

For example, we should avoid steep-sided, nearly rectangular passband profiles, if we have a choice. Instead, we should use smooth functions that look like Gaussians, or cosine-squared profiles. One can design systems with gently-sloping rather than steep passband edges, so that a given error in edge placement contributes less to the mean-square difference integral. Notice that this condition is independent of spectral resolution, and applies to spectrophotometric samples as well as to broadband photometry.

Furthermore, small errors in band placement produce differences from a nominal band profile that look like its derivative. Therefore, we can achieve good linear combinations if some of the bands look like the derivatives of others. In particular, we should try to space bands so that their peaks fall at the steepest parts of neighboring bands.

These prescriptions for band shape and spacing are similar to what I have advocated on other grounds, namely, satisfying the requirements of the sampling theorem (Young, 1974, 1988). Evidently both approaches lead to similar conclusions about the design of photometric systems. This is not surprising, as a central feature of both is to obtain complete, rather than partial, information about stellar spectra at low resolution.

If these requirements are observed in the future, we can look forward to considerable improvements in not only photometric accuracy and reliability, but also in the amount of useful astrophysical information we obtain by multicolor photometry. If they are not, we can anticipate continued transformation problems, and continued degradation of the status of photometry.

Acknowledgements:

I thank P. Zvengrowski and K. Salkauskas, of the Univerity of Calgary Mathematics Department, for helpful discussions of Hilbert space and transformation problems. Part of this work was supported by NSF Grant AST-8913050.

References:

Borucki, W.J., ed., 1988, Second Workshop on Improvements to Photometry (NASA CP-10015), Moffett Field, NASA Ames Research Center.

Borucki, W.J., and Young, A.T., eds., 1984, Proceedings of the Workshop on Improvements to Photometry (NASA CP-2350), Moffett Field, NASA Ames Research Center.

Gilliland, R.L., Brown, T.M., Duncan, D.K., Suntzeff, N.B., Lockwood, G.W., Thompson, D.T., Schild, R.E., Jeffrey, W.A., and Penprase, B.E., 1991, Astron. J., **101**, 541.

Hampel, F.R., Ronchetti, E.M., Rousseeuw, P.J., and Staehl, W.A., 1986, *Robust Statistics*, Wiley, New York, p. 4.

Hawes, R.C., 1971, Appl. Opt. **10**, 1246.

Johnson, H.L., 1952, Astrophys. J., **116**, 272.

King, I., 1952, Astron. J., **57**, 253.

Kolmogorov, A.N., and Fomin, S.,V., 1961, *Measure, Lebesgue Integrals, and Hilbert Space*, Academic Press, New York.

Menzies, J.W., Marang, F., Laing, J.D., Coulson, I.M., Engelbrecht, C.A., 1991, M.N. **248**, 642.

Mielenz, K.D., Eckerle, K.L., Madden, R.P., and Reader, J., 1973, Appl. Opt. **12**, 1630.

Oden, J.T., 1979, *Applied Functional Analysis*, Prentice-Hall, Englewood Cliffs, NJ.

Pel, J.W., 1991, in: Philip et al., p. 165.

Philip, A.G.D., Upgren, A.R., and Janes, K., eds., 1991, *Precision Photometry: Astrophysics of the Galaxy*, L. Davis Press, Schenectady.

Quinn, T.J., 1992, Meas. Sci. Technol. **3**, 141.

Sterken, C., and Manfroid, J., 1987, in: *Observational Astrophysics with High Precision Data*, Proc. 27th Liège International Astrophysical Colloq., p. 55.

Young, A.T., 1974, in *Methods of Experimental Physics*, vol. **12**, Part A, *Astrophysics: Optical and Infrared*, N. Carleton, ed., Academic, New York, Chapter 3.

Young, A.T., 1988, in *Second Workshop on Improvements to Photometry*, (NASA CP-10015) W. J. Borucki, ed., p. 215.

Young, A.T., Genet, R.M., Boyd, L.J., Borucki, W.J., Lockwood, G.W., Henry, G.W., Hall, D.S., Smith, D.P., Baliunas, S.L., and Epand, D.H., 1991, Pub. A. S. P. **103**, 221.

Young, A.T., 1992, Astron. Astrophys. **257**, 366.

Discussion

R.M. Genet: *What is the best shape for filter band passes? How should filters be spaced?*

Young: You need to have a filter that looks like the *derivative* of each neighbouring pass-band. To avoid introducing higher-order terms in the Fourier transform of the spectrum, a sinusoidal profile is ideal. Then you need to place the neighbouring bands so their peaks fall on the steepest part of the profile of the central band. Thus this vector-space approach leads to the same conclusion as the sampling theorem does.

L.E. Hawkins: *How significant are ambient temperature and polarization effects when doing precision photometry?*

Young: Polarization effects are significant. A good example of ambient temperature effects can be seen in Park's poster paper. It is an easily measurable effect, particularly with sharp cut-off filters. The Geneva photometers have always been temperature controlled for this reason, and failure of the temperature controller is noticeable in the transformations.

D. Crawford: *I have three questions. Many who have developed a photometric system have chosen filters to isolate and to avoid strong features in the spectrum. Your approach seems to imply something different; use many filters and space them to sample the spectrum. Will you comment on these two different approaches?*

What about $H\beta$ or $H\alpha$ photometry, as such?

It seems to me (as we proposed many years ago) that an excellent photometer would be a well chosen spectrograph with cross-disperser to put the spectrum onto a CCD, filling the chip with all the spectrum to which it is sensitive (sky included). What do you think?

Young: Isolating a feature is fine if your instrumental system never changes, and if no one else tries to reproduce your results. But if you need to transform between two instruments, you need local spectral derivatives. That requires overlapping passbands to measure those local slopes accurately, because there are many stars with strong local spectral features at any wavelength.

On your second question; these are both very strong features. The wide and narrow passbands used represent basis vectors at a large angle, about 70°.

On your third question; spectral dispersion helps define accurate wavelengths, though there is still the problem of relative calibration of the red and blue sides of a passband, Jaap Tinbergen will have more to say about this in his talk later in the Colloquium.

S.B. Howell: *With regard to your comment on non-uniformities in filter glass over large areas, how will this be important for large filters to be used with large format CCD's?*

Young: Laboratory measurements in the optics literature show spatial variation of several percent in a distance of a few millimetres across filters.

Irriducible Elements of Quality in High-Precision Photometry

C. Sterken

University of Brussels (VUB), Pleinlann 2, 1050 Brussels, Belgium

Abstract

A consise overview of the errors related to transformations of data from one photometric system to another is given.

1. Introduction

In the previous paper, Young considers the conditions that allow accurate transformations of measurements from an instrumental photometric system to a standard one for any object spectrum. As he points out, simple photometric systems (like the UBV system), do not have enough bands to capture the astrophysical information needed to distinguish between e.g. an unreddened G star and a heavily reddened B star. Also, systems with more bands (and even intermediate-bandwidth systems) show such deficiencies. This situation leads to complications when measuring binaries, as was already pointed out in 1955 by M. Ovenden at the IXth IAU General Assembly:

> Since the distribution of energy in the combined radiation of an unresolved double star is not necessarily the same as that in the spectrum of a single star of the same mean colour index, the transformation from one photometric system to another on the basis of colour index is not permissible. The problem is particularly serious for eclipsing binaries, where the relative contribution of the components to the total light varies throughout the eclipse. The comparison of different series of observations of the same binary in different photometric systems was possible only when the photometric systems are accurately defined.

The question simply is: how accurate can one perform photometric transformations in real life, that is, in the case where one combines data obtained in many (slightly) differing instrumental systems (eventually at different sites) into large, homogeneous datasets that extend over many months or years of time.

As an example, we reproduce in Fig. 1 the phase diagram of HD 46407, an eclipsing-binary Barium star with period 452.5 days and eclipse depth that barely exceeds $0\overset{\mathrm{m}}{.}02$ in V. The data were obtained in different *uvby* "systems" at a single site (see Jorissen et al. 1992).

[1]Belgian Fund for Scientific Research (NFWO)

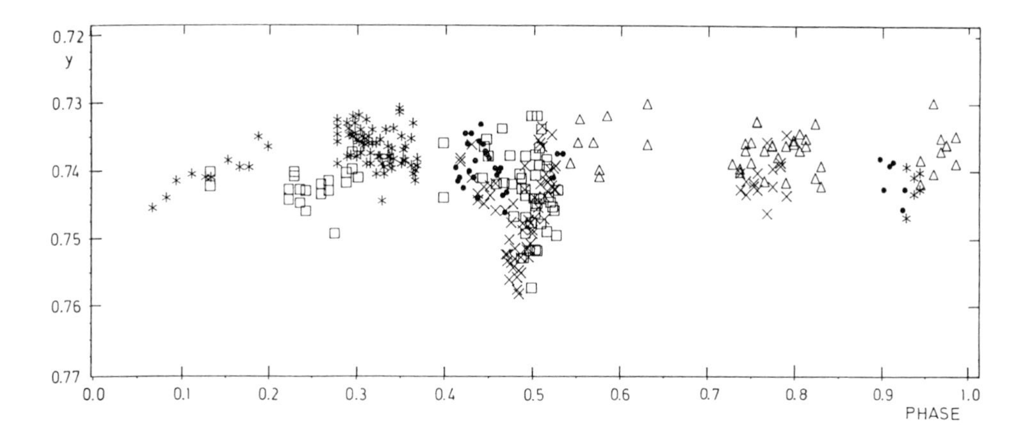

Figure 1: Phase diagram for HD 46407 in y for the $452^{\rm d}\!.5$ eclipsing-binary Ba star. The different symbols refer to different cycles (from Jorissen et al. 1992).

The analysis of data of objects where the light curve exhibits intrinsic scatter, and where the measurements are obtained over a very long time-baseline—eventually at more than one site (networks)—can only yield conclusive results if the error budget is under control and if this budget is well known. The accuracy with which the transformation can be performed depends on

- the degree of excellence of the site(s)
- the quality of functioning of the observer(s)
- the compatibility of the detector(s)
- the unisonance of the photometric system(s)
- the congruence of the reduction schemes

The first item, the photometric quality of the site, seems a trivial factor. However, few observers do realize how crucial this element is for obtaining results of the highest quality. A *necessary* condition for a good site is that the atmosphere above that site is of high transparency. Fig. 2 gives a rough idea of the diversity of occurence of this qualification, but it is obvious that the prerequisite of high transparency must be more than occasionally fullfilled. But, as is well known, good sites may also yield periods of highly-variable (low) transparency, as is the case during several weeks or months following a major volcanic eruption.

The second point is one that is mostly not considered, since it is usually assumed that observers are qualified people. Though it is generally accepted that the replacement of an observer by a robotic system tends to eliminate personal errors, manual

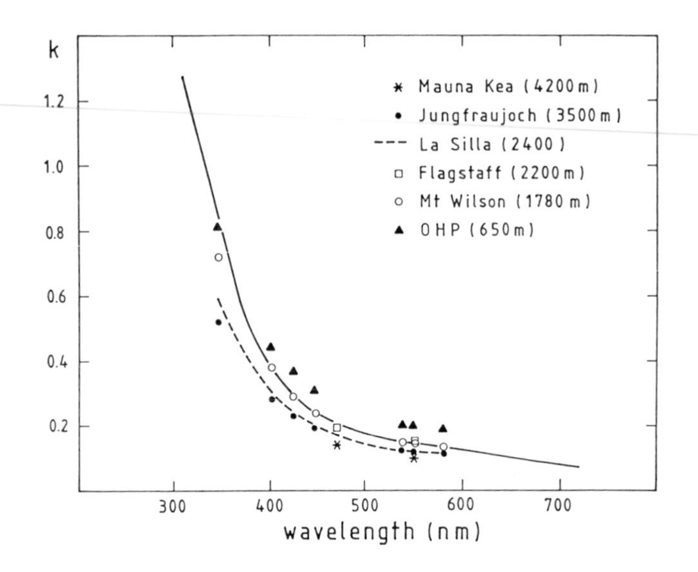

Figure 2: Atmospheric extinction coefficient in function of wavelength for different sites (from Sterken and Manfroid 1992a). The dashed curve is based on 7-color data from Rufener (1986), the full line covers data from Melbourne (1960).

observers can outperform robotic systems when it comes to efficiency and planning (see Sterken and Manfroid 1992b, where it is shown that human observers worked at average airmasses that were systematically lower than when automatic mode was used).

And when it comes to the third point—that is, detectors—one should realize that, through the years, the changes in detectors have drastically modified the original photometric systems (the *UBV* system, for example, has been used as well with photographic emulsions, as with photomultiplier tubes, as with CCD detectors), and one must make sure that such changes (or even changes from one detector of a kind to a second one of a same type) do not reflect into the final data.

For discussing the fourth and fifth point, in what follows we shall assume that we have a perfect site, a perfect detector and photometer, versatile software, and a good observer.

2. Compatibility of systems and congruence of reductions

Manfroid and Sterken (1992) discern *conformity errors* and *reduction errors*. The former arise from the fact that the photometric systems have mutually different passbands, and that there is no way to evaluate the corrections needed to properly transform data from one system to another. The latter are of a purely methodological nature.

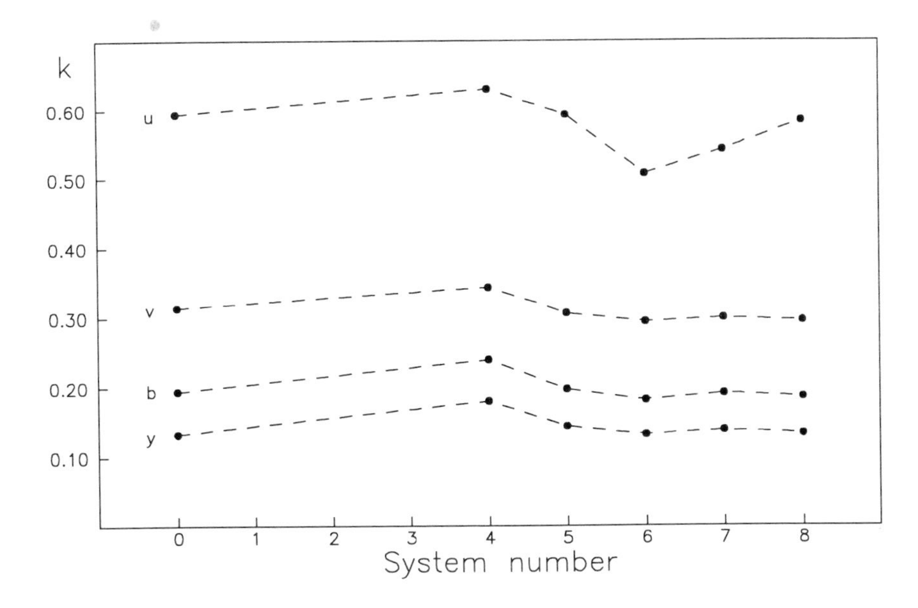

Figure 3: Atmospheric extinction coefficient in u, v, b and y for 6 instrumental *uvby* systems (based on data from Table 2 of Sterken and Manfroid 1992c).

One must not forget that conformity errors are often unavoidable, since prescriptions of a purely practical origin (such as the availibility of a given photometric system at one observing site) may force the investigators to rely on data coming from different such systems. It happens, however, that large passband-mismatches cause only small errors, whereas small passband-mismatches may invoke large discrepancies. As an example of the former situation, we refer to the observing campaign on BW Vulpeculae (a large-amplitude β Cephei star) by Sterken et al. (1992), where very different photometric systems have been put to use, with very little effect on the final astrophysical conclusions. As an example of the second situation, we refer to Sterken and Manfroid (1987), who illustrated that small passband mismatches (in *uvby*) lead to astrophysically contradictory results concerning the evolutionary state of pulsating B stars in the young open cluster NGC 3293. Manfroid (1992) demonstrates that conformity errors have a detrimental effect on the reddening vector, and consequently on the reddening-free indices, and that such is also the case when color indices of composite objects (binaries) are transformed. Let us also point out that deviations from conformity—even if they are small—will reflect in the derived extinction coefficients (see Fig. 3), and that such errors may strongly bias any interpretation of variations in atmospheric extinction.

Reduction errors can be of two kinds: one class is due to the limited range of stellar types used in the color-transformation procedure, and the other category are those errors that result when different transformation schemes are applied (see Manfroid

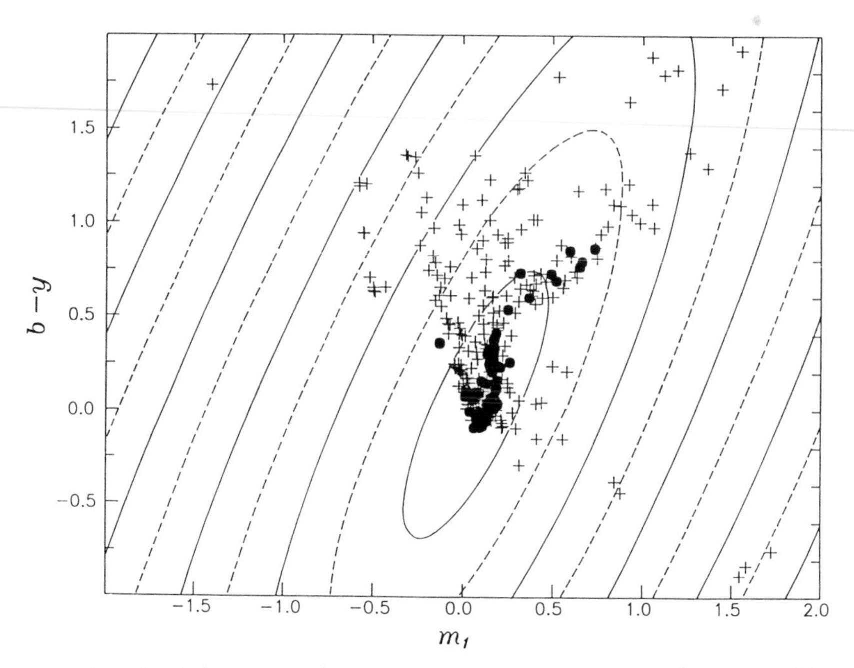

Figure 4: Standard deviation of inter-run variations in m_1 as a function of $b-y$ and m_1. Adjacent contours are separated by $0\overset{\text{m}}{.}005$. Crosses are variable stars, dots are standard stars (from Sterken and Manfroid 1992a).

et al. 1992).

Reduction errors of the first category are typical for batches of data that are treated with a consistent method of reduction, as is the case in long-term and network projects. Some of the parameters in the reduction schemes have larger errors than others (for example, in *uvby* photometry the ratio of the uncertainties of the coefficients in the transformation equation of m_1 to the coefficient related to the $b-y$ transformation may amount to a factor of five), and the resulting errors are appreciably large for stars with extreme color indices. Such effects are random shifts that affect all measurements of a given star by a same amount (during a specific observing run). In Fig. 4 we show, as an example, the standard deviation of the inter-run variations in m_1 as a function of $b-y$ and m_1 for discrete observing runs of several weeks duration each. Adjacent contours are separated by $0\overset{\text{m}}{.}005$. The figure makes clear that the application of differential photometry will not help, unless one compares (exotic) stars that are located very close to each other in that diagram.

Reduction errors of the second type are *extrapolation errors* that occur when different schemes of transformations are applied. Such situations typically occur when data, obtained and reduced by individual observers, are being taken from the literature and are combined in quasi-homogeneous datasets. These errors are of the order of several tenths of a magnitude (see Manfroid et al. 1992) and appear as method-dependent shifts, and show up for stars having color indices that fall outside the range of standard values, where the color-transformation relations are necessarily extrapolated. Again, differential photometry does not help, since the effects do not show up for the comparison stars (if their color indices belong to the range of indices of standard stars). Since usual schemes of color transformation do not adequately represent the effects of interstellar reddening (Manfroid 1992), the application of a variety of differing color transformation schemes must lead to problems.

3. Conclusions

The further deployment of long-term projects and multisite networks is imposing a stringent requirement of homogeneity. The importance and usefulness of any photometric system will depend on the activity of the system in terms of numbers of users, the rate of collection of data, and the geographical spread of the sites where the system is implemented, and on the extent of the ever expanding field of astronomy that must be covered by any system put to use. All of this urgently calls for the elaboration and publication of uniform and solid reduction procedures, not only for the sake of global and long-term campaigns, but also for the small batches of data that, on an individual basis, are submitted for publication.

References:

Jorissen, A., Manfroid, J., Sterken, C. (1992) *Astron. Astrophys.*, **253**, 407

Manfroid, J. (1992) *Astron. Astrophys.*, **260**, 517

Manfroid, J., Sterken, C. (1992) *Astron. Astrophys.*, **258**, 600

Manfroid, J., Sterken, C., Gosset, E. (1992) *Astron. Astrophys.*, in press

Melbourne, W.G. (1960) *Astrophys. J.* **132**, 101

Rufener, F. (1986), *Astron. Astrophys.*, **165**, 275

Sterken, C., Manfroid, J. (1987) *Proc. 27th Liège International Astrophysical Coll.*, 55

Sterken, C., Manfroid, J. (1992a) *Astronomical Photometry, a Guide.* Kluwer, Dordrecht

Sterken, C., Manfroid, J. (1992b) in *Automatic telescopes for photometry and imaging*, eds. S.A. Adelman et al., A.S.P. Conference Series **28**, in press

Sterken, C., Manfroid, J. (1992c) *Astron. Astrophys.* in press

Sterken, C., Pigulski, A., Liu Zongli (1992) *Astron. Astrophys. Suppl. Ser.*, in press

Discussion

T.J. Kreidl: *It certainly seems that not only the tightening of uniformity of filters and the resulting better transformations, but also the economic advantage and improved uniformity of a large number of filter sets could be realised if many observatories pooled their orders for filters.*

Young: Large filter orders do not completely eliminate variations from one filter to the next. Manufacturing processes, even, can produce significant differences between individual filters made in large batches.

Sterken: Sure, your are right, but Kreidl is right too in the sense that the differences between filters will be a magnitude less than those to be expected when each of us separately orders filters to be manufactured. You should not forget also that some of us are somewhat forced to place filter orders with manufacturers inside our own country, a procedure that adds to the confusion, whereas a bulk-purchase should avoid such problems. So the truth lies somewhere in the middle.

W. Tobin: *Do you have any comment concerning whether in Strömgren reductions it is better to use* m_1 *and* c_1 *directly, or reduce in (u-b) and (v-b), which seems simpler, and then compute* m_1 *and* c_1*?*

Sterken: We work with (b-y), m_1 and c_1 for all-sky photometry, and the transformations are established for all stars observed. Reductions with (u-b) and (v-b) instead of m_1 and c_1 are often applied to subsets of stars (e.g. only B, or B and A type stars).

Lessons from Very Long-Term, Very High-Precision Photoelectric Photometry

G. W. Lockwood, B. A. Skiff, and D. T. Thompson
Lowell Observatory, Flagstaff, AZ 86001 U.S.A.

Abstract

Traditional photoelectric techniques can achieve millimagnitude precision if a few simple rules are scrupulously followed. We illustrate the practical limits of differential photometry obtained with a dedicated telescope and photometer. Using a highly homogeneous set of measurements of bright F, G, and K solar analog stars, we present the distribution of observational noise and intrinsic stellar variability over a typical observing season and over the entire 8-year program. Finally, we list Strömgren *uvby* standards which we found to vary slightly.

1. Introduction

Drawing upon experience gained in a 20-year effort to monitor the brightness of solar system objects and solar analog stars as precisely as possible, we show that stellar variability can be found at millimagnitude levels over intervals ranging from days to nearly a decade. Our techniques are neither unique nor remarkable. Rather, they confirm the benefits of a straightforward, albeit steadfastly systematic, approach to differential photoelectric photometry using a dedicated reflecting telescope.

The basic procedures follow a scheme initiated by M. Jerzykiewicz in 1971, and we have found no reason to change them. In brief: (1) observe within one hour of the meridian; (2) use at least two tightly spaced comparison stars and measure them and the program star with equal weight, on the assumption that any or all may be variable; (3) complete an entire "cycle" of differential observations in one filter before moving on to another; (4) use a large star diaphragm--for our bright star program, we use 50 arc-seconds; (5) if observations don't repeat within ~1%, quit observing; (6) leave equipment energized and cooled year round.

2. An example of precision differential photometry

A program of differential, intermediate-band, Strömgren *b* (472 nm) and *y* (551 nm) photometry carried out using Lowell Observatory's 0.5-m reflector illustrates the practical consequences of these recommendations. Since 1984, we have monitored three dozen ordinary main-sequence sunlike stars, measuring the rotational brightness modulations by starspots and long-term luminosity variations (e.g., Skiff and Lockwood 1985; Lockwood and Skiff 1988; Radick *et al.* 1990; Lockwood *et al.* 1992). We find detectable variability at levels as low as 0.0015 mag rms in an 8-year time series of annual mean differential magnitudes. Conversely, the differential magnitudes of several star pairs were stable to 0.0006—0.0010 mag rms over 8 years. Long-term measurements of planets and satellites made with the same instrumentation have produced results of similar precision over nearly two decades (Lockwood and Thompson 1991; Thompson and Lockwood 1992).

For differential observation, the stars are organized into trios or quartets, closely spaced on the sky, each containing one or two program stars and two or three comparison stars. Typically, we obtain 10—20 nights per season and four cycles of observation (y,b,b,y) per night. Figure 1 is a sample light curve for two of the six differential magnitude combinations drawn from the quartet of stars featuring the program star HD82885.

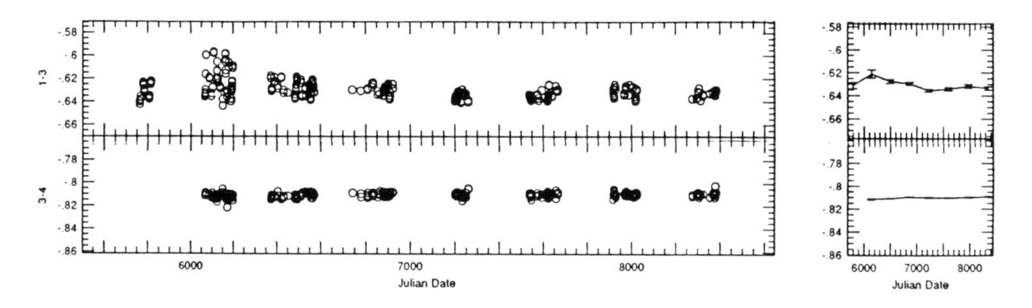

Figure 1. Differential light curves for the program star HD82885 (SV LMi, G8 IV-V, 5.4 mag, a *uvby* standard!) minus the comparison star HD83951 (F3 V, 6.1 mag) and, below, the comparison star HD83951 minus HD83525 (F, 7.0 mag). The b and y magnitudes are averaged in this display. Individual observations are on the left; annual mean differential magnitudes with 95% confidence interval error bars are on the right. SV LMi showed a strong rotational modulation in the second season (Skiff and Lockwood 1985) and an overall, possibly cyclic, long-term variation with an 0.015 mag peak-to-peak amplitude in the annual mean. The amplitude of variation for the comparison star pair was 0.002 mag (0.0008 mag rms).

The y,b,b,y sequence typically requires about 35 minutes of observing time for a quartet (program star, three comparison stars). Thus, the differential magnitudes within a given b or y cycle are determined in less than 10 minutes; this is therefore the time interval over which sky transparency must remain constant.

3. Variability among some *uvby* standard stars

Sixteen of the more than 100 stars on our program incidentally happen to be *uvby* standard stars (Crawford and Barnes 1970). In the upper part of Table 1, we list ten stars that we found to be variable over the 8-year interval, 1984—1992.

The rms deviations of their annual mean magnitudes (b and y averaged) range from 0.0015—0.0053 mag. Seven of the ten are persistently variable, year after year, on a timescale of days to weeks. The remaining six, in the bottom section of the table, display fluctuations from 0.0015—0.0029 mag rms but we were unable to detect variability. HD117176 and HD111812 are, however, variable on a seasonal timescale.

We determine the variability or constancy of a given time series of differential magnitudes according to the significance (99% or better) of the correlation between the magnitudes in pairs of light curves having one star in common. For example, among the six combinations of stars within a quartet group, if the light curves for *star 1 minus star 2* and *star 1 minus star 3* are correlated, we presume *star 1* to be variable. Often, more than one such combination of light curves is available, in which case we require a consistent statistical conclusion. Thus, our determination of variability is more likely to err on the side of declaring a star to be constant when in fact it may be slightly variable. Typically, short-term (night-to-night) variability repeats at about the same level year after year and accompanies long-term variations.

Table 1. Observed Variability of Strömgren *uvby* Standard Stars

Standard stars which are long-term variable stars

HD	HR	Name	Sp. Type	rms 8-yrs	rms-season	Notes
10476	493	107 Psc	K1V	0.0015	0.0017 (const)	NSV 600*
114710	4983	β Com	F9.5V	.0020	.0021 (const)	
201092	8086	61 Cyg B	K7V	.0021	.0021 (var)	NSV 13546*
101606	4501	62 UMa	F4V	.0021	.0018 (var)	
82635	3800	10 LMi	G7.5III	.0021	.0049 (var)	SU LMi
201091	8085	61 Cyg A	K5V	.0022	.0022 (var)	V1803 Cyg
115383	5011	59 Vir	G0Vs	.0033	.0033 (var)	
81997	3759	τ^1 Hya	F6V	.0033	.0025 (const)	
39587	2047	χ^1 Ori	G0V	.0053	.0053 (var)	
82885	3815	11 LMi	G8 IV-V	.0053	.0053 (var)	SV LMi

* these stars are variable, but not for the reasons given in the NSV catalog

Standard stars which appear to be constant

HD	HR	Name	Sp. Type	rms 8-yrs	rms-season	Notes
143761	5968	ρ CrB	G0+Va	0.0015	0.0015 (const)	
103095	4550	Grmb 1830	G8 VI	.0017	.0015 (const)	
13421	635	64 Cet	G0 V	.0020	.0020 (const)	
185144	7462	σ Dra	G0 V	.0020	.0020 (const)	
117176	5072	70 Vir	G4 V	.0024	.0018 (var)	
111812	4883	31 Com	G0 IIIp	.0029	.0030 (var)	

4. Detection of variability on seasonal and longer time scales

Using the correlation properties of pairs of light curves as a basis for variability assessment--whether within a single season or over much longer time intervals--we find that we can usually identify variable stars down to the level of several millimagnitudes without ambiguity unless a majority of stars in a group are variable at about the same level. In this section we describe the properties of our data.

Within the brief interval of a few minutes required to make a differential measurement, the dispersion of observed values is dominated by photon noise, scintillation noise, and, depending on the quality of the night, an additional error due to variations of sky transparency. From a very large set of homogeneous data on bright stars obtained over several years (cf. Lockwood and Skiff 1988), we show in Figure 2 the distribution of cycle-to-cycle repeatability of *b* and *y* differential magnitudes (that is, how well measurements repeat over a time interval of about ten minutes). The distribution is strongly skewed to higher values, probably due to the failure to discard a few marginal nights. While the median values are nearly identical for *b* and *y*, 0.0022 mag rms and 0.0024 mag rms, respectively, the small difference is statistically significant. This may be caused by inadvertently including faint red background stars in the sky measurements. Alternatively, it may be due to transparency fluctuations during the *y,b,b,y* sequence of observation, but we suspect not.

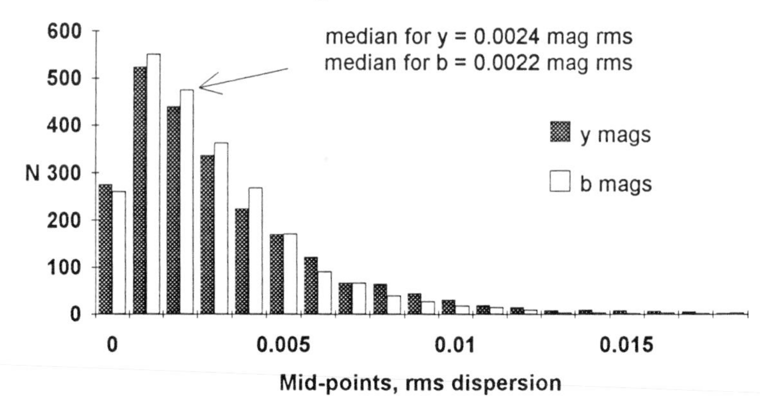

Figure 2. Distribution of cycle-to-cycle repeatability of *b* and *y* measurements made in a *y,b,b,y* sequence. Values larger than about 0.005 mag are probably caused by transparency variations, but the difference in the medians for the two filters may not be (see text).

Over the longer time intervals present in an observing season, intrinsic stellar variability begins to emerge and is generally distinguishable from observational noise on the basis of the correlations available within trios and quartets of stars. Intrinsic stellar variations are slightly greater in *b* than in *y*, sure evidence of an astrophysical rather than an observational effect. Figure 3 shows the distribution of observed night-to-night variation of variable and non-variable F, G, and K, mostly main sequence, bright stars. Observational error is, of course, included in these data, dominating the distribution of values for the non-variable stars. The overlap between non-variable and variable stars is admittedly a region of some uncertainty, but it is fairly safe to say that an observed dispersion greater than 0.005 mag indicates intrinsic stellar variability.

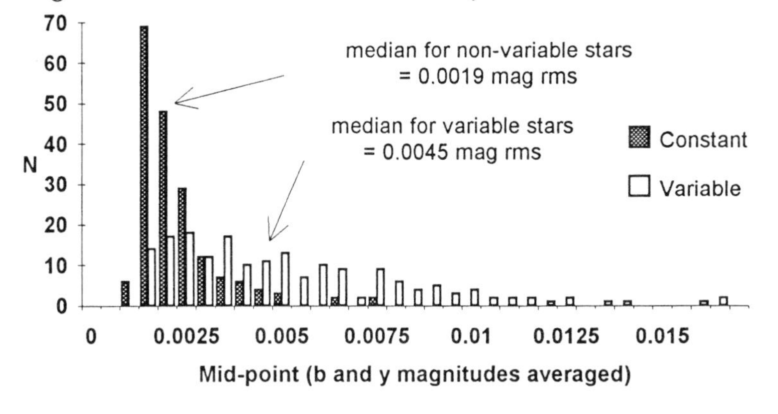

Figure 3. Distribution of the observed magnitude dispersion over an observing season for non-variable and variable stars.

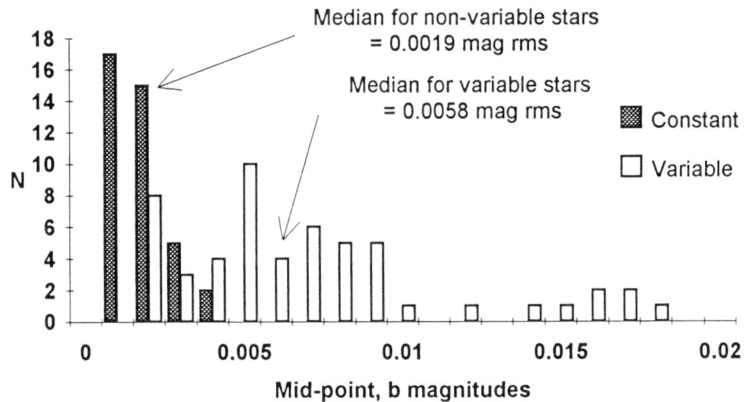

Figure 4. Distribution of the dispersion of annual mean b magnitudes over 8 seasons for non-variable and variable stars.

Figure 4 shows, similarly, the observed distribution of rms fluctuations for annual mean b magnitudes, with a similar distinction between non-variable and variable stars. In this particular sample of stars, the median observed variability is about three times larger than the observational noise.

The small but astrophysically significant difference between the amount of variability in b and y is revealed in Figure 5, where we show the distribution of *differences* between the two filters. Here the non-variable stars center on zero, indicative of color-independent observational error, while the variable stars are displaced toward an excess of b variability relative to y variability.

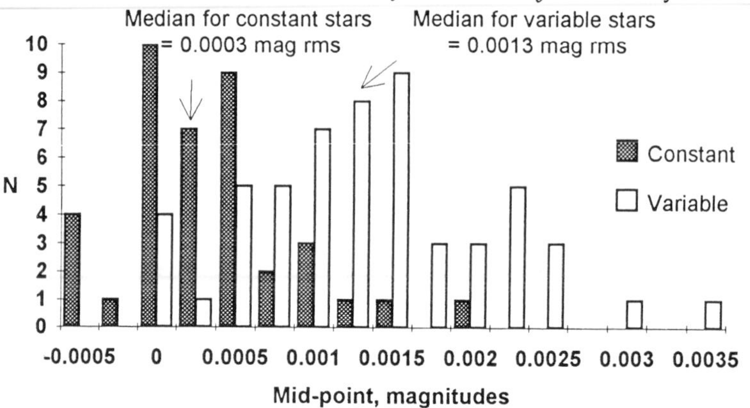

Figure 5. Distribution of the differences between the rms variability in b and the rms variability in y. The distribution for the non-variable stars centers on zero, as expected for a dispersion arising mainly from observational error.

5. Conclusions

Careful, systematic differential photometry can be used to study the variability of bright stars over time intervals from days to many years down to the level of a few millimagnitudes. At this level of precision, a substantial fraction of ordinary solar analog dwarf stars, including some well-known photometric standard stars, exhibit variability. We are continuing to investigate variability in hopes of determining whether cyclic luminosity cycles are a common feature of cool dwarf stars.

The Division of Atmospheric Sciences of the National Science Foundation and the Lowell Observatory have steadfastly supported this work since 1972.

References:

Crawford, D. L., Barnes, J. V. , 1970, Astron. J. **75**, 978.

Lockwood, G. W., Skiff, B. A., 1988, *Luminosity Variations of Stars Similar to the Sun*, Final Report, AFGL-TR-88-0221, Air Force Geophysics Lab, Hanscom Air Force Base.

Lockwood, G. W., Skiff, B. A., Baliunas, S. L., Radick, R. R., 1992, submitted to Nature.

Lockwood, G. W., Thompson, D. T., 1991, Nature **349**, 593.

Radick, R. R., Lockwood, G. W., Baliunas, S. L. ,1990, Science **247**, 39.

Skiff, B. A., Lockwood, G. W. ,1985, Publ. Astron. Soc. Pacific **97**, 904.

Thompson, D. T., Lockwood, G. W., 1992, JGR Planets, in press.

Discussion

R.M. Genet: *Do you control the temperature of the filters? Have you seen filter aging effects?*

Lockwood: No we've looked for temperature effects (nightly, seasonal) but have found none. We see a slow change in colour terms over 20 years, but since it occurs in both filters we believe it's due to the aging of the photomultipliers (an EMI 6256s tube).

D.S. Hall: *How often do you determine your transformation coefficients: nightly, seasonally, annually or what? I ask because transformation coefficients can be affected directly by temperature and so can be used to compensate for temperature change.*

Lockwood: We compute transformation second-order coefficients in batches of 10 nights or so. However we've not noticed a seasonal variation related to temperature effects, (probably because the colour differences in our differential groups are small). The colour range of our standard stars is from (b-y) $\sim$ 0.0 to (b-y) $\sim$ 0.4; the color difference in differential pairs seldom exceeds half this range, i.e $\Delta(b-y) < 0.2$.

E.F. Milone: *CF UMa is not a variable star?*

Lockwood: We've not been able to confirm its variability.

E.F. Milone: *How do you maintain the stability of your PMT power supply and of PMT cooling?*

Lockwood: The PMT is thermo-electrically cooled and maintained at -15°C. The HV power supply is never turned off and we don't know how stable it is. We do observe a small change in absolute sensitivity of the photomultipliers in the first hour or so each night as the equipment responds to changing ambient temperature. This amounts to $\sim$ 1%. We suspect a change (drift) in the HV or the amplifier/discriminator gain, and/or band-width.

A.J. Penny: *Has the reflectivity of the telescope mirror changed?*

Lockwood: Yes — when we aluminized it in the mid-eighties, (probably for the first time since Harold Johnson used the telescope in the 50s), we gained 30% more photons.

Accuracy in variable star work: The three-star single-channel technique

Michel Breger

Institute of Astronomy, University of Vienna, Austria

Abstract

The three-star technique for photoelectric photometers is discussed. The technique is used to study millimag variations of variable stars with periods in the range of 30 minutes to a few days. It is emphasized that for highest precision two comparison stars should be used and observed as often as the variable star. The different types of numbers used to express the precision of measurement are discussed together with possible misinterpretations.

Finally, the successful use of an extinction coefficient derived from differential extinction between the three stars is illustrated with measurements showing an extreme amount of instrumental sensitivity drift.

1. The standard 3-star technique

For over two decades the three-star observing technique used with standard single-channel photolectric photometers has been the primary method to study short-period variable stars with periods longer than 30 minutes (f $<$ 50 cycles d^{-1} or f $<$ 0.5 mHz) up to a few days. Although a precision of 2 mmag per single measurement had already been achieved more than 25 years ago (Breger 1966), the improvement since then has been minor. The present limit to the photometric precision is still just slightly under 1 mmag with good photoelectric photometers under excellent weather conditions. In fact, considerable observational care has to be taken in reducing the observational errors sufficiently in order to obtain the 'old' limit near 1 mmag precision per single measurement.

The three-star observing technique consists of observing the variable star, V, together with two comparison stars, C1 and C2, using the same instrumental equipment. Cycles of [C1 –V – C2] – [C1 – V – C2] – etc. are chosen. In order to keep scintillation errors to 1 mmag or less, for telescopes in the 0.5 to 1-m class integration times between 30 and 60 s are usually selected. Furthermore, with such long integration times, photon statistics is rarely a problem for relatively bright stars. After considering the additional time required for sky measurements, telescope setting and centering of the star, a complete cycle can be completed in about five minutes. Note that all three stars are observed for identical lengths.

Care must be taken to select the comparison stars in such a manner that the properties of the three stars are as similar to each other as possible. In particular, the stellar brightnesses, positions in the sky and spectral types are important. Since all three conditions cannot usually be fulfilled, reasonable compromises must be made. Of course, close comparison stars of spectral type K would not be used to study the variation of B stars! To reduce the errors of measurement, only one (or possibly two) filters are usually used.

During the data reduction, the measurements of star V are reduced relative to both comparison stars. Care needs to be taken to choose the best interpolation scheme and not to put undue weight on single measurements. Some numerical experimentation to choose an optimum interpolation scheme is recommended.

The three-star technique relies on the cancellation of most sources of error and avoids the drift problem experienced with multichannel instruments. The price paid for this success lies in the low duty cycle and the restriction to periods of 30 minutes or longer.

An example of the technique is shown in Figure 1, where unpublished measurements of the δ Scuti variable HR 7222 obtained at McDonald Observatory are shown. Two comparison stars, HR 7263 and HR 7286, were used. The measurements were made with the B filter, extinction corrections were applied, and the average magnitudes subtracted for clarity. The data were reasonably free of systematic drifts.

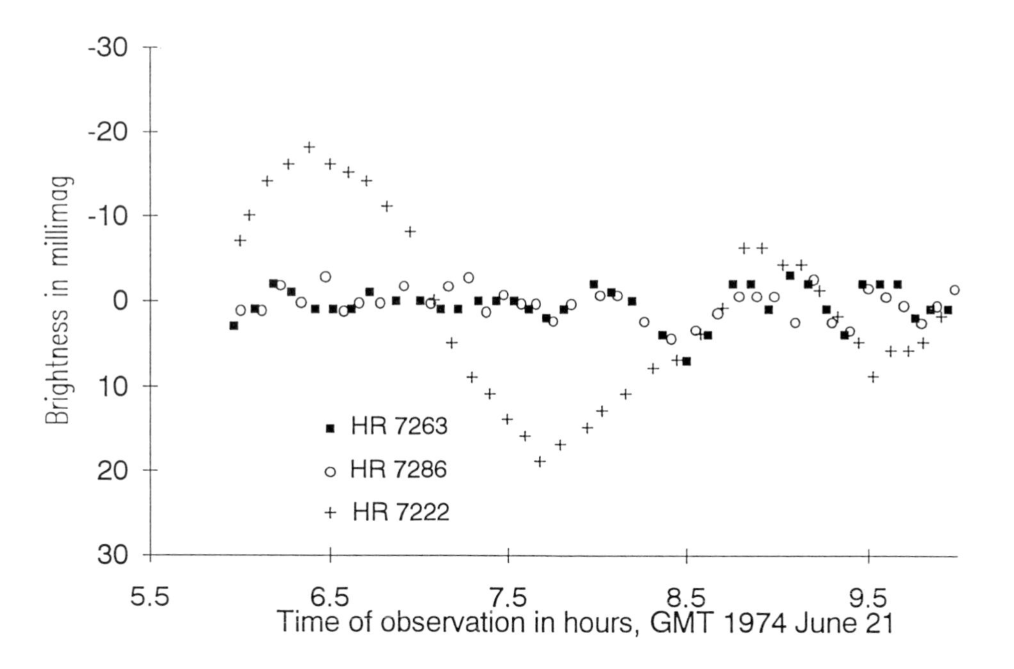

Figure 1 A typical example of the three-star technique. HR 7222 is a δ Scuti variable. The nonrepeatability from cycle to cycle is normal for these stars due to the presence of multiple periods. The comparison stars agree with each other to ± 2 mmag.

2. The variable-comparison-check star technique

A popular variation of the standard three-star technique involves measuring the second comparison star less often, possibly only once every half hour. This star, now called a check star, is therefore used only to provide a rough check of the constancy of the primary comparison star. A motivation for applying this variation lies in the fact that the duty cycle of the variable star measurements can be increased up to 40%, which might be important for very short periods under one hour.

For variable stars with periods of one hour or longer the standard method with two comparison stars is to be preferred. Some of the advantages of the standard method are:

* The possibility of variability of comparison stars cannot be excluded. The check for constancy is much more reliable when both comparison stars can be compared every five minutes rather than every half hour.

* If one of the comparison stars turns out to be variable, the use of two comparison stars might permit the observations to be saved. When all three stars are observed equally well and often, inspection of the three power spectra of the reduced magnitude differences C1 – C2, V – C1, V – C2 might reveal which star is constant. Note here that one of the three differences should contain the variability information of both variable stars! The method works best if the measurements are of high quality and the time-scales of variability of the two variable stars are different. The latter is often the case.

* Inspection of the light curve of the difference (C1 – C2) reveals those times at which the precision is low and the data should be discarded. This is especially useful for the beginnings and ends of the nights.

* The precision of measurement for each night can be determined from a comparison of the two comparison stars. Knowledge of this number is important for the subsequent analysis of the variable star measurements.

* If all three stars have similar spectral types, it might be possible to derive a differential extinction coefficient from the two comparison stars. If there exists instrumental sensitivity drift, this method to determine extinction becomes extremely useful.

3. What kind of precision are they talking about?

With discussions of millimag or even micromag precision, it is important to define the different types of precision. In the variable-star field, the uncertainties of measurement are usually expressed in one of two ways:

Precision per single measurement. This description of observational quality can be understood as the precision to which the difference between two or more constant stars can be measured, i.e. the repeatability in the instrumental system. (Note here that the precision of the magnitude difference between two stars is not numerically the same as the precision of the individual measurements.) Since tranformation errors can be avoided (no transformations), the measurements can be quite precise.

In fact, a precision of 2 mmag per single measurement had already been obtained more than 25 years ago. Today, the typical precision obtained during multisite campaigns with a variety of telescopes and photometers is about 3 mmag. The best photometry published has uncertainties between 0.5 to 1 mmag. The reason for this relative lack of improvement lies in the observational difficulties in the millimag range. For an excellent discussion of the errors of measurement and their reduction we refer to Young et al. (1991).

Noise in an amplitude, frequency diagram. The relatively new field of asteroseismology requires that for pulsating stars amplitudes of a millimag or smaller be extracted from photometric data. This requires long data strings and stable pulsations. A different description of precision is often used to describe the ability to find such small pulsations, viz. the noise in an amplitude, frequency (Fourier) diagram near the frequency of interest. This noise is the amplitude of a sine curve at a hypothetical frequency forced through data which contain no such variation.

Gilliland and Brown (1992) in their analysis of CCD measurements conclude that the detection of amplitudes as small as 15 μmag should be possible.

An enthusiastic and possibly misleading comparison of the precision obtained with different techniques is given by Kjeldsen and Frandsen (1992, hereafter called KF). In this very informative paper they compare high-precision CCD photometry with standard photoelectric techniques, including the three-star technique described in this paper. In their Table 1 the noise at 3 mHz is compared. They list $\sim$ 1 mmag/min for CCD techniques, $\sim$ 2 mmag/min for single-channel work with no comparison stars and $\sim$ 8 mmag/min for single-channel work with two comparison stars.

Are measurements with comparison stars really so much worse than those without comparison stars? After noticing that the evaluation was made at periods of 5 mins (3 mHz), we can explain most of these numbers without having to examine the potential of CCD techniques or comparing the observational quality of the different papers used in the comparison. Since one of the papers cited contains data familiar to the author (HR 2724, see next section), we can use these data to examine the analysis of KF. The SAAO observations of HR 2724 used in the table are $\sim$ 200 hours of single-channel measurements of a cycle of up to 4 stars measured in 4 colors each. (Some averaging was undertaken before calculating an amplitude-frequency diagram.) We note:

* The observations of HR 2724, as listed, are spaced 20 mins apart. Hypothetical amplitudes of sine curves at periods $\sim$ 5 mins (3 mHz) are not meaningful. The original publication, naturally, does not present a power spectrum in this range of periods. This and other data obtained with the three-star technique *should not be used for frequencies beyond the Nyquist frequency*, as was done in the table of KF.

* The duty cycle of the data < 10%. This can partially explain the fact that in the table of KF observations made without comparison stars appear to be more accurate. Nevertheless, in the range of periods intended for the three-star technique (and discussed in this paper), comparison stars must be observed.

We conclude that the seemingly poor performance of the three-star method as listed by KF is caused by using its data for purposes for which they had not been intended and should never be used. At periods of one hour or longer the game is completely different than at 5 mins. Any measurements made without comparison stars would probably show additional errors due transparency variation and equipment drift. (Of course, here the situation is reversed: these measurements without comparison stars are not intended to be used to search for such long periods.) Furthermore, a realistic analysis of the noise needs to include the effects of aliasing (caused by day-time observing gaps) of the multiple periods.

Finally, it needs to be emphasized again that the 'best' observing method strongly depends on the length of the periods one wants to determine.

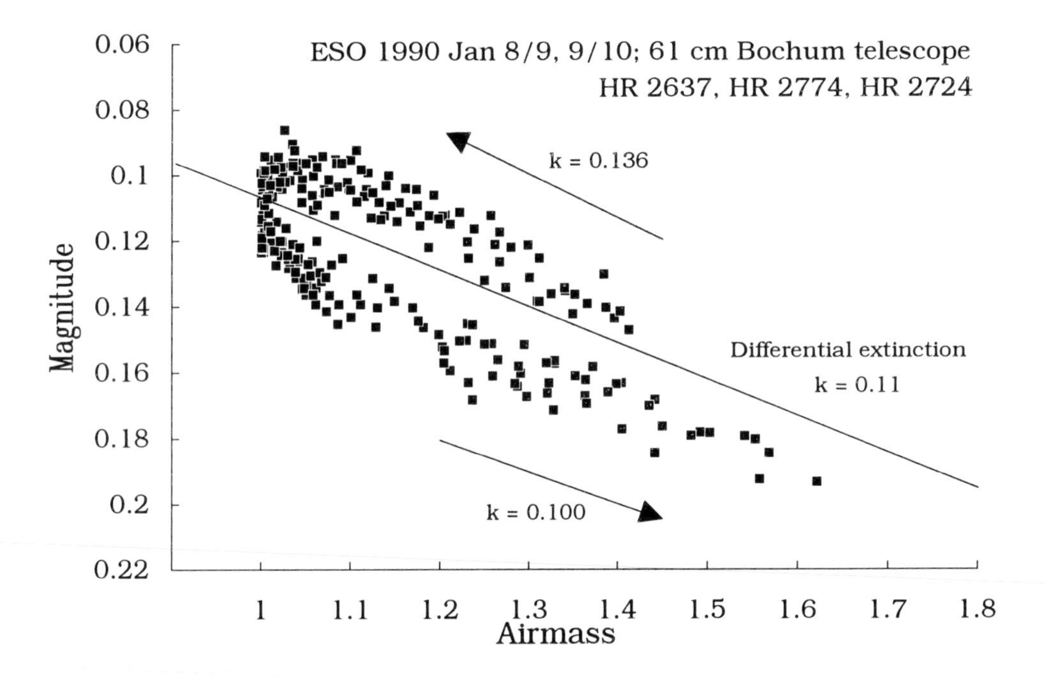

Figure 2 Bouguer plot of measurements (squares) of three stars obtained during two nights during which the instrumentation with instrumental sensitivity drift. The feature at small air masses is typical of drift. A solution for the extinction coefficient from differential extinction shows k = 0.11, in agreement with values obtained on subsequent nights with less or no instrumental sensitivity drift.

4. Some comments on extinction

The use of close comparison stars eliminates uncertainties in the extinction coefficients, at least to the first order. However, the remaining small errors are important since they are systematic. Different recommendations in the literature on how to correct for extinction include the use of

* mean seasonal extinction coefficients
* nightly extinction coefficients
* extinction coefficients changing throughout the night.

Which of these reduction schemes should be adopted? We cannot give a universally valid recommendation. Since the present technique results in a large number of measurements obtained at different air masses, a solution might be to determine the extinction differentially between the stars, i.e. to examine the relation between Δ(mag) and Δ(air mass) between the stars. This assumes, of course, that the stars are situated several degrees apart and have similar spectral-energy distributions.

Figure 2 gives an example of extreme instrumental sensitivity drift. One of the three stars shows millimag variations, which can be disregarded in this context. The measurements were made by H.-G. Grothues (see Breger, Balona and Grothues 1991). This example shows clearly that solving for time-variable extinction coefficients would not be suitable for these data.

References:

Breger, M. 1966, ApJ 146, 958
Breger, M., Balona, L. A. and Grothues, H.-G. 1991, A&A 243, 160
Gilliland, R. L. and Brown, T. M. 1992, PASP 104, 582
Kjeldsen, H. and Frandsen, S. 1992, PASP 104, 413
Young, A. T., Genet, R. M., Boyd, L. J., Borucki, W. J., Lockwood, G. W., Henry, G. W., Hall, D. S., Pyper Smith, D., Baliunas, S. J., Donahue, R., Epand, D. H. 1991, PASP 103, 221

Discussion

C.L. Sterken *I recall that on one occasion we saw that the non-differential signal was more precise than the differential result: we were observing Spica (α Vir) during a fortnight of continuous, excellent, weather conditions at La Silla, with the Danish 50 cm telescope in 1985.*

Breger: If your measuring errors are purely random in nature, the difference between two stars will be $\sqrt{2}$ times the individual errors. Such excellent observing conditions, with excellent equipment, I have not seen personally so far, but one should always check that reductions of the variable relative to the comparison stars does not increase the scatter.

R.M. Genet: *We have made observations with a precision of 0.0005 magnitude. These observations matched predictions by Andy Young of observational error versus air mass. Thus reducing errors further requires a larger telescope or longer integration times.*

Breger: The present limit with standard photoelectric photometers seems to be ± 0.0005 mag, which has been obtained by several astronomers. But can we measure more precisely, or are all stars variable at that level (with periods > 24 hours)?

R.R. Shobbrook: *Scintillation requires some 50s - 100s integration times on telescopes of $D \sim 0.5$ m. One can improve precision slightly by fitting a polynomial (in time, during the night) to the magnitudes of the comparison stars. The first term is the gain drift and the other terms account for the occasional $\sim 1\%$ transparency variations.*

Breger: In practice, we also have found it useful to fit polynomials through the data or to draw a curve by eye. In this respect it is important for the fitted curve not to follow every little wiggle.

C.L. Sterken: *Differential Bougner method: Observers of δ Scuti stars, and other variables with periods of the order of hours, frequently monitor such stars throughout a complete night. If, as you suggest, both comparison stars are equally frequently observed, and if the stars are at an appropriate declination, the determination of extinction coefficient is often done using the original Bougner method, viz., from a plot of observed magnitude versus air mass X. Sometimes, however, the so-called "differential Bougner method" is used, viz a plot of differential magnitude (of the comparison stars) versus associated differential air mass. Provided that the ΔX-range remains substantial, that method should yield the same result as the classical Bougner method when atmospheric conditions are good, but the method may break down when this is not the case. The classical method is more likely to show that conditions are not optimal.*

Atmospheric Intensity Scintillation of Stars on Milli- and Microsecond Time Scales

D. Dravins, L. Lindegren, E. Mezey

Lund Observatory, Box 43, S-221 00 Lund, Sweden

Abstract

Stellar intensity scintillation on short and very short time scales (≈100 ms – 100 ns) was studied using an optical telescope on La Palma (Canary Islands). Photon counting detectors and real-time signal processing equipment were used to study atmospheric scintillation as function of telescope aperture size, degree of apodization, for single and double apertures, in different optical colors, at different zenith distances, times of night, and seasons of year. The statistics of temporal intensity variations can be adequately described by log-normal distributions, varying with time. The scintillation timescale (≈10 ms) decreases for smaller telescope apertures until ≈5 cm, where the atmospheric 'shadow bands' apparently are resolved. Some astrophysical sources may undergo very rapid intrinsic fluctuations. To detect such phenomena through the turbulent atmosphere requires optimized observing strategies.

1. The challenge of high-speed astrophysics

The present study is part of a broader program in high-speed astrophysics, an exploratory project entering the domains of milli-, micro-, and nanosecond variability. Possible future studies include:

- Plasma instabilities in accretion flows onto white dwarfs and neutron stars
- Magneto-hydrodynamic instabilities in accretion disks around compact objects
- Radial oscillations in white dwarfs, and in neutron stars
- Optical emission from millisecond pulsars
- Emission fine structure ('photon showers') from pulsars and other objects
- Photo-hydrodynamic turbulence ('photon bubbles') in extremely luminous stars
- Stimulated emission from magnetic white dwarfs ('cosmic free-electron laser')
- Non-thermodynamic-equilibrium photon statistics from exotic sources

To study such rapid phenomena is not possible everywhere in the spectrum. For high photon energies (e.g. X-rays), a limit is set by the observable photon count rates with current spacecraft, and for low energies (e.g. radio), counting of individual photons is not yet feasible, precluding studies of certain quantum effects. The most promising domain therefore seems to be the ground-based optical, the region for which we have constructed a dedicated instrument, '*QVANTOS*' ('Quantum-Optical Spectrometer'). Its design considerations included, in particular, the following points:

Data glut: 1 ms resolution means 3.6 million points an hour, and 100 Mb in three nights. However, 1 μs gives 100 Gb, and a mere plot of the light curve (at laser printer resolution) would subtend $\simeq$100 meters per second. Thus, there is a need for real-time data analysis, and a reduction of the data to manageable statistical functions only.

Faint sources: Many interesting sources are faint. To study variability on time-scales *shorter* than typical intervals between successive photons, a statistical analysis of their arrival times is required to test for deviations from randomness.

Time resolution: The highest time resolution that is meaningful, is set by quantum-optical properties of light, e.g. the bunching of photons in time. Such properties are fully developed on times equal to the inverse frequency bandwidth of light ($\simeq 10^{-14}$ s), but may be detectable also on much longer (nanosecond) timescales.

The terrestrial atmosphere causes rapid fluctuations of the source intensity. Accurate determinations of astrophysical fluctuations require a correspondingly accurate measurement, calibration, understanding, and correction for atmospheric effects.

The first version of our instrument (*QVANTOS Mark I*) has been extensively used to study atmospheric scintillation using the Swedish 60-cm telescope, located at 2400 m altitude at the observatory on Roque de los Muchachos (La Palma, Canary Islands). In this paper, we describe some of the results from these observations.

2. Observations, instrumentation, and data reduction

Systematic observations of stellar intensity scintillation on timescales between $\simeq$100 ms and $\simeq$100 ns were made. About 25 full nights were used to study the dependence on telescope aperture size, degree of apodization, for single and double apertures, for single and binary stars, in different colors, using different passbands, at different zenith distances, at different times of night, and different seasons of year. Scintillation properties were recorded as temporal auto- or cross covariance functions, and intensity probability distributions; sometimes supplemented by simultaneous video recordings of the stellar speckle images, as well as seeing disk measurements in an adjacent telescope.

The photometer optics and detectors (very fast photon-counting photomultipliers) were placed at the Cassegrain focus, with cables leading to digital signal processors and real-time data displays on the observing floor. Computations are made of the temporal autocovariance (ACO), cross covariance (CCO) and photon count distribution (probability density, PDE). The time resolution (sample time) is selected in software to between 20 ns and 1 s. With 64 registers for data storage, the covariance measurements are made with 64 different time delays simultaneously. In PDE mode, the number of occasions when $0, 1, \ldots, 63$ photons are recorded during one sample interval, is measured.

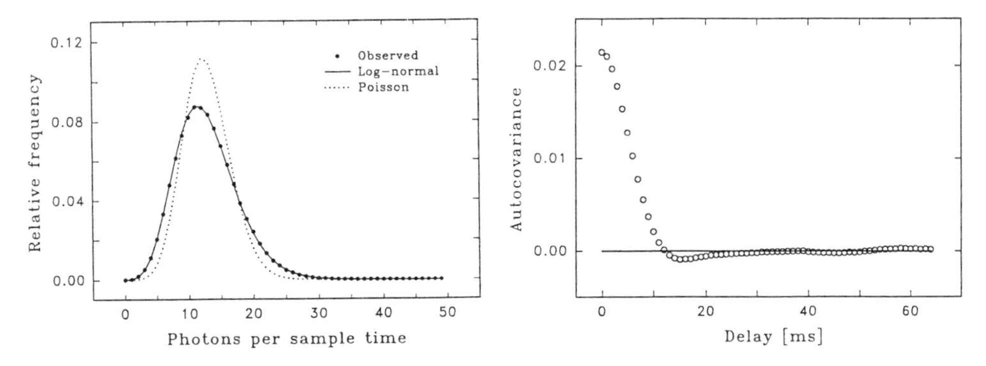

Figure 1. To the left a typical observed photon-count distribution is shown. A log-normal distribution is fitted to the data. The Poisson distribution corresponding to zero atmospheric intensity fluctuation is also shown. To the right is a typical autocovariance function of the starlight intensity, measured at time delays 1, 2, 3, etc. milliseconds, here integrated for 100 s.

The function $ACO(\tau)$ measures the strength of the intensity fluctuations for different delays τ. It satisfies $ACO(0) = \sigma_I^2$ and $ACO(\infty) = 0$, where σ_I is the root-mean-square value of $(I - <I>)/<I>$. The Fourier transform of $ACO(\tau)$ yields the scintillation power spectrum. The autocorrelation is the normalized function $ACO(\tau)/ACO(0)$.

The PDE is a measure of the statistics of the intensity fluctuations. The observed distribution is a convolution of the atmospheric fluctuations with the Poisson distribution (inherent in the detection process) of the photon counts. We have fitted log-normal distributions to all our data (several hundred measurements under different conditions), and found it to be an adequate distribution for $\sigma_I^2 < 0.4$.

Elementary theory predicts a functional dependence of σ_I^2 on aperture diameter D, wavelength λ, and zenith distance z, with the limits for large and small apertures:

$$\sigma_I^2 = \begin{cases} \text{const} \cdot D^{-7/3} (\sec z)^3 \int_0^\infty h^2 \, C_N^2(h) \, dh, & D >> r_0 \quad \text{(1a)} \\ \text{const} \cdot \lambda^{-7/6} (\sec z)^{11/6} \int_0^\infty h^{5/6} \, C_N^2(h) \, dh, & D < r_0 \quad \text{(1b)} \end{cases}$$

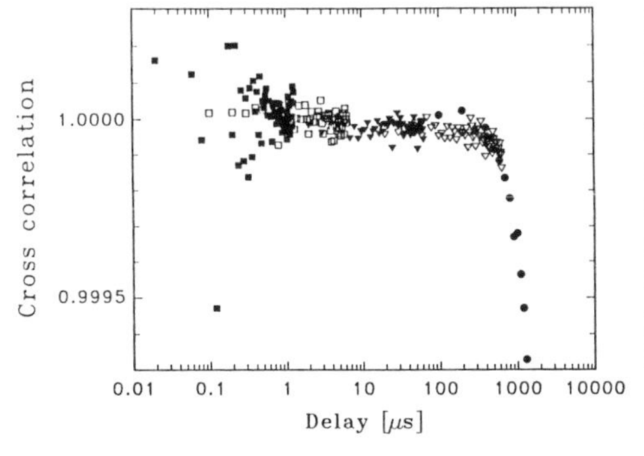

Figure 2. Stellar scintillation on very short time-scales. This plot shows the small deviations from unity close to the origin. This particular measurement (λ 550 nm, 60 cm aperture) was made as a cross correlation between the signals from two detectors, each measuring the same star through a beamsplitter. This arrangement minimizes effects of photomultiplier after-pulsing. Photon noise causes a spread of data points for the shortest delays.

(e.g. Roddier 1981), where r_0 is Fried's parameter and $C_N^2(h)$ the refractive index structure coefficient as function of height h above the telescope.

Several properties can be understood in terms of the illumination pattern caused by diffraction in inhomogeneities in high atmospheric layers. These structures are carried by winds, resulting in 'flying shadows' on the ground (e.g. Codona 1986).

3. Scintillation on extremely short timescales

Figure 2 shows an example of the autocorrelation for very short time delays. On longer timescales, the curve is smoothly decreasing (cf. Fig. 1b), but at shorter delays there is largely a lack of structure. In this domain one may expect to find signatures of the *inner scale* of atmospheric turbulence, i.e. the smallest eddies with temperatures, etc. distinctly different from their surroundings. From other studies, that scale is expected to be $\simeq$3 mm. With the 'flying shadow' patterns carried across the telescope by windspeeds of $\simeq$10 m/s, a linear size of 3 mm corresponds to characteristic times of $\simeq$300 μs.

4. Aperture dependence

Rapidly changeable mechanical masks in front of the telescope were used to study the dependence on different types of aperture. Figure 3 shows how the fluctuations change with aperture size until $D \lesssim 5$ cm, where the structures in the 'flying shadows' on the ground appear to be resolved. The spatial properties were also studied with *apodized* (i.e. 'unsharp') telescope apertures, achieved by placing suitably airbrush-painted thin mylar films in front of the telescope. Such apodized apertures are expected to generate

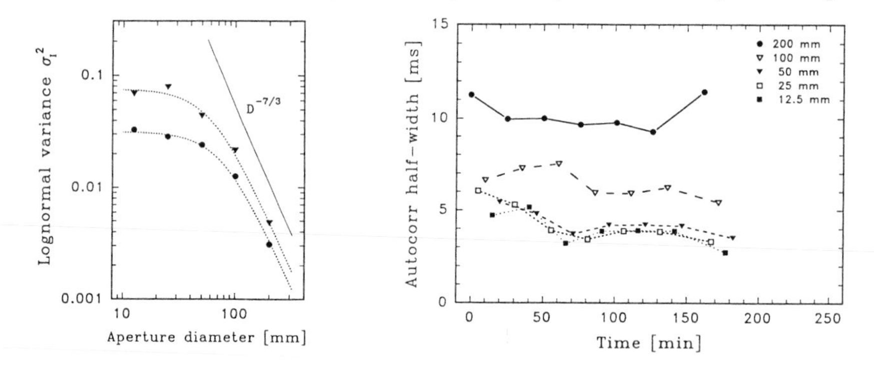

Figure 3. Aperture dependence of the scintillation variance σ_I^2 (obtained by fitting a log-normal distribution to the observations; left), measured at two different times. The slope $-^7/_3$ corresponds to Eq. (1a), the expected behavior for large apertures. At right is the aperture dependence of autocorrelation timescale [width at half-maximum of ACO(τ)], as well as typical changes over a few hours. Both the half-widths and the amplitudes become essentially independent of aperture size for $D < 5$ cm, where the structures in the 'flying shadows' on the ground appear to be resolved.

less scintillation power at high temporal frequencies, since the bright and dark bands in the 'flying shadows' are then subject to a more gradual intensity cutoff at the edge of the telescope aperture (Young 1967).

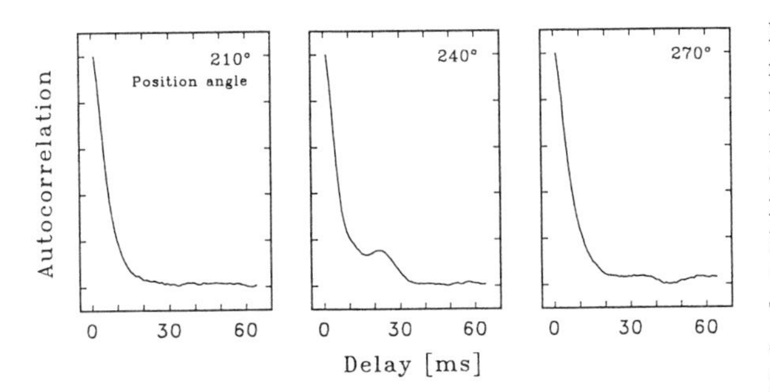

Figure 4. Intensity autocorrelation, measured through a mask with *two* holes. If the same 'flying shadows' pass both holes, a secondary peak appears. From the spacing (here 30 cm), and position angle of the holes, the speed and direction of the 'flying shadows' can be determined. Typical delays indicate a speed of $\simeq 15\,\mathrm{m/s}$, apparently the wind-speed in the upper atmosphere.

5. Wavelength dependence

This was studied with different broadband filters and different aperture sizes. For apertures smaller than $\simeq 10\,\mathrm{cm}$ there is a measurable difference for different colors. For small apertures, our data agree well with the theoretical slope $-7/6$ in Eq. (1b), but the wavelength dependence rapidly diminishes for larger apertures. At shorter optical wavelengths, the fluctuations are thus more rapid, and show a greater variance.

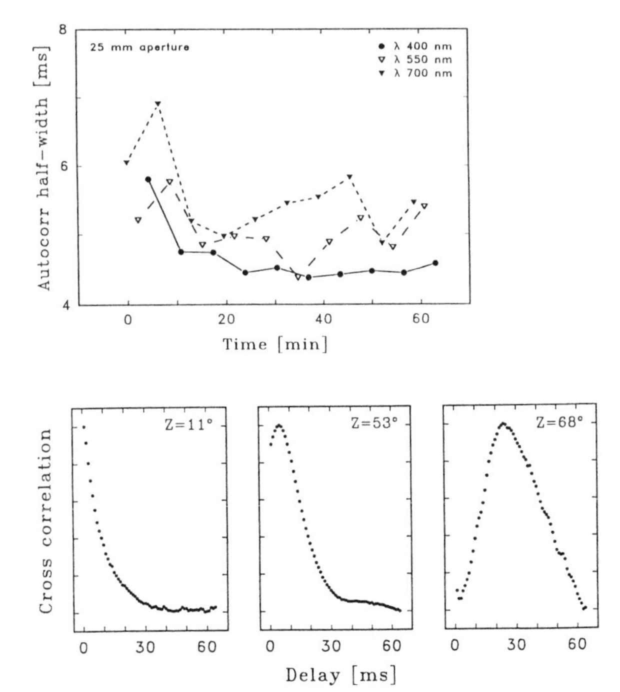

Figure 5. Wavelength dependence of the scintillation timescale, measured with a 25 mm aperture, also showing typical atmospheric changes during one hour.

Figure 6. *Cross correlation* between fluctuations at λ 400 and 700 nm, and its zenith-angle dependence. Near zenith the fluctuations are simultaneous, but with increasing zenith angle a time delay develops. This is due to atmospheric dispersion, which causes chromatic displacements of the 'flying shadows'.

6. The search for rapid astrophysical phenomena

The analysis of these and other data is in progress with an aim of better understanding the properties and physical origins of stellar scintillation (e.g. Jakeman et al. 1978; Stecklum 1985), and for comparing the La Palma conditions to those at other observatories, such as Mauna Kea on Hawaii (Dainty et al. 1982).

The final aim, however, is to utilize this understanding in order to detect rapid fluctuations in astronomical objects. From laboratory experiments, Figure 7 shows an example of how microsecond scale fluctuations in light intensity can be detected. For astronomical sources, such fluctuations will be superposed on the most rapid atmospheric scintillations (cf. Fig. 2), and a careful segregation of atmospheric effects must precede any astrophysical conclusions. It appears that simultaneous observations of a calibration star will be required, whereupon variability in the target may be identified from differences of the intensity statistics between it and the calibration star.

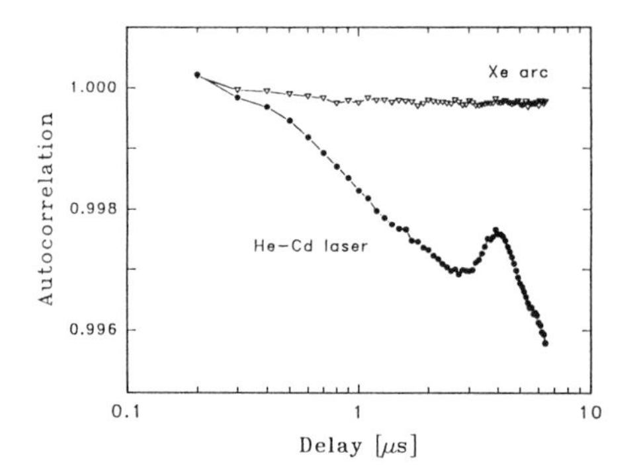

Figure 7. Rapid fluctuations in laboratory light sources. Autocorrelation functions are shown for a largely stable xenon arc lamp, and a more unstable He-Cd laser. The gradual falloff of the latter curve indicates a slower variability, while the peak around $4\mu s$ is a signature of plasma oscillations at $\simeq 250$ kHz in this low-pressure source. Fluctuations this rapid have not yet been detected in any astronomical object.

Acknowledgements

The high-speed astrophysics project is supported by the Swedish Natural Science Research Council and the Swedish Council for Planning and Coordination of Research. At Lund Observatory, we thank in particular research engineers H.O.Hagerbo and B.Nilsson for their highly competent help with the sophisticated electronic units. Likewise, the staff at The Research Station for Astrophysics on La Palma (Royal Swedish Academy of Sciences) are thanked for their valuable help during our several visits there.

References

Codona, J.L.: 1986, *Astron. Astrophys.* **164**, 415
Dainty, J.C., Levine, B.M., Brames, B.J., O'Donell, K.A.: 1982, *Appl. Opt.* **21**, 1196
Jakeman, E., Parry, G., Pike, E.R., Pusey, P.N.: 1978, *Contemp. Phys.* **19**, 127
Roddier, F.: 1981, *Progr. Opt.* **19**, 281
Stecklum, B.: 1985, *Astron. Nachr.* **306**, 145
Young, A.T.: 1967, *Astron. J.* **72**, 747

Discussion

R.M. Genet: *Would a very close comparison star or fast, on-line computation of some sort, allow more accurate differential photometry? This is of interest to us as scintillation is the main limit to the precision of our measurements.*

Dravins: Yes, real-time monitoring of a close enough comparison star should indeed permit more accurate photometry. Ideally, one could think of a grid of close comparison stars, surrounding the target object. Note, however, that the comparisons should be observed in the same colour as the target, since scintillation is colour-dependent.

A. T. Young: *The area of sky over which scintillations of stars are correlated is the angular size of the telescope pupil as seen from the upper troposphere where scintillation comes from, (about 10 Km). For an 8-metre telescope, this angular scale is* $8 \cdot 10^{-4}$ *radians, or about 2 arc minutes. So stars separated by less than 1 or 2 arc minutes will have partly correlated scintillations.*

The Deputy Lord Mayor of Dublin, Councillor Brendan Brady with Denis Sullivan, Whetu Tirikatene-Sullivan and Patrick Wayman at the Civic Reception in the Mansion House

Photometric and Spectrophotometric Data Required for the SUSI Programme

R.R. Shobbrook

Chatterton Astronomy Department, School of Physics, University of Sydney, NSW, 2006, Australia

1. Introduction

The Sydney University Stellar Interferometer, SUSI, a modern version of a Michelson stellar interferometer, will be capable of measuring the angular diameters of stars down to V $\approx$ 8.0, for all spectral types. The resolution of the 640 m baseline — <0.1 milliarcsec at optical wavelengths — includes reasonable sample sizes of all spectral types and luminosity classes to this magnitude limit.

The angular diameter data require complementary photometric, spectrophotometric and also spectroscopic data for the determination of fundamental parameters of single and multiple stars. This paper discusses the accuracy required of these data for some of the main problems to be tackled by SUSI. A selection of the programmes planned for this unique instrument is listed below, where the need for complementary data is also indicated.

- The measurement of the changes of angular diameters of Cepheids and other pulsating stars such as Mira variables. Comparison with the linear changes determined from radial velocity curves will enable their distances and luminosities to be determined by an essentially geometric method.

- The measurement of the angular sizes of the orbits and of one or both components of spectroscopic binary stars. This effectively makes them 'visual' binaries so that the orbital inclination, i, may be determined. When these data are combined with the velocity curve solutions which include $(mass) sin^3 i$ and $(semi-majoraxis) sin\ i$, we may determine masses and linear radii of one or both components and also the distances to the systems. Also, the light curve of an eclipsing binary provides information on the ellipticities and radii of the components and on the eccentricity of the orbit — information which may be used to aid the analysis of the SUSI data. In addition, the determination of the light curve (some bright systems have not been observed for up to one or two decades) is the quickest method of finding the current phase of its orbit.

- The establishment of the total-flux-based temperature scale of all spectral types and luminosity classes from O to M. This requires accurate photometry and spectrophotometry over as wide a wavelength range as possible; it therefore also requires observations from space observatories, as was also necessary for the analysis using the Narrabri intensity interferometer data (Code *et al.*, 1976).

- The measurement of diameters of stars with shells or extended atmospheres, such as Be and Wolf Rayet stars, at different wavelengths — in particular, in the emission lines.

It will be appreciated that the value of an angular diameter alone is of little use. To obtain the distances of Cepheids, high resolution spectra are required throughout the pulsation cycle. To obtain stellar temperatures, spectrophotometry must be obtained over as much of the spectrum as possible — especially in the UV for hot stars and in the IR for cool stars. For determining the masses, radii and distances of stars in binary systems, radial velocity and photometric data are also required. In addition, parameters such as the period and the time of maxima of variable stars are useful for reducing the number of variables for which the raw SUSI data — consisting of time variation of fringe visibility at different resolutions — has to be solved. Many of these parameters are more accurately and easily obtained by standard techniques.

The aim of the SUSI programme is to obtain angular diameters to an accuracy of $\leq 2\%$. This accuracy will be approximately equal to that to which distances of, for instance, spectroscopic binary systems can be measured by relating the angular and linear dimensions of the orbital semi-major axes. For the Cepheid programme, the distances (and therefore the luminosities) determined for individual Cepheids will necessarily have a somewhat lower accuracy. Since we relate the change in diameter during the cycle determined from the velocity curves to the change in angular diameter, the errors will certainly be greater than 2%. However, since more than 40 Cepheids are available to SUSI, the zero point of the period/luminosity relation should be well determined.

2. Requirements of Photometry and Spectrophotometry

2.1 Determination of stellar temperatures and fluxes

Davis (1985) argues that the figure of $\pm 2\%$ accuracy for angular diameters is necessary if we are to increase substantially our knowledge of the surface temperatures and emergent fluxes of stars, especially for the purpose of improving the accuracy of theoretical stellar models. Davis refers to the relation:

$$F = 4f/\theta^2 = \sigma T_e{}^4$$

where F is the flux from the star, f is the observed flux at the surface of the earth, θ the angular diameter, σ the Stefan-Boltzmann constant and T_e the effective temperature. So that errors in the diameters should not contribute significantly to the error in the derived temperatures, the fractional error in θ should therefore be less than one half of the fractional error in f, which Davis found to be of the order of 4% for well measured stars with $B - V > -0.15$. Such errors lead to 1% errors in T_e. Reversing this argument, if the angular diameters are to be determined to 2% and temperatures to 1%, then the spectrophotometric errors, including calibration errors and extrapolation to unavailable regions of the spectrum, should be no more than 4%. Thus, if the ground-based UV to infrared spectrophotometry is internally accurate to 1%, the space observations and the interpolated and extrapolated stellar flux must contribute no more than 3%.

We intend to investigate the accuracy of determining the total flux, f, from Johnson, Strömgren and/or Geneva photometry of programme stars. The catalogues of data, at least in the latter two systems, are accurate to about 1% (often less for bright stars) so that if they can be satisfactorily calibrated with respect to the absolute flux standards, the desired accuracy might be expected. That this is likely to be the case is indicated by the results of Petford *et al.* (1985). They compared their Reticon spectrophotometry with fluxes computed from integrating appropriately normalised Johnson 13-colour photometry for 216 bright stars and found agreement to a standard deviation of 0.55%.

2.2 Determination of temperature variations

Mean stellar temperatures must be determined from spectrophotometry (or many-wavelength-band photometry) over as wide a wavelength range as possible. However, small temperature *differences* may be accurately determined from *changes* in only one or two of the standard Johnson $UBVRI$ or Strömgren uvbyβ colour indices. Such temperature differences are of interest, for instance, in the study of the changes of atmospheric conditions during cycles in pulsating or eclipsing variable stars.

An important factor here is the sensitivity of the colour parameter used as the temperature discriminant. It is easy to show that provided the optimum discriminant is chosen (depending upon the spectral type of the star) colours should be measured to 0.003 to 0.01 mag (0.3–1%) if temperature changes of 1% are to be discerned. For instance, between F0 and K0 stars there is a temperature difference of about 2500K and a difference in $B-V$ of about 0.5 magnitude. Therefore 0.005 mag (0.5% accuracy) corresponds to 25K, or about 1% in temperature. For B stars, $(U-B)$, c_1 or $(u-b)$ are better temperature discriminants, where for a ~1.0 mag change in colour, there is a ~15,000K change in temperature. Here, 0.01 mag corresponds to 150K in temperature, or about 1% on average.

In general it may be true that the Strömgren $(b-y)$, $(u-b)$, c_1 and β values are more readily obtained to such accuracy than are the broad band indices $(U-B)$, $(B-V)$, $(V-R)$, $(R-I)$ and $(V-I)$. However, when only accurate colour *changes* are required (needing only the *slopes* of transformation equations) it is likely that observations of the broad band indices will suffice.

Usually, for bright stars at least and excepting many stars in clusters or in crowded fields, higher accuracy (<1%) is obtained by using a photometer having one or more photomultiplier tubes than by using a CCD. It is not only more accurate but also faster, both for data collection and reduction. With a computer-controlled filter wheel in front of a single photomultiplier tube, bright stars can be measured to two or three millimags in a few minutes through several filters. There are two main advantages to this system. First, fast rotation of the filter wheel enables all filters to be selected for short integration intervals (of about 1s) several times for each 'observation'. The signals through all filters are therefore measured at essentially the same time, which is important at larger air masses and at poorer sites. High precision is often maintained for the colours even through thin cloud, as long as the dwell times are short. A second advantage is that faster measurement means that more frequent observations of comparison or standard stars is feasible.

The procedure is even faster with a photomultiplier tube behind each filter, such as in the photometer used by the Danish observers at ESO (*e.g.* Grønbech & Olsen, 1976 and many later publications). In this case the analysis is more complex since it involves the sensitivity variations of several photomultipliers.

2.3 Low amplitude variable stars and 'millimag' photometry

With low amplitude variable stars such as δ Scuti stars, β Cephei stars and many ellipsoidal binaries, precision differential photometry to a limit of one to three millimag is desirable. For these stars, the variation may be no more than 0.01 to 0.1 mag, often due to more than one period. Such precision has been routinely accomplished by several observers since the δ Scuti and β Cephei work of Michel Breger and the present author independently in the late 1960s and the 1970s (see Shobbrook *et al.*, 1969, where the term 'm.mag' was first coined!).

With small telescopes on which much of the high precision work on bright stars is done, a primary consideration is starlight scintillation. On one night of fairly good seeing on the Siding Spring 16-inch telescope, extended tests showed that 100 successive one-second integrations on α Vir (with over 200,000 counts/sec) gave an r.m.s. scatter of over 0.02 mag. This indicated that integrations of 100s were required to approach 2 mmag precision. Since scintillation decreases approximately linearly with aperture, one does gain significantly by using neutral density filters with larger 1m class telescopes even on very bright stars.

It will not be possible to measure the fractional change in size and shape of variable stars to the same accuracy as that to which we measure the diameter itself. However, for those with short periods we should be able to use differential techniques over one to two hour intervals on each night. Such measurements promise to be of great interest, since the variations will not be spherically symmetric in the case of non-radial pulsations and ellipsoidal variations, and such phenomena have not been directly measured before.

Conclusion

It has been demonstrated that a precision of $\pm 2\%$ in the angular diameters determined by SUSI will require photometry and spectrophotometry of stars to an accuracy of from 1 to 10 millimag. As we can see from discussions in many papers presented at this conference, this precision is certainly attainable, provided that sufficient attention is paid to details of observational procedure and to internal (instrumental) and external (standard star) calibration.

References:

Code, A.D., Davis, J., Bless, R.C. & Hanbury Brown, R., 1976. Astrophys. J., **203**, 417.

Davis, J., 1985. IAU Symposium No. 111, *Calibration of Fundamental Stellar Quantities*, eds. Hayes, D.S., Pasinetti, L.E. & Davis Philip, A.G., p 193.

Grønbech, B. & Olsen, E.H., 1976. Astron. Astrophys. Suppl. **26**, 155.

Petford, A.D., Leggett, S.K., Blackwell, D.E., Booth, A.J., Mountain, C.M. & Selby, M.J., 1985. Astron. Astrophys., **146**, 195.

Shobbrook, R.R., Herbison-Evans, D., Johnston, I.D. & Lomb, N.R., 1969. Mon. Not. R. Astron. Soc., **145**, 131.

Discussion

M. Cohen: *Can we request that you remeasure Sirius and, if possible, Vega with SUSI? If you did so what kind of accuracy could you get on these two stars?*

Shobbrook: Sirius has been measured by the SUSI prototype and will soon be measured again by SUSI. We cannot expect a result much better than 1% for temperature, though. Vega is too far north for SUSI's declination limit of $< +30^o$.

E. Budding: *You mentioned that the actual determination of Cepheid diameters was likely to be more complicated in detail than it might, at first sight, seem. I would guess that the assigned stellar limb-darkening, for example, will have a bearing on your derived diameter values, and probably that can be well taken care of, to within your 2% target, by available model atmosphere values. However, what about the role of rotation — in relation to gravity darkening, say, or physical distortion? How will these factors affect things?*

Shobbrook: Limb and gravity darkening theory is known well enough for normal stars to be corrected for, within the 2% accuracy. We shall also be observing at a range of wavelengths, so that such effects will be able to be determined where the effects are large — this is one of SUSI's proposed programmes. Change of shape due to rotation or pulsation can be measured as the star's axis changes its orientation with respect to the baseline during a night.

T.J. Kreidl: *Have you considered that should non-radial pulsations be present in any of the target objects, the geometrical orientation of the pulsation axis will possibly have a large influence on the measured change in angular diameter?*

Shobbrook: The stars change their orientation with respect to SUSI's north/south baseline during the night. Observations over a sufficient number of cycles will enable the change in shape to be determined.

Secondary Spectrophotometric Standards: Results and Future Observational Programme

I.N. Glushneva

Sternberg Astronomical Institute, 13 Universitetskij prospect, 119899 Moscow, Russia

Abstract

Three sets of spectrophotometric standards are presented : eight bright B and A stars of $0^m_.0$ – $2^m_.65$, fourteen B – A stars of 3^m–4^m from the catalogue of 238 secondary spectrophotometric standards and 24 A0-G0 stars of 7^m–8^m. For all of these stars UBV photometry was produced and brightness variations exceeding $0^m_.01$were not found.

Spectrophotometric standards are used widely for a comparison with the variable and pecular stars when information on spectral energy distribution is necessary. Two main properties of spectrophotometric standards must be taken into account: the reliability of energy distribution data and the absence of brightness variations at least within the limits of the accuracy of measurements. As stars with reliable energy distribution they may be used for comparison with model atmospheres, for obtaining mean energy distribution for stars of different spectral types and creation of the new observational techniques.

Secondary spectrophotometric standards are produced as the result of comparison with stars provided with reliable energy distribution data, *i.e.* primary standards, compared with black body model or lamps calibrated in absolute monochromatic fluxes. In the process of creating two large spectrophotometric catalogues containing data on the energy distribution in the spectra of the more than 2000 stars, eight stars of the spectral types B and A and brightness interval $0^m_.03$–$2^m_.65$ were chosen as standards: α Lyr, γ Ori, α Aql, β Tau, α Leo, η UMa, α Peg, β Ari. Each star of both catalogues (Voloshina, Glushneva *et al.* 1982; Kharitonov *et al.* 1988) was compared with one or several standard stars by means of the method of equal altitudes taking into account differential extinction.

At first spectral energy distribution was obtained for each standard star as the mean for all the data of different authors published up to that time. The energy distribution of Vega, the main spectrophotometric standard, was obtained also on the base of spectral calibration. In 1976, the energy distribution were changed as a result of the new comparison of standard stars to Vega which were done independently at the Fessenkov Institute and Sternberg Institute. Newly published data were used also (Kharitonov and Glushneva, 1978).

Table 1. Mean square errors of the energy distribution data for standard stars (%)

λ	σ
3200–3300	2.1
3300–3700	1.2
3700–4000	1.5
4000–5000	1.0
5000–6000	1.0
6000–7000	1.0
7000–7600	1.3

The next step for more exact data for the system of standards was a reduction to the new scale of the Vega energy distribution based on data obtained by D.S.Hayes (1985). This reduction was done for all the stars of both catalogues. Energy distribution data of eight standard stars in the range 3200-7700 A may be taken from Kharitonov *et al.* (1988) or Glushneva *et al.* (1992). For all standard stars, including α Lyr, UBV-photometry was produced during three observational seasons at Mount Maydanak near Kitab (Uzbekistan, about 3000m above sea level). Observations did not show short time variations (during several hours or days) exceeding the errors of measurements ($\sigma = 0\overset{m}{.}005$). Spectrophotometric standard α Aql is suspected of long time variability with the variations of the brightness about $0\overset{m}{.}02$ (Sparauskas, private communication).

The set of 238 stars was proposed as secondary spectrophotometric standards in the range 3200-7600 A (Glushneva *et al.* 1992). These stars are common to the catalogues of the Sternberg Institute and Fessenkov Institute and the differences between spectral energy distribution data of the two catalogues do not exceed 5%, while the mean inner accuracy of both catalogues data in this range are about 3.5%. For 99 stars the energy distribution data in the near infrared (6000-10800A) obtained at the Sternberg Institute are also presented. Table 2 demonstrates the consistency Δ of the data obtained at the Sternberg and Fessenkov Institutes.

Table 2. Mean values of Δ as a function of S_p and λ

Number of stars	76	68	25	19	47
Spectral type	O-B	A	F	G	K-M
3200–3700	0.012	-0.004	-0.019	0.016	0.020
3700–4000	0.004	-0.003	-0.003	0.010	0.001
4000–5000	0.013	0.006	0.004	0.015	-0.008
5000–6000	0.013	0.009	0.013	0.019	-0.005
6000–7000	0.001	-0.004	0.004	0.020	-0.010

$\Delta = 2(E_A - E_M)/(E_A + E_M)$, where E_A and E_M are monochromatic fluxes of stars in the Alma-Ata and Moscow catalogues.

Good agreement of the results obtained fully independently may be considered as indirect evidence that these 238 stars are not variable within the limits of a few percent. However, many of them are variables or suspected variables, i.e. the members of GCVS or NSV. At the same time many suspected variables are in fact of constant brightness within the limits ≈1%. In any case photoelectric observations of the brightness variations are very important. Fourteen B and A stars of 3^m–4^m which are not members of GCVS or NSV and the 67–70 Name-Lists were chosen. The Bright Star Catalogue numbers of these stars are: BS 269, 580, 1165, 1520, 2540, 3690, 5107, 5867, 6396, 7236, 7710, 8335, 8597, 8634. For all these stars we have energy distribution data in the near infrared, obtained at the Sternberg Astronomical Institute Crimean Station. These data were combined with the mean of the results of the Sternberg and Fessenkov catalogues in the range 6300-7600 A common for all the measurements. In this range the differences between these sets of observations (Sternberg visual, Fessenkov visual and Sternberg near-infrared) also do not exceed 5%, and in many cases the differences between $1/2\ (E_A + E_M)$ (mean of Alma-Ata and Moscow data) and Sternberg near-infrared data are even less (for example, for BS 8634 mean $\mid \Delta \mid = 0.01$). Such good agreement makes it possible to obtain the mean of data presented in the Table 3 and 5 of the paper by Glushneva *et al.* (1992) in the range 6300-7600 A.

Energy distribution data in the range 3200-10800 A for 14 stars are presented in the paper, submitted to a Joint Meeting of Commissions 9 and 25 at the 21-st General Assembly of the IAU, Buenos Aires, Argentina *Automated Telescopes for Photometry and Imaging* (Glushneva, 1992, in press). These data are based on the Vega energy distribution obtained by D.S. Hayes (1985). Mean value of differences between averaged Alma-Ata and Moscow data in the range 6300-7600A and Moscow infrared data in this range for 14 stars is $\mid \Delta \mid = 0.034$. All these stars were observed photometrically at the Sternberg Institute Tyan'-Shan' expedition and brightness variations exceeding 0.01 were not found for the W, B, V, R bands (Kornilov *et al.*, 1991). Now the main task is a creation of a set of fainter standards, connected with the eight bright ones. For this aim a new programme of spectrophotometric standards observations for the space project 'Lomonosov', planned at the Sternberg Institute was begun. In the addition to the great amount of astrometric measurements, this project (Nesterov *et al.* 1990) will include photometric and spectrophotometric observations of stars of 10^m–12^m. For these stars it is necessary to have spectrophotometric standards and now observations of 7^m–8^m stars located at the zone +40 deg relative to the ecliptic are produced at the Sternberg Institute Crimean Station.

Eighty A0–G0 stars spread across the sky more or less uniformly were chosen. These stars were not mentioned as variables or suspected variables, however U, B, V photometric measurements were started as a test of possible variability. As a first step in the spectrophotometric investigations of these stars energy distribution data for 24 stars were obtained in the range 3400-7500A (Biryukov *et al.* 1992, in press). Eight bright stars were used as standards. To decrease their flux only one half of the main mirror aperture remained opened and for the brightest standards the Cassegrain mirror was stopped down also. A Czerny-Turner grating spectrophotometer with discrete scanning was used. A photomultiplier with photon counting electronics was used as a detector. Table 3 contains the data on the internal accuracy for the energy distribution of the 24 stars investigated.

Table 3. Mean square errors of 24 A0–G0 stars of 7^m–8^m (%)

λ	σ
3400–3700	2.2
3700–4000	1.6
4000–5000	1.1
5000–6000	0.9
6000–7000	1.0
7000–7550	1.6

The future programme of spectrophotometry of standard stars includes continuation of observations of about 60 7^m–8^m stars with parallel U, B, V photometry which on completion willbe followed by the programme for stars of 9^m–10^m.

References:

Biryukov V.V. *et al.* 1992, Crimean Journ. of Astrophys. in press.

Glushneva I.N. *et al.* 1992, Astron. Astrophys. Suppl. Ser. **92**, 1.

Glushneva I.N. 1992 in: Proceedings of a Joint Meeting of Comm. 9 and 25, 21st General Assembly PASP Conference Ser., Ed. S.J. Adelman and J. Dukes, Jr., in press.

Hayes D.S. 1985, *Calibration of Fundamental Stellar Quantities*, in IAU Symposium 111, Eds. D.S. Hayes, L.E. Passinetti and A.G. Davis Philip, p.247.

Kharitonov A.V. and Glushneva I.N. 1978, Astron. Zh **55**, 496.

Kharitonov A.V. *et al.* 1988, *Spectrophotometricheskij Katalog Zvezd*, Alma-Ata, Nauka.

Kornilov V.G. *et al.* 1991, Trudy Gos. Astron. Inst. Stern. **63**, 399 pp.

Nesterov V.V. *et al.* 1990, *Inertial Coordinate System on the Sky* in: IAU Symp. 141, Eds. J.H. Lieske and V.K. Abalakin, p.355.

Voloshina I.B. *et al.* 1982, *Spectrophotometry of Bright Stars*, Ed. I.N.Glushneva (Nauka Publ., Moscow), 255 pp.

Discussion

A.T. Young: *Have you compared your results with the Gunn-Stigler atlas? There must be many stars in common.*

Glushneva: Yes, we did it for separate stars. But in general the comparison of two catalogues as a whole is not a simple task. At first it is necessary to discover and exclude systematic errors which may be connected with the standards also. In the process of the creation of the Moscow and Alma–Ata catalogues common stars were compared several times and systematic errors $\sim$ 2–4% for the Moscow catalogue and up to 10% for several stars of Alma–Ata catalogue were found and excluded. I presented the result of comparison of the corrected data.

Session 3

New Techniques

New Techniques

Jaap Tinbergen

Sterrewacht Leiden, Kapteyn Sterrenwacht Roden, Netherlands

Abstract

Routine millimagnitude photometry may require a new approach to reduction of photometric errors. Such an approach is outlined in this paper; it stresses elimination of each error as close to its source as possible. The possibilities provided by modern technology are reviewed in this light. An engineering design group dedicated to photometry is a prerequisite and an on-site photometric technician may be necessary. In this concept, observers are mainly remote users of a database. Implied is the idea of *accurate* photometry necessarily developing into a single but multi-site astronomical facility (cf. VLBI) and the communal discipline that goes with it.

Introduction

"Techniques" is a pleasantly vague term. Reviewing 'New Techniques', I have chosen to capitalize on 3 decades of varied experience as a hybrid astronomer/engineer and I intend to put up for discussion a model for a different kind of photometric instrument system than you will be used to. I do not claim uniqueness for this model; the 'conversation piece' I shall present will be there to provoke further thought. It should be knocked down if it can't stand up, and major surgery may be required. But I hope there will prove to be some virtue in at least some parts of it, to help make some real progress towards *routine* millimagnitude photometry, which we all claim should be possible. 'New Techniques', therefore, I shall take as 'technology which does exist, but which most of this audience may not associate with the aims and practice of photometry'. My assumed role will be that of the systems engineer, translating system aims into requirements for the component subsystems. Development of those subsystems to the requirements is something I shall leave to an imaginary staff of specialized engineers. Hopefully, these imaginary specialists have kept me sufficiently well-informed of the state of their respective arts.

My own experience started in radiopolarimetry, then shifted to optical photometry and polarimetry with small telescopes, and finally included 10 years of the more hi-tech environment of equipment design for large telescopes. Technological areas which I judge to be of interest include telescope construction, autoguiding, filter design, detectors, on-line calibration facilities, optical fibre techniques and computer control.

Actual implementation of most of the concepts discussed will require an engineering group covering optics, mechanics, electronics and software. This should not discourage us, at least not at this point: world-wide photometry is a large enough community to support 'somewhere' a central development laboratory for the benefit of all; it is a matter of wanting it badly enough.

The Story So Far

The design of many classical photometers has been dictated by a requirement of minimal cost. As a consequence, such photometers are extremely simple, consisting of little more than an entrance aperture, a filter wheel, a Fabry lens, a photomultiplier and the bare minimum of electronics. Recently, a computer has been added to this basic setup, and much effort has gone into automating and standardising the data collection and analysis. As a result, we now know a great deal about systematic errors of various kinds in our photometry and in many cases we know the basic causes. They are discussed in Young's papers entitled 'Improvements to Photometry' and in the proceedings of the two workshops with the same name. Examples of error mechanisms are (e.g. Young et al. 1991):

— Image motion translates into photometric errors at the photometer aperture, at the filter and at the detector; the details differ between photometers and depend mainly on where pupil and focal plane are re-imaged and on the type of component used.

— Change of ambient temperature affects filter passbands and detector properties.

— Filter sets in most cases do not fully sample the spectrum, so that transformations to other photometric systems depend on information which has not actually been measured.

— Filter bandpass shapes are generally suboptimal for successful transformations.

— Atmospheric dispersion causes relative image displacement as a function of wavelength and elevation, leading to elevation-dependent errors in measured photometric colours.

— Atmospheric extinction is more complex than is often assumed; the only good way to reduce its influence on the observations is to disentangle its spatial, temporal and wavelength dependences routinely by a suitable standard star strategy *applied systematically.*

— Humidity may affect the behaviour of optics (including those of the telescope), filters and possibly detector.

— Variable flexure of telescope and photometer affect their response.

— Magnetic fields (of the Earth and of local motors and solenoids) affect the response of photomultipliers.

— Neutral-density filters are generally not sufficiently neutral for the wide bands used in photometry, therefore affect passbands.
— Stray light from the Moon or from within the dome can enter photometers that have been sealed or baffled insufficiently, or via insufficiently clean telescope optics.

In the 'classical' photometer I have sketched, there is often not much one can do to eliminate these defects, and the trend has been to take the photometer for granted and determine its properties by comparison with standards (stars, systems, seasons, etc). This approach seems to have reached its limit at an accuracy of perhaps 0.01 magnitude (in the best cases). Young (1992a and earlier papers in the series) points out possible improvements, but these are not all really simple to implement and one wonders whether *routine* improvement by an order of magnitude or more can in fact be achieved by simple means at all. Scientifically speaking, improvement by much less than an order of magnitude is not of interest: imagers and low-dispersion spectrographs, used in standardized fashion, will supersede classical photometers for such purposes; this trend has started already and it will continue now that the detectors are good enough and as spectroscopic observers become aware of photometric basics.

Imagers using array detectors are in many ways *arrays of classical photometers* and to apply imagers properly we shall need most of our 'classical' experience, possibly augmented by such things as an accessible pupil for some of their filters and for shutters, by a telecentric beam for other filters and for detectors, and by compensation for field rotation during the exposure. Imagers are receiving attention from others and I shall not mention them often in this review. This is not to imply that they are unnecessary or 'not photometry'; imagers and spectrophotometers are complementary and several hybrids may also be necessary.

Plan of Attack

If it is not going to be simple to improve the 'classical' photometer, let us examine another possible approach, which is not simple either but has the virtue of being step-by-step rather than having to tackle many interacting effects at the same time. Let us try to specify a photometric installation, in which all the known causes of error are eliminated at source or as close to it as we can get, therefore as a first-order effect rather than as a second-order systematic error in the data. Let us attempt to reduce each cause of errors by a large fraction without introducing new errors and without interacting with the mechanisms that cause other errors. Isolate, decouple and eliminate at source will be the watchwords. At the end of that exercise, one should again use standards of all kinds, but hopefully to eliminate much smaller errors than in the classical case.

My feeling is that, *for spectrophotometry as in imaging, we must learn to live with the array detectors,* both for the higher accuracy of colour information allowed by simultaneous measurements and for the multi-channel advantage which helps in observing standards sufficiently frequently to tackle the extinction problem systematically. We have learnt to live with photomultipliers, which at first sight are not particularly promising, either.

Image motion in the focal plane may serve as an example of how I intend to proceed. In the classical photometer, photometric errors may arise from light being spilt outside the aperture, from non-uniformity or angle-dependence of the filter passband, and from non-uniformity or angle-dependence of the detector characteristics. Rather than accepting image motion as a fact of life and spending great efforts and expense on reducing the (thoroughly mixed-up) secondary effects, let us see how well we can stabilise the position of the light source that serves as input to the photometer, using everything modern technology can offer and if necessary sacrificing a modest amount of light to gain much higher accuracy. The focal plane is the interface between 'telescope' and 'photometer' and the next section of this paper will concentrate on making the telescope deliver to the focal plane the kind of input a photometer works best with. Similarly, the section after that will concentrate on specifying a photometer that delivers an output signal suitable for recording accurately with an array detector such as a CCD.

Arguing backwards from the astronomical goals, we can make the following, logically more or less connected, series of statements:

- For astrophysical reasons, we require a system accuracy of 0.1% in N bands.
- To obtain this accuracy, measurements in these N bands must be simultaneous.
- If N is large, we are forced to use array detectors.
- For photometric stability, we need M pixels per band.
- Array detectors require that bands are defined *by optical means* and are spatially separated. The latter can best be implemented by a prism spectrometer, the former by a filter in the white beam (cf Walraven and Walraven 1960).
- Photometric stability demands a laboratory environment for the photometer.
- The photometer requires stable illumination of its entrance aperture.
- Such stable illumination requires compensation of atmospheric dispersion, it requires autoguiding and autofocus, and it requires an optical (fiber?)scrambler.
- Taming atmospheric extinction will require frequent short observations of standards all over the sky and a means of monitoring extinction *variations* near the object being observed.

These are the basic system requirements I have had in mind. The rest of this paper concerns the techniques for possible solutions. These solutions look suspiciously like very-low-resolution spectrometry. They are, in fact; the emphasis, however, will be on quantitative flux determination within a high-purity spectrum.

Telescopes

Since photometry is our business and photometers work best in a laboratory environment, we should attempt to satisfy that condition. We can either use a fibre to couple the telescope to a laboratory photometer, or design the telescope to give a stationary output. Fibres tend to have variable losses depending on bending or mechanical stress, so it may be best to avoid them for this application (but see Ramsey 1988 p. 285 and Heacox 1988 p. 300). A Coudé mirror train has atrocious photometric properties, partly due to the polarizing power of successive oblique reflections in a variable configuration. A feasible compromise is the Nasmyth configuration; this introduces only 1 oblique reflection and has the highly convenient platform, on which we can install our photometers horizontally and have the space to seal, thermostat and screen them. My idea of heaven is a 1-metre altitude-driven telescope on, say, a 10-metre azimuth platform. I would reserve one side for imagers, which must be mounted on large rotary bearings (almost of telescope-mounting precision); the other side would be reserved for the other main type of future photometer, the point-object spectrophotometers.

The oblique Nasmyth reflection will cause photometric errors with polarized objects. The solution is to depolarize the light before it strikes the Nasmyth flat. A continuously rotating superachromatic halfwave plate is the right kind of depolarizer for this application (Tinbergen 1974); unfortunately, such components are small, so that we must bring the Cassegrain beam to a premature focus above the Nasmyth flat (cf. the design of LEST: Andersen et al. 1984, p. 15) and limit the focal plane field to a few cm (4 arcmin per cm for a 1-metre telescope and an F/10 beam, which is about the fastest beam such a depolarizer can handle; with 1-arcsec seeing, a 10-arcmin field will be about what even a very large CCD can handle for good imaging photometry). After the Nasmyth flat, we shall need re-imaging optics to produce the final, real, image in an accessible position on the platform. New UV-transmitting optical glasses make it worth considering refractive optics for this. A second depolarizer will be needed near this focus.

Frequent observations of standards all over the sky demand a fast-slewing and fast-settling telescope; of order 5 deg/sec is certainly possible and seems adequate for slewing, the transition from full slew speed to dead stop must not take more than a few seconds and modern computer drives can achieve that. The requirement of high slewing speed may determine the size of the Nasmyth platform (as may dome cost, of course).

We shall require atmospheric-dispersion-compensation prisms to produce a white focal-plane image. The best position for these is probably just in front of the secondary mirror, using them in double pass; this position will minimize polarization effects and consequent photometric errors with polarized stars.

We shall need a moving secondary mirror, under full computer control (Milone and Robb 1983). I envisage at least 2 functions for it: rapid switching from object to comparison stars and sky, and as a component of the autoguider system. Detailed considerations of allowable photometric errors will determine the maximum sweep allowed in each application.

With the telescope which we have just 'constructed' we can deliver a white, unpolarized image in the second focal plane. Unfortunately, this image moves around as the atmosphere does its worst, and our photometer is not going to like that. We shall have to install an autoguiding system. Excellent CCD autoguider cameras exist and can be expected to handle 15th, or even down to 17th, magnitude on a 1-metre telescope; they can also double as a monitor of 'local' extinction variations. The error signals from such a sensor will be separated into 2 or 3 frequency ranges. The slowest and largest errors will be corrected by driving the entire telescope, faster and smaller error components will drive the secondary mirror, and finally a very small optical element at the photometer input can be made to correct the highest frequencies and smallest amplitudes. Avoiding oscillations in such a system may not be easy, but a similar problem has been tackled successfully for the Mt Wilson interferometer delay lines (Shao et al. 1988, p. 360). Most of the time the autoguider will use an offset star to guide on, but for initial impersonal centering, and for cases where no suitable offset guide star exists, we need an option to use the object itself. This option can use all the flux for a fraction of the time, or it can use a 'neutral' fraction of the flux or a restricted wavelength range all of the time. Extracts from guide star catalogues must be on-line for quick selection of the most suitable star in the field. Clearly, the autoguider system is not trivial.

Since best photometric practice with fibre scramblers may be to defocus the stellar image by a controlled amount (Heacox 1988, p. 299), the CCD autoguider system should be constructed to auto*focus* the telescope as well (e.g. just before an integration); this function could in fact just be a computer algorithm: a 'focus run' to achieve best focus on the autoguider CCD, while the photometer input is permanently defocused by a fixed amount with respect to the autoguider.

Photometers

The prime requirement for the photometer is a laboratory environment. On the Nasmyth platform out in the open, this implies a sealed thermostated enclosure. Flexure is not a problem on the platform, but screening from local or the Earth's magnetic field may be necessary. The detector could be outside the thermostated enclosure, but will have its own temperature control. The entire instrument should be flushed with clean dry air at all times.

None of this poses any special problems and the list of requirements reads like the actual operating conditions of a recent large Nasmyth spectrograph. And indeed our photometer is going to look like a spectrograph; the fact that we do photometry will show in our detailed use of the detector pixels. For photometric stability in the face of any remaining instrumental drifts, each elementary channel will in fact be recorded on M pixels; M could be of order 100. In general we can expect to use a finite evenly-illuminated input aperture, which we shall image on to the detector (perhaps slightly out of focus) through a filter that defines the passbands and a prism that spatially separates them. Each pixel records a finite wavelength range; conversely each wavelength is recorded on a finite number of pixels. Stray light will have to be kept under stringent control; in this kind of system, it is like a red leak in a filter photometer, viz. it is light recorded in the wrong place for its wavelength, hence represents an error.

Since the input aperture and output light patches are of finite size, we cannot use detector windowing for the superior band definition we need. The solution is to use a filter in the white beam to produce a so-called channel spectrum, in which bright and dark bands alternate. If the filter is of the birefringent variety, there will be 2, 4, or even 8 beams with complementary bands; the resultant spectra are recorded side-by-side on the detector, and the edges of the detector readout windows are put within the dark bands; in this way, band definition is almost exclusively a filter function, and obtaining stable passbands is mainly a question of good temperature control rather than some uncertain mix of residual instrument flexure and spectral smoothing by the finite input aperture. The filters are the key to well-defined and stable passbands and are discussed in more detail in a separate section.

For stable results, the photometer requires stable, preferably broadly peaked, illumination of the input aperture. The telescope/autoguider system delivers a central 'point' source. This is surrounded by a field of much lower brightness, which nevertheless is essential to photometric precision. The central source has a stable average position, but a size determined by the momentary seeing. To match this signal to the photometer, we require a scrambler of some sort, a device that rearranges this changing light distribution into something more constant with time. The level of photometric accuracy which we can attain routinely may depend critically on the scrambler and some hard thought and experiment will be needed in this area. The best performance would probably be given by an integrating sphere, but its efficiency is prohibitively low. Fiber-bundle scramblers exist, but they transform small residual image motions into large (random) ones as well as the other way around, so they may not be the answer. A single fiber scrambles azimuthally but not radially (Heacox 1988). However, if one moves the fiber input back and forth along a diameter, radial scrambling is achieved, *in the time-average sense.* The attractive aspect is that, by controlling the pattern of motion, one has control over the shape of the light patch that serves as input to the photometer; one could thus compensate on-line for slow variations of seeing as detected by the autoguider . Such a device could be integrated into the final actuator of the autoguider system, merely by adding a periodic signal to one of the error signals driving the fiber input. Another way of introducing radial scrambling is to use 2 lengths of single fiber, linked by pupil-to-image conversion optics (Ramsey 1992 and Barwig et al. 1988). This provides a very uniformly-illuminated patch, but with sharp edges which we may not want and may have to remove by defocusing.

Another possible feature of the photometer is an option to observe several sources simultaneously, along the lines of Walker 1988. A really fast telescope reduces the need for a multi-object input, but for some programmes it will not be adequate to use the entire telescope or the secondary mirror to chop between object and comparison star, or one may need the optical efficiency of true simultaneity. Walker's data indicate that 0.1% precision can be obtained in practice, and there are likely to be enough pixels on the detector for multiple beams. Whether a multi-object input option is feasible will depend on the details of the birefringent filter. *Use* of the option may possibly conflict with the moving-fibre type of scrambler.

Detectors

To me, it is an article of faith that we must press the more sophisticated versions of the array detectors into service for photometry; their parameters have improved enormously. A very encouraging differential accuracy of 0.004 mag over a 3-magnitude range is reported by Penny and Griffiths (1991; see also Penny et al. 1992), for an imager; for a properly designed spectrophotometer using many pixels for a single output quantity, it must be possible to do much better than that. Gilliland et al. (1991; see also Gilliland and Brown 1988) reach a *precision* of 0.002 mag in *time variations* of each of over 100 objects within a repeated CCD frame.

The most troublesome property of CCDs for accurate photometry is likely to be their high reflectivity; recent trends to coat them for anti-reflection (to improve blue sensitivity) will help, but some hard thinking will be needed on how to stop the reflected light getting back to the detector a second time, often in a position where it does not belong (in other words: red and blue leaks!!). One of the problems is that the reflection is specular, leading to structured ghosts in instruments equipped with CCDs (e.g. Gilliland et al. 1991, fig 1), thus to highly local photometric errors. If a birefringent filter is used, the light striking the detector is likely to be 100% polarized and one may eliminate excessive reflections by a polarization trick as is used for computer terminal screens. If the light is not strongly polarized, this approach will lead to loss of half the light; in that case, a diffusing surface extremely close to the detector might be worth considering, to scatter the reflected light as widely as possible and thus to reduce the part that gets specularly reflected. Since reflection coefficients of spectrometer optics will be only a few percent per surface, reduction by a factor of 10 may be enough to make any structured ghost harmless, so that scattering over an angle of 3 times the angular extent of spectrometer optics may be very effective. Very interesting 'holographic diffusers' have been announced, which reportedly can be tuned during manufacture both to a specified wavelength range and to a specified angular width of the scatter diagram.

One thing we must **NOT** do is to focus the spectrum sharply on the CCD and match the optical resolution to the size of one pixel. In such a case, we would produce both undersampling by a factor of at least 2 *and* the worst possible band shape (Young 1992b). Instead, we should spread the light for one channel (predefined in wavelength content by the filter) into a fuzzy patch of M pixels by judicious means such as size and shape of the photometer input light distribution, or defocusing the spectrometer part of the photometer. A value of M of 100 might be sufficient: reduction of errors by a factor of 10 by averaging, from the 0.004 mag quoted above for an imager (which did use some degree of pixel averaging itself).

In my other paper I shall argue that we may need some 1500 photometric channels, each of about 10 Angstrom width, to cover the optical range. This would mean 150000 pixels for recording the channels themselves and large chips will be needed to accommodate some dark space in which to put the *spatial* band limits. Such large chips may take of the order of a minute to read out, so arrangements must be made to read out only those pixels that contain photometric information. The maximum signal in 100 pixels will be of order 25 million electrons; with several tens of 10-Angstrom channels finally going into one intermediate-bandwidth photometric quantity, it seems we can accommodate a dynamic range of order 1000 with photon noise of 0.1% or better throughout and with a single exposure. Given that narrow bands will allow extensive use of neutral-density filters and that exposure times can vary from 1000 seconds down to where scintillation noise becomes prohibitive, such a dynamic range seems ample.

A point of some concern with CCDs is of course the stability of the gain and of the departures from linearity. We have lived with deadtime corrections for photomultipliers and can live with similar corrections for CCDs, as long as they are stable. In photometers such as I have sketched, a particular pixel will always be exposed to a narrow spectral range only, and very much the same mixture for all sources (ignoring stray light for the moment); this will help to obtain the desired *stability* of any imperfections that may be present. Actual data are hard to find; since for most purposes 1% is considered highly linear, measurement noise tends to be at that level. However, ignoring the noise in the published data, linearity better than 1% and stability better than 0.1% seem to be realistic. Linearity can be tested on-line by timed exposures to a well-stabilized lamp in the calibration system (see separate section) and by observing sources of different flux levels (can be lamps, too) both with and without a neutral density filter. Such tests will have to be done in extenso and reduced both in an absolute and a relative way. Gain stability may be more difficult to test. Comparison with calibrated diodes illuminated by the same stabilized lamp should allow sufficiently accurate interpolation between reliable standard-star sequences. The problem is likely to be much less serious than it ever was with photomultipliers and their nth-power dependence on supply voltage; we now have a million detectors all tied to the same supply and we are mainly interested in relative gains, since absolute detector gain variations mimic things we have little control over, such as thin cirrus, aging of mirrors and minor dewing of optical surfaces (telescope mirrors, for instance; KenKnight 1984).

We should realise that, by using array detectors to record our data in much more detail than we shall finally need, we have extra design freedom to improve our 'instrument'. For instance, by reducing the weight of data from outlying pixels, we can trim the wings of the passbands to achieve the log-concave-downward condition expressed by Young (1992a). At another level of resolution, we have the freedom to shape our scientific passbands as we like (my other paper). Yet another example: the recorded data are a first-order approximation to the spectrum of the source; if we have calibrated, for each wavelength, the level of scattered light into other channels than intended, we can correct for such scattered light afterwards, as long as the scattering function is smooth and the level is sufficiently low for the iterative correction procedure to converge. These examples emphasize that it is premature to predict the limit of CCD performance in such untried applications, but there does seem to be plenty of room for optimising to something practicable. Needless to say, the technology of CCD applications will need years of accumulated experience before it can be called mature; photomultipliers took decades. But at least the basic building blocks are available and we know more or less what we wish to find out about their detailed performance; the engineers can take over.

At my request, an instrument designer experienced in CCD optimisation has commented: *"A) for a thinned CCD, 450 to 850 nm is the best range and pixel-to-pixel response variation will be of order 1%; possibly more in the UV (surface cleanlinesss etc). B) temperature stability of gain and Q.E. can be made 0.1%. C) linearity is routinely better than 1%, while 0.1% is probably attainable with care. D) an LED internal to the cryostat, used with multiple standard flashes is excellent for checks of stability and linearity to better than 0.1%"* (Jorden 1992).

Filters

A simple prism spectrometer is sufficient to disperse a white beam into a spectrum fit for a CCD. In order to avoid problems of residual wavelength shifts in this spectrometer, it is very desirable to incorporate into the white beam a multi-band filter producing alternating bright and dark regions in the spectrum of that 'white' beam. The spectrum on the detector will then consist of isolated bright patches; these patches may move around slightly as conditions change, but their wavelength content remains constant, as defined by the filter passbands. To the extent that the detector spectral response is constant from one pixel to another, no photometric errors will result, as long as the boundaries of the readout window for a particular patch remain within the dark region. A Fabry-Perot etalon is an example of such a multi-band filter, but for several reasons it is not suitable. Filters based on the spectral dispersion of birefringence are much more promising. Applications known to astronomers are by Lyot, Šolc and Walraven; each of these designs is capable of further development. What such filters have in common is the possibility to conserve all the light by using beamsplitting polarizers: the light that is rejected by one beam is shifted into the other beam, of opposite polarization. One can pass both those beams through the spectrometer and record them both with one array detector (Wizinowich 1989); by using the correct polarization and passing the prisms at the Brewster angle, one can construct extremely efficient systems (Walraven and Walraven 1960). The passbands are all of the same shape (on a linearized birefringence scale), all of them being determined in the same way by material constants, thickness of crystal slices and position angle of the components. Since one hardware filter determines all the passbands, care in stabilizing the operating conditions of this filter is well-spent indeed and calibration is a matter of determining a small number of parameters. By changing the position angle of components in the filter, one can move all the passbands in synchronism; this feature allows creation of passbands separated by half their FWHM, the separation that is required for fully sampling the spectrum at the resolution determined by the bandwidth.

Two developments of the basic filters are of great interest. Ai Guoxiang and Hu Yuefeng (1985, p. 10, 11) describe adaptations of the Lyot filter that can produce, in 4 or 8 beams, multiple passbands separated by wide dark regions. The efficiency is not far from 100% if one uses both beams from the entrance polarizer. In a series of papers, Ammann and associates (Yarborough and Ammann 1968 is the best reference to start with) present a synthesis technique (with experimental verification!) for arbitrarily-shaped passbands by filters similar to Šolc's. Again, the efficiency is close to 100%. For bandwidths of order 10 Angstrom, these two types of filter are more suitable than Walraven's; the choice between them must be resolved by detailed engineering considerations of passband shape, sidelobe level, maximum size of input aperture, adaptability to multiple output beams, etc. It seems likely that Young's (1992a) conditions for good transformability can be met well enough by either type (see also section on Detectors: 'trim the wings' etc).

Calibrations

Calibration is going to be much more important than with classical photometers and the main reason is that there is much more opportunity for it. For classical photometers with relatively wide passbands, calibration with one source will not be of much use for a source with a noticeably different spectrum. If, however, our instrumental passbands are narrow compared to spectral variations of some of our sources, we can usefully observe those sources to monitor relative channel sensitivities with time; for short timescales we can use stable lamps, for longer timescales we may have to refer to groups of selected standard stars. The aim of calibration is to relate the changing channel sensitivities and central wavelengths of our actual instrument to those of a virtual instrument with absolutely constant properties; this virtual instrument may be a similar instrument somewhere else or the same instrument at some other time. If we can do that consistently, we shall no longer be plagued by transformations from instrumental to standard systems, at least not at the stage of using the data for a scientific purpose. The traditional difficulties of absolute calibration remain, but fortunately most of our work can be done without such absolute calibration.

Several kinds of calibration will be necessary. Some of these can be carried out once in a while, others could be quick online checks or necessary monitoring. An extremely important matter is that beams from lamps should resemble beams from the telescope. They should therefore have the same focal ratio, and proceed from a pupil at the same distance and with the same central obstruction. The most likely way to achieve this is to inject the light by a flat mirror introduced just downstream of the secondary mirror and mount the pupil and beam-defining optics on the side of the telescope. This optical system should be fed by fiber (bundle) from a sealed and screened 'lamp laboratory' on the Nasmyth platform. To ensure that the lamps and fibers have constant, or at least known, properties, there will need to be calibrated diodes inside the photometer; these diodes will most probably be the best final reference components (Borucki et al. 1988). With lamps and diodes protected from environmental effects and the transmission path calibrated out, we have probably gone as far as we can with terrestrial sources. To include the telescope mirrors and for the very highest accuracy over long time periods, nature probably still provides the best reference; if a select group of stars (cf. Lockwood and Skiff 1988, p. 201) all give the same results, one must conclude that those stars are stable or be very sure of an alternative explanation.

The most basic calibration will be to scan through the passbands with a monochromator source of considerably higher resolution, as a check that they are indeed of the shape intended. This very time-consuming procedure will need to be speeded up by multiple output slits in the monochromator. We also need to know scattered-light levels. These can be determined by tuning the monochromator to the central wavelength of each channel in turn and reading out *all* channels every time. For this purpose, we shall have to be very sure that the monochromator delivers just one wavelength; order-sorting and other blocking filters will be needed, even with a double monochromator.

A continuum source, stabilized by reference diodes as above, will be needed for routine (several times per night) calibration of channel sensitivities. This takes very little observing time, since the lamp is a bright source and there is no scintillation.

For routine checking of wavelength stability, a temperature-stabilized solid Fabry-Perot in the beam of this continuum source may be the best component. The multiple transmission bands of this filter will lead to a characteristic pattern of outputs of the photometer; any wavelength drift of the photometer will lead to an equally characteristic change of that pattern. Fewer than 10 parameters determine F-P, transmission optics, birefringent filters and detector, whereas measurements will be available in hundreds of channels. By methods similar to determining a pointing model of a telescope from measured apparent positions of standard stars, it should be possible to disentangle drifts in the several component systems. This type of measurement will also take very little time from the observational programme.

Any neutral-density filters used in the photometer will have to be calibrated regularly. With narrow passbands, this is an accurate and not very time-consuming procedure. It can be combined with the detector linearity check and could be carried out on the continuum lamp in the daytime, as a check on occasional determinations on standard stars.

Computers, Automation and Organisation

The installation I have sketched contains many automatic processes that are essential to its functioning. These will have to be computer-controlled. The trend is to use dedicated computers for single well-identified functions. One expects to have at least the following computers in some sort of hierarchy:

— System has overall control, serves local observer and engineer
— Telescope drive includes on-line extracts from star catalogues
— Autoguider reads autoguider CCD, computes errors, checks focus
— Photometer controls temperature, ND filter wheel, etc
— Detector controls shutter, reads out CCD windows
— Communications serves remote observer and remote archive

There are two separate reasons to organize actions through computers. One is ease of operation, or impossibility of doing the same thing manually. Examples are autoguiding, autofocus, auto-guidestar-selection and CCD readout. The other reason is that some activities are too important to leave to individual observers. Examples of this type: scheduling of standard star observations sufficiently frequently for good extinction handling, and calibration by suitable lamp exposures. Except for very specialized programmes, it is to be expected that the observer will be remote, entering his wishes into a scheduler programme running on the system computer and receiving messages that tell him when his observations have arrived at the archive. To achieve good photometry, standardized operation is essential and cannot be left to observers' social conscience: one's own programme is always more important than someone else's calibrations. For the benefit of all, the scheduler (or whoever controls its parameters) must firmly resist any attempt by observers to get more than their fair share of programme star observations.

Conclusion

I have outlined what kind of instrument development seems necessary to me, if we are to profit from our knowledge of the causes of photometric errors. The approach has been that of modularity, allowing gradual buildup and enhancement of the installation, and effective detection and diagnosis when things go wrong. Once a satisfactory instrumental system has been achieved, standardization of the observational process and routine calibration will be basic to the approach. As I argue in my other paper, instrumental bandwidths could be as narrow as 10 Angstrom. The standardized observations will be stored in a data-base. The data-base contains the raw measurements, with all auxiliary data that could be relevant; it also contains a version of the photometric data corrected for *known* instrumental drifts and scattered light, and transformed to outside-atmosphere by standard extinction handling. The individual observer will take observations from this standardized observational machine and put them to his own creative use by synthesizing his own scientific passbands and creating his own photometric system from scratch, using the standard star observations in the data-base which are immediate public property. It is by such methods that I feel we can *routinely* achieve millimagnitude photometry and perhaps beyond; the classical method of 'build it cheap and see what it does' will only achieve such accuracy now and then, by luck or extreme hard work.

The amount of modern technology implied may seem frightening, particularly to those who have used similarly complex installations at large telescopes and have lived through numerous partial breakdowns during observations. Experience shows that such installations become spectacularly more reliable, the more they are left undisturbed and used only in standard ways. Designing for maximum reliability rather than for maximum flexibility will also help. However, an on-site 'photometric assistant' may be a necessary condition for success.

References:

"**Second Workshop**": Second Workshop on Improvements to Photometry, 1988, ed. W.J. Borucki, NASA Conference Publication 10015, NASA, Moffett Field, California

"**My other paper**": Transformations and Modern Technology. This conference, session 5

Ai Guoxiang and Hu Yuefeng 1985, LEST Technical Report no 14 (Institute of Theoretical Astrophysics Oslo)

Andersen T.E. et al. 1984, LEST Technical Report no 7 (Institute of Theoretical Astrophysics Oslo)

Barwig H. et al. 1988, Second Workshop, p. 35

Borucki W.J. et al. 1988, Second Workshop, p. 47

Gilliland R.L. and Brown T.M. 1988, Pub. Astron. Soc. Pacific, vol 100, p. 754

Gilliland R.L. et al. 1991, Astron. Jour., vol 101, p. 541

Heacox W.D. 1988, Second Workshop, p. 289

Jorden P.R. 1992, Private communication (summarised by JT)

KenKnight C.E. 1984, p. 222 in 'Proc. Workshop on Improvements to Photometry', eds W.J.Borucki and A.Young, NASA Conference Publication 2350, NASA, Moffett Field, California

Lockwood G.W. and Skiff B.A. 1988, Second Workshop, p. 197

Milone E.F. and Robb R.M. 1983, Pub. Astron. Soc. Pacific, vol 95, p. 666

Penny A.J. and Griffiths W.K. 1991, RGO Gemini Newsletter, no 33, p. 4

Penny A.J. et al. 1992, This conference, poster session 2

Ramsey L.W. 1988, Second Workshop, p. 277

Ramsey L.W. 1992, to be published in Pub. Astron. Soc. Pacific

Shao M. et al. 1988, Astron. Astrophys., vol 193, p. 357

Tinbergen J. 1974, p. 175 in 'Planets, Stars and Nebulae studied with Photopolarimetry', ed. T. Gehrels, Univ. Ariz. Press, Tucson, Arizona

Walker E.N. 1988, Second Workshop, p. 57

Walraven Th. and Walraven J.H. 1960, Bull. Astron. Inst. Neth., vol 15, p. 67 (no 496)

Wizinowich P.L. 1989, Optical Engineering (SPIE), vol 28, p. 157

Yarborough J.M. and Amman E.O. 1968, Jour. Opt. Soc. America, vol 58, p. 776

Young A.T. 1992a, Improvements to Photometry V; Astron. Astrophys., vol 257, p. 366

Young A.T. 1992b, Improvements to Photometry VI; private communication

Young A.T. et al. 1991, Pub. Astron. Soc. Pacific, vol 103, p. 221

Discussion

S. C. Russell: *In your ideal photometric telescope, where you are intending to focus within the telescope anyway, could you not do away with the mirror and depolarizer and collect the light near the focus with the fibre image scrambler?*

Tinbergen: Yes, that could work for the single object photometers. I envisage that the telescope might be needed for several different jobs - uncrowded fields, medium crowded fields and crowded fields for instance; imagers and single-object spectro-photometers are the two extremes. For imagers the Nasmyth flat is needed.

S.C. Russell: *So you envisage telescopes that can do many jobs rather than one job well?*

Tinbergen: Yes, if we are to ask for dedicated and expensive photometric telescopes to be built, they would have to be able to perform many jobs. It is hardly likely funding would be approved for five telescopes to do five particular jobs. Besides, we don't *know* yet quite how well fibres can perform; with a Nasmyth flat, conventional optics remain an option.

D.L. Crawford: *We definitely want to go in this direction, most likely as fast as we can, and where we can. We will also be using 'classic' photometers (all we can afford, or what we need). I have been promoting a 'law' or 'goal' where the telescope costs less than the instrumentation which cost less than the software for analysis.*

Tinbergen: In easy steps, most of this is possible, I'd say. The telescope and drive mechanics would be exceptions, but are receiving attention (see Genet's review). With reference to your second point: compared with solar physics; we are growing up, too!

D.L. Crawford: *Please also add the Site (atmosphere) to your list, at the top end. Quality of the site (clear, photometric, stable, dark skies) is a quite important aspect to all this, of course, including site preservation.*

Tinbergen: I know. La Silla (photometric), versus La Palma (spectroscopic), criteria were paramount in their selection.

A. T. Young: *A problem you did not mention with fibres is that of coupling light, in and out. With a 100 μ fibre, a defect or dust speck just a few microns across is photometically important, at the millimagnitude level. I think this is one reason why people have had difficulty in using fibres photometrically, in addition to the bending-loss problem you mentioned.*

Tinbergen: Is a 1mm fibre more difficult to keep clean than a 1 mm diaphragm? I am hoping that multi-fibre components with low input loss can be produced. Mantel (MEKASPEC project) mentions 85% transmission for uncoated end faces - about 7% short of perfection. The trick is precisely that of removing the cladding at the input end. However, I do not wish to pretend that engineering development is unnecessary.

R.M. Genet: *Many of the telescope features, 1-metre alt-azimuth autoguide, fast slewing, autocentreing, fast secondary motion, etc., are being applied at Autoscope. We are currently building a 1-metre alt-azimuth telescope for the University of New Mexico to be placed at* *the Apache Point Observatory.*

Tinbergen: I am very pleased to know I have companions along this route and engineers, too. I can't believe my luck.

E. F. Milone: *On an aspect you haven't fully covered, but to which you alluded, — time resolution; there is the binning problem, which presumably requires additional time to correct. The read-out time is another problem. On another point - the limited size of CCD chips may well limit the precision attainable if there is no suitably bright (and colour-matched) comparison star in the frame (see Schiller and Milone, 1990).*

Tinbergen: I was referring to software re-binning on conversion to wavelength scale. This is a transformation in the photometric sense, with rectangular bands under sampling the spectrum by a factor of two; the worst of all, according to Young (these proceedings). It does not concern my main argument, but was an aside on an illustration not reproduced here. Your second point concerns imagers, which are not the optimum way of using CCDs for accurate photometry. In the narrow-band spectrophotometry I propose, in my other talk, colour-matching is not necessary. A fast telescope is a more important point (cf. comment after Genet's review).

A. J. Penny: *In addition to your dedicated 1-metre telescope, it would also be cost effective to have an additional 0.5-metre telescope dedicated to monitoring the extinction over the sky during the night.*

Tinbergen: Yes, an extinction monitor will always be useful. Considering Young's comment, and Genet's figures for the slewing speed and settling time of a 1-metre telescope, primary extinction data could best be derived from observations by the prime instrument. Monitoring extinction variations of stars very close to the object might be done in say 3 broad bands. Perhaps the autoguider CCD could perform this task.

Crowded Field Photometry using Post-Exposure Image Sharpening Techniques*

R.M Redfern[1], A. Shearer[1], R. Wouts[1], P. O'Kane[1], C. O'Byrne[1], P.D.Read[2], M. Carter[2], B.D. Jordan[3] and M. Cullum[4]

[1]*Physics Department, University College Galway, Ireland*
[2]*Dublin Institute for Advanced Studies, Ireland*
[3]*Rutherford-Appleton Laboratory, United Kingdom*
[4]*European Southern Observatory, Germany*

Abstract

The technique of image sharpening which allows high resolution images to be produced from ground-based telescopes is applied to the problem of photometry in crowded field regions - such as close to the cores of globular clusters. The conditions for image sharpening are discussed and the technique is demonstrated using simple objects (close double stars). Preliminary results from image sharpening of M15 are presented.

1. Image Sharpening

Plane waves (from distant objects) become distorted by refractive index fluctuations while passing through the atmosphere, limiting the resolution which can be obtained by even small telescopes at sea level and modest-sized telescopes at the best high altitude sites. The resolution of a large telescope is entirely determined by atmospheric effects and by its optical quality rather than by diffraction. Refractive index fluctuations are generated by temperature fluctuations in the atmosphere arising (a) from convective motion and/or turbulence locally and in the atmospheric boundary layer, and (b) the dissipation of turbulent energy arising from high-altitude atmospheric wind shears. Collectively the effects are known as seeing. It has proved possible recently, by site selection, careful attention to local thermal effects and by improving optical performance, to produce substantial improvements in average seeing. Nevertheless, atmospheric distortion remains the limitation to high resolution imaging.

---

* This article describes work being carried out with the William Herschel Telescope in the Spanish Observatorio del Roque de los Muchachos of the Instituto de Astrofisica de Canarias and the New Technology Telescope of the European Southern Observatory.

The quality of the atmospheric seeing is usually expressed by a characteristic scale R_0 (Fried's parameter) and time scale τ_0. One definition of R_0 is the diameter of a diffraction-limited telescope producing seeing-limited resolution, ω_{eff} ;

$$R_0 \propto 1.22\lambda/\omega_{eff}$$

R_0 has a value of a few cm to perhaps 0.5 m, and τ_0 lies in the range of a few milliseconds to 50 milliseconds for the best high altitude sites. The probability of an image having less than 1 rad^2 rms phase error is; (Fried[1978]):

$$prob \approx 5.6\ \exp[-0.1557\,(D/R_0)^2]$$

Strikingly different behaviour can be observed through small ($D \simeq R_0$) and large ($D > 6R_0$) aperture telescopes. For a large aperture there is little image motion, and the image is always blurred - many bright "cores" can be observed moving about within the image, if the star is sufficiently bright, and only occasionally coalescing into a single "core". For a small aperture one may frequently observe a single bright "core" in constant motion. Figures 1 and 2 express this graphically, and underline the importance to image sharpening of the correct choice of pupil.

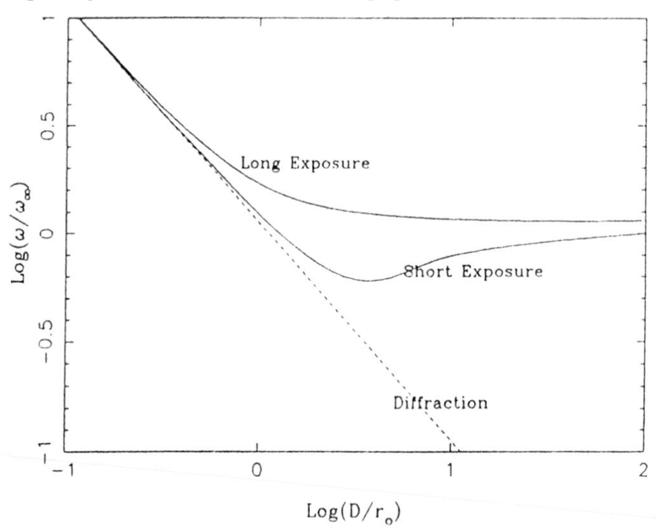

Figure 1: Image width, ω, normalised to that of a large aperture, ω_∞, as a function of aperture, for long and short exposure images (Roddier[1981]).

ω corresponds to the diffraction limit only for small apertures. The short exposure image width (where short means that all image motion is frozen, say 1 millisecond) follows the diffraction limit more closely and shows a minimum at $D/R_0 \sim 3.7$ which is a factor 2 smaller than the long term limit. What is happening is that for values of

$D/R_0 \leq 4$ the dominant term in a polynomial expansion of the wavefront distortion is the linear one (tilt). A gain of a factor 2 can be achieved by removing tilt alone. For values of $D/R_0 > 10$ the dominant terms are of a higher order - removing tilt does not produce any improvement. Figure 2 predicts that it ought to be possible to make large improvements in resolution by using choosing those (few) moments of diffraction-limited seeing for a large aperture.

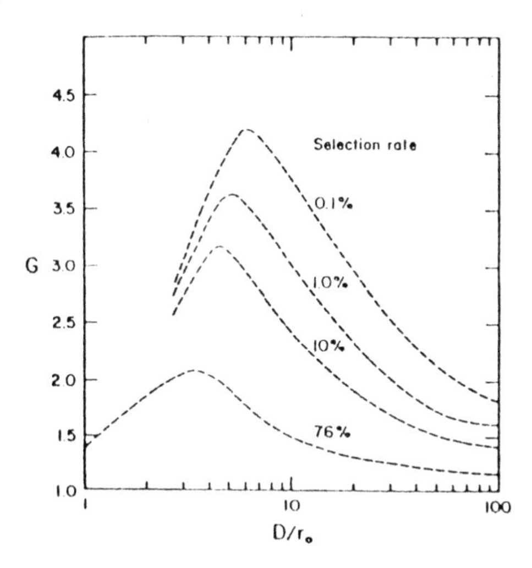

Figure 2: Expected gain in resolution, G, using selection in addition to re-centring, as a function of D/R_0 for different selection rates. (from Hequet & Coupinot [1985].

The technique of image sharpening exploits these effects, and in post-exposure image sharpening image data (usually single photon addresses) are collected in "frames" which are short compared with τ_0. Image motion and quality may be assessed by considering the image of a bright unresolved star or stars in the field of view. **The minimum required reference brightness directly affects the fraction of the sky which is accessible and hence determines to usefulness of the technique.** The crucial trade-off in image sharpening is between small apertures which allow a large fraction of the images to be selected, but have a limited maximum resolution and produce few counts in the reference star(s), and larger apertures which require more severe selection criteria but do, in principle, produce sharper images, and do produce more reference counts.

So, the recipe for (post-exposure) image sharpening is:

1. Choose a site with excellent seeing, a telescope and system resolution with resolution $\ll$ the resolution of a $3.5R_0$ telescope, and focus the telescope to a similar order
2. Choose a telescope sub-pupil $\approx 3.5R_0$, and fill the telescope aperture with sub-pupils to make efficient use of the total area.

3. Record photon address data with a time resolution $\ll \tau_0$
4. Choose a field with identifiable star(s) producing $\approx$ several x 1000 cps - for a field containing many unresolved stars this may mean $>$100,000 cps for a detector field containing many sub-pupils.

2. TRIFFID[2]

The image sharpening camera built in Galway and Dublin (TRIFFID) has:

1. An 18.2 mm collimated beam (for f/11 telescopes) within which are placed colour filters, polarising filters, etc., and in which a mask defines sub-pupils in a pupil-plane conjugate to $\approx$5 km altitude (roughly within the high altitude turbulence causing the seeing.)
2. 2 photon counting detectors imaging the same set of sub-pupil images, each sub-pupil image separately. The light is split up using a dichroic filter to allow wide (B) band imaging on an event counting detector[3] (to produce the image sharpening reference signals) and V, R, or narrow band (Hα, for example) imaging onto a framing detector[4] unsuitable for image sharpening by itself.
3. (Remote) data collection system capable of framing photon event into short ($>$500 μs) frames and storing 450 Kbytes/second directly onto optical disk from up to 3 photon counting detectors of different types.
4. Absolute time-tagging of events to 10 μs to enable phase-resolved high resolution images to be produced.
5. Hardware and software to enable (a) the telescope to be focused to $<$0".1 and (b) R_0 to be determined.
6. Software to enable image sharpening to be performed with $<$1000 reference star photons per sub-pupil per second (each sub-pupil being essentially independent)

3. High Spatial Resolution Photometry in Crowded Field Regions using TRIFFID

TRIFFID was used for the first time in June 1992 in the GHRIL Laboratory (Noordam [1985], Redfern[1991]) on 4.2m Herschel Telescope in a programme of study of the post-core-collapse globular cluster M15. A MAMA camera, belonging to ESO, was used as the straight through event-counting detector and the RAL-PCD was used as the sidearm detector. The RAL-PCD was used mostly for narrow band (Hα) imaging. Figure 3 shows both raw and sharpened images of a small region, almost central, to the core of M15, resulting from 5 minutes data accumulation with the ESO MAMA detector. The pixel scale was 0".074 per pixel so that the displayed region is $\approx$4" square, taken from a 14".0 diameter original image. The raw data, 3(a), displays excellent resolution of

---

[2] **TR**ansputer **I**nstrument **F**or **F**ast **I**mage **D**econvolution

[3] MAMA (Timothy et al [1985]), PAPA (Papaliolios et al [1985]), IPD (McWhirter et al [1982])

[4] RAL-PCD (Carter et al [1990])

0".65. The positions and sharpnesses of 12 stars (in the whole image) were used to form a composite reference and ≈25% of the data was selected and accumulated to form image 3(b). The numerical algorithm used for image sharpening is still under development and image 3(b) represents a preliminary analysis only. Nevertheless, it reveals information which cannot be seen in 3(a). Features can be distinguished down to a resolution of ≈0".25. At least 50% of the dataset displays a final resolution as good or better than this.

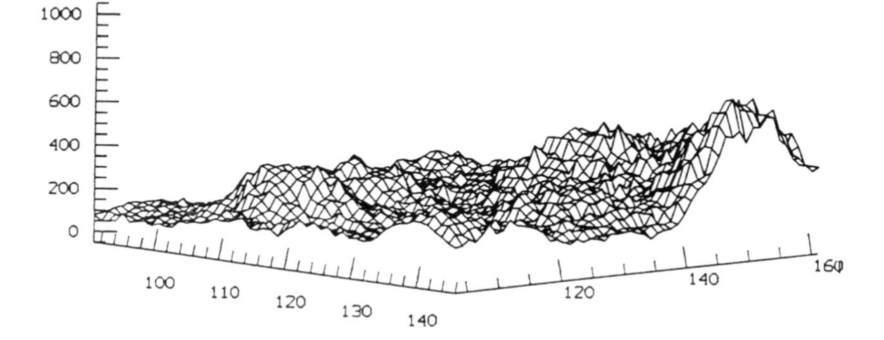

3(a) Raw Image

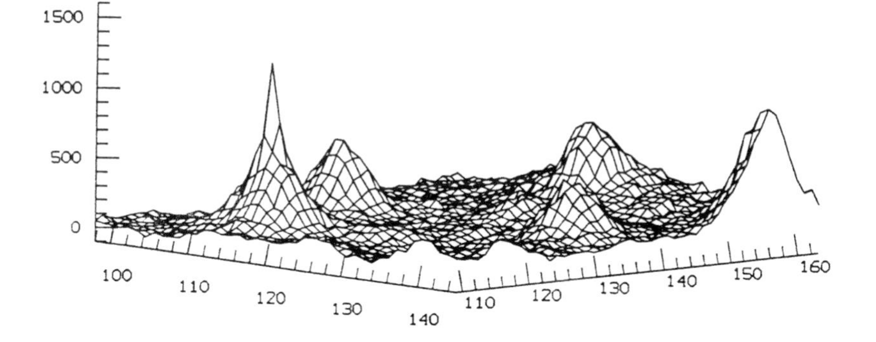

3(b) Sharpened Image

Figure 3: Image Sharpening of the Core of M15

One of the several objectives of this particular study was to obtain a detailed light curve of AC211, the optical counterpart of the binary X-Ray source X2127+11, which displays an 8.5 hour periodicity (Ilovaisky et al [1987]). Photometry of AC211 has difficult heretofore because it is close to a much brighter star and superimposed on a steeply sloping background ≈ 2" from the core of the cluster. The position of AC211 is indicated in figure 3(b) - it is clearly resolved, and photometry can be performed by normal means on this image. Whilst in its bright phase AC211 is known to be bright in Hα (Naylor et al [1988]), so that narrow-band imaging in Hα is likely to be particularly useful.

References:

Carter,M.K.,Cutler,R.,Patchett,B.E.,Read,P.D.,Waltham,N.,& van Breda,I.G.,1990. *Instrumentation in Astronomy VII, Proc.S.P.I.E.*,**1235**,644
Fried,D.,1978.*J.O.S.A.*,**68**,1651
Hequet.J.,Coupinot.G.,*J.Optics*,**16**,21
Ilovaisky,S.A.,Aurière,M.,Chevalier,C.,Koch-Miramond,L.,Cordoni,J-P., & Angebault,L.P.,1987.*Astr.Astrophys.*,**179**,L1
McWhirter,I.,Rees,D.,& Greenaway,A.H.,1982.*J.Phys.E:Sci.Instrum.*,**15**,145
Naylor,T.,Charles,P.A.,Drew,J.E.,& Hassall,B.J.M.,1988.*Mon.Not.Roy.Astr.Soc.*,**233**,285
Noordam,J.,1985.*Proposal to Establish the GHRIL Facility, under the La Palma Instrumentation Announcement of Opportunity.*
Papaliolios,C.,Nisensen,P.,& Ebstein,S.,1985.*Applied Optics*,**24**,287
Redfern,R.M.,1991.*Vistas in Astronomy*,**34**,201
Roddier,F.,1981.*The Effects of the Atmosphere in Optical Astronomy,* In: *Progress in Optics XIX (E.Wolf,Ed.)*,283
Timothy,J.G.,Bybee,R.L.,1985.*Proc.S.P.I.E.*,**687**,1090

Discussion

W. Tobin: *How many big telescopes actually focus well enough for your technique?*

Redfern: The number may be small, but the next generation of telescopes should be better. We are very happy to have access to the NTT which has an outstanding measured image width of 0″.18 (80% energy enclosed). The Nordic Optical Telescope in La Palma is also reputed to be very good.

The problem is, of course, alleviated by taking independent sub-pupils since many abberations reduce as the square of the pupil diameter.

T.J. Kreidl: *Can you quantify to what precision the sharpening process will conserve flux, hence photometric precision?*

Redfern: There is no reason to suppose that the image sharpening process produces any systematic change in the relative intensities of the reference stars and other parts of the field. In the double star images shown the relative intensities remained constant within statistical accuracy.

P.A. Wayman: *Your M15 example is of how many apertures, and how may apertures were you using altogether?*

Redfern: The data which I showed is very preliminary - we have had access to the data for less than three weeks in all. In this data one aperture only, out of four, was used and only 5 minutes total from more than 10 hours of data. In order to produce final images we must address the problems of flat-fielding and image rotation in addition to co-adding the four images.

Photometry with Infrared Arrays

I.S. Glass

South African Astronomical Observatory

Abstract

Infrared arrays have been in use at a number of observatories for several years. They are more complicated in their construction than optical ones and more problems arise in obtaining good photometry from them. The types of arrays currently available are described together with the observational techniques and the problems encountered in obtaining accurate photometric results.

1. The available arrays

Most infrared arrays used in astronomy are 'hybrids', *i.e.* they are made like a sandwich whose bottom layer is a silicon read-out multiplexer chip and whose upper layer is an array of infrared detectors. The layers are contacted together, pixel by pixel, by microscopic pillars of indium which form cold welds under pressure. Unlike optical arrays, the infrared ones are not usually read out by CCDs, but the charge is accumulated in the self-capacitance of each photodiode and read out by a network of switches in the silicon.

There are three types of detector material in use in astronomical arrays at present. The most popular is HgCdTe, whose proportions can be tailored to make band gaps corresponding to a large range in cut-off wavelength. The well-known NICMOS arrays, developed for an infrared camera to be carried by the Hubble Space Telescope, have a cutoff at about 2.5μm, while the 64×64 pixel Philips' arrays used by ESO and SAAO cut off at 4.1μm so that they can cover the JHKL$'$ bands (1.25 to 3.8μm effective wavelength).

Somewhat less tractable as a detector material is InSb. Arrays of 58×62 elements, manufactured by SBRC, were until recently the most popular sensors in infrared cameras but have been overtaken by the NICMOS chips. InSb has a fixed cut-off wavelength at about 5.5μm. Sensitivity at L$'$ (3.8μm) and M (4.8μm) is highly desirable for locating objects with cool shells by means of their high K-L$'$ or L$'$-M colours, especially in galactic plane fields which are crowded at K.

PtSi is a technology which is much more reliable than either of the two compounds already mentioned. Most chips manufactured to date are monolithic, *i.e.* the PtSi is formed by implanting Pt on a Si chip and the whole process is thus much simpler. However, a 256×256 sandwich-type PtSi chip has been used with considerable success at KPNO and CTIO (Fowler *et al.*, 1989). Unfortunately, the quantum efficiency of PtSi detectors is at most a few percent, tailing off at a little beyond 5μm. This problem is to some extent offset by the very large formats in which they can be constructed — detectors of 512 × 512 have been used in Japan and others of 1024 × 1024 are being developed.

Table 1 Some currently available infrared arrays suitable for astronomy:

Manufacturer	Rockwell	SBRC	Mitsubishi	Hughes
Material	HgCdTe	InSb	PtSi	PtSi
Format (pixels)	256 × 256	256 × 256	512 × 512	256 × 256
Pixel size (μm)	40 × 40	30 × 30	20 × 26	30 × 30
Eff. fill factor	100%	~100%	71%	~100%
Operating temp.	60-77K	40K	57-60K	40-60K
Dark current	$< 5e^-/s$	$< 10e^-/s$	$< 1e^-/s$	$< 10e^-/s$
Read noise	$< 60e^{-*}$	$< 60e^-$	$< 30e^-$	$< 60e^-$
Well capacity	$2.5 \times 10^5 e^-$	$6 \times 10^5 e^-$	$2 \times 10^6 e^-$	$6 \times 10^5 e^-$?
Quant. efficiency	>50%	>80%	3% to 6%	3% to 6%
Defective pixels	<2%	<1%	<0.01%	<0.5%
"Entry" price	$90,000	$180,000	Y10M (in camera)	$30,000

* $27e^-$ using multiple non-destructive read-outs (Hodapp *et al.*, 1992)

2. Problems in the arrays themselves

Bad pixels

As can be seen from the tables, it is not unusual to find that about 1% of the pixels in an array are faulty — they may be of low sensitivity, noisy, dead or have high dark current. This problem may be overcome by multiple exposures with the array displaced by one pixel at a time and subsequent re-alignment before median averaging during image processing (see Fig. 1). At the very least, three exposures are necessary. Davidge (1992) reports using nine exposures per field in an attempt to obtain high accuracy observations of the globular cluster NGC 4147.

Quantum efficiency variations

Most array detectors need some form of flat fielding to reduce the pixel-to-pixel variations in sensitivity. For example, Hodapp *et al.* (1992) report that a 256 sq NICMOS3 array in use at the University of Hawaii has average q.e. of 32% at J, 41% at H and 45% at K with a factor of 1.6 variation from the worst to the best area of the array. The quantum efficiency and its variations are found to be highly temperature sensitive, necessitating careful thermostating.

Read noise

The read noise of the hybrid arrays (see table) is usually quite high compared with optical CCDs (the latter have typically 3-10e^-), although it is small compared to the fluctuations in the background photons from the sky in broadband exposures. However, for low-background work such as spectroscopy, read noise is often excessive and it is usual to resort to multiple non-destructive read-outs. Several authors have found that read noise can be reduced by up to a factor of ~4 in this way (Persson *et al.* (1992), Hodapp *et al.* (1992).

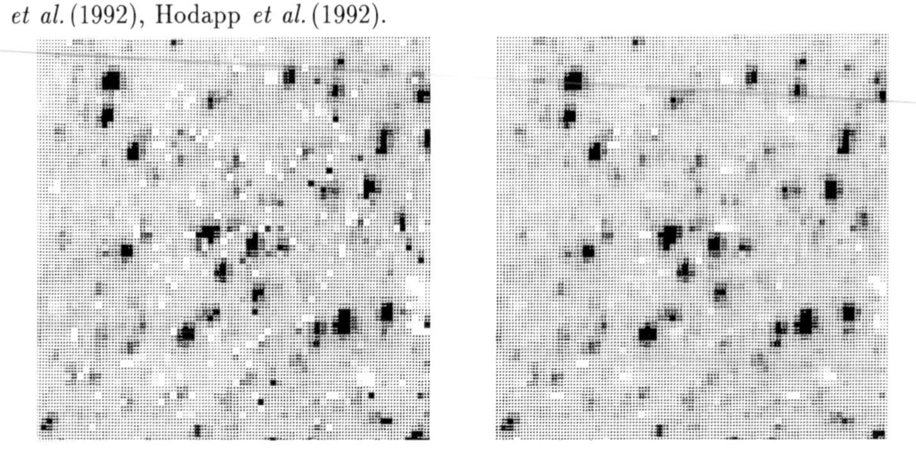

Figure 1 Eliminating bad pixels by median averaging: the left frame is one of three (the other two of which were exposed with offsets of one and two pixels respectively) median averaged to make the right frame, after image processing to remove the offsets. The isolated bad pixels have been removed but extended areas of bad pixels remain in part. This exposure is a K image of the field of the X-ray source 1E 1740.7 -2942 and it covers about 64 × 64 arcsec. Each of the three images is formed from four on-source exposures and four sky exposures of one minute.

Non-linearity

Non-linearity in direct read-out arrays arises from the use of the photodiodes as integrating capacitors. Unfortunately, the capacitance of a solid-state diode changes with voltage so that, as charge accumulates, the developed voltage (the measured quantity) is not strictly proportional to photon flux. A deviation of 4% has been reported for the SBRC 58 × 62 InSb array by Guarnieri *et al.* (1991) and about 1% at 2/3 full well for a NICMOS3 array by Hodapp *et al.* (1992). Persson *et al.* (1992) mention a non-linearity of 5% for a NICMOS2 (128 × 128) array. Such non-linearities must be corrected before any other processing of the array data can be undertaken.

Thresholds and charge-transfer problems

Threshold problems, *i.e.* a reluctance to transfer charges properly, particularly at low light levels, were a common defect in early optical CCDs and had to be overcome by such techniques as pre-flashing. Although this is not usually a problem with direct read-out arrays, it seriously affects the operation of the HgCdTe Philips' arrays at J,H & K. These arrays are unusual in that they have a CCD readout layer. By ensuring that exposures are always sufficiently long that the background level acts as a 'pre-flash', these effects can be overcome partially. However, difficulties remain with flat-fielding in the neighbourhood of 'hot' pixels (*i.e.* those with very high dark current) due to their 'tails' which tend to be background-dependent. The duration of a single exposure is limited by the fact that more pixels become saturated by dark current as the exposure time increases.

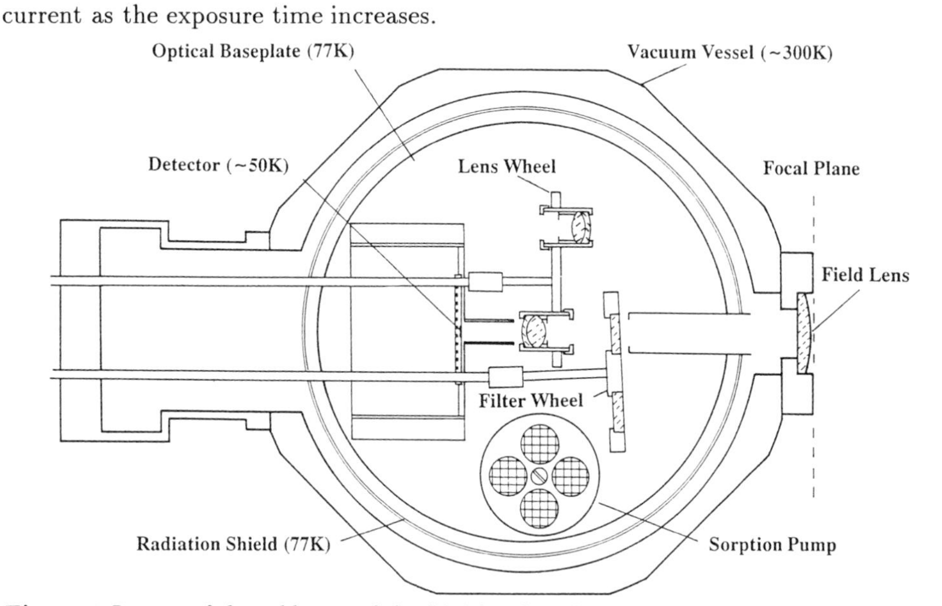

Figure 2 Layout of the cold part of the SAAO infrared camera. The entrance of the baffle forms a cold field stop and the pupil stop is a cold aperture close to the relay lens. The detector is a Philips' 64 ×64 HgCdTe array with a cut-off wavelength of 4.1μm.

Well capacity

Especially when observing at wavelengths beyond 2.5μm, the background from grey- or black-body radiation from the telescope becomes overwhelming except for a cooled instrument in space. The limited well capacity of arrays restricts exposure times in most situations to a small fraction of a second. In order to obtain long effective exposure times, an array must be read out many times and the frames co-added. Even the Philips' array at SAAO, although it has an exceptionally high well capacity of 10^7e^-, can integrate for only about 1s if used at L′ (3.8μm) and 1 arcsec^2 per pixel. Other arrays are much more limited. An extreme example is the 10μm camera described by Gezari *et al.* (1992), based on a 58 x 62 SBRC Ga:Si hybrid chip with 0.25 arcsec square pixels, which requires to be read out 30 times per second and the frames co-added. This requires high-speed electronics and data processing. Larger format arrays for long wavelengths rapidly become impractical.

Residual Excess Dark Current

Hodapp *et al.* (1992) report an excess dark current which follows the detected pattern of a previous strong exposure. Although its effects are negligible after a few complete reads of the device for normal background-limited exposures, it can remain a nuisance for up to an hour for exposures with low backgrounds.

3. External Problems

Extraneous Background

Needless to say, every effort must be made to reduce the extraneous radiation falling on the detector. This is done by cooling the detectors and their surroundings as well as the filters (to reduce their emission out-of-band) and by the use of cold field and pupil stops. Fig. 2 shows the design of the camera used at SAAO.

Undersampling of stellar images

The small size of arrays tempts the observer to use a scale which maximizes the sky coverage but which leads to undersampling. This is particularly serious when there is dead space between pixels or uneven sensitivity within pixels such as in the Philips' and many single-layer arrays. The problem can be reduced by multiple exposures with shifted images and by fitting the stellar profiles with models tailored to the characteristics of the array in question. Uneven response within individual pixels has been found in a NICMOS2 array (Allen, private communication).

Flat fielding methods

Ideally, a flat field should be an object in space with colour similar to the objects being measured. In practice, the methods used are to compare the sky with the interior of the dome, to observe a white sheet illuminated by incandescent light, to observe a card similarly illuminated at the top of the Cassegrain baffle and to use the median average of long sky exposures. At L', a sky exposure can be subtracted from the interior of the dome! Although some of the more stable infrared arrays can be used in almost the same manner as optical CCDs, *i.e.* without frequent observation of background frames, the more usual procedure seems to be to 'chop', *i.e.* to observe blank areas of sky. Because of the threshold problem of the Philips' chip, at SAAO we observe at K by making one exposure on the object followed by two on the sky, two more on the object and so on as necessary in ABBA pattern. The duration of the individual exposures is kept constant at 30s. At L', we take typically 40 0.6s exposures for each of the As and Bs and co-add the data before recording.

Standard Stars

Most infrared standard star lists contain objects far too bright for array exposures of reasonable length, *i.e.* comparable to those used for observation of programme objects. Some observers report de-focussing the telescope to reduce the flux per pixel. To solve this problem, Brian Carter of SAAO has been observing about 64 stars of around 8th mag at JHK in collaboration with the IRIS group at AAO and others at Mt Stromlo. The programme comprises 64 stars scattered widely around the southern sky. So far, 54 have been observed to the necessary accuracy ($\sim$.03 mag). Each programme star is referred to a nearby star from Carter (1990).

Sky background and special filters
Sky backgrounds of J=15.9, H=14.0 and K=11.8 mag/arcsec^2 have been reported by Persson *et al.* (1992). The H-band flux (mainly due to airglow) varies by almost a factor of two from night to night. In order to achieve higher sensitivities at K, Wainscoat & Cowie(1992) advocate the use of a K′ filter which has shorter cut-on and cut-off wavelengths than the standard K band, by about 0.1μm. This reduces the contribution from the thermal radiation of the telescope and atmosphere, at the expense of introducing a new broad-band colour. However, the ability to make measurements on a standard system should not be abandoned lightly, as experience with differing J filters has shown in single-detector photometry.

Photometric Performance
With great care, including proper attention to the quality of the standard stars and the transformations between photometric systems, it appears that photometry of stars bright enough to be well above the background is possible at about the 1% error level (Guarnieri *et al.* ,1991; Hodapp *et al.* ,1992; Persson *et al.* ,1992). Difficulty in obtaining perfect flat fielding, rather than the intrinsic sensitivity of the arrays, seems to be the limiting factor for faint mags.

Acknowledgement

I would like to thank D.A. Allen, J. Elias and A. Moneti for sending me information on various arrays in use at their institutions.

References:

Carter, B.C., 1990. MNRAS **242**, 1.

Davidge, T.J., 1992. A.J., **103**, 1259.

Fowler, A., Joyce, R., Gatley, I., Gates, J. & Herring, J., 1989. SPIE **1107**, 22.

Gezari, D.Y., Folz, W.C., Woods, L.A. & Varosi, F., 1992. PASP **104**, 191.

Guarnieri, M.D., Dixon, R.I., & Longmore, A.J., 1991. PASP **103**, 675.

Hodapp, K.-W., Rayner, J. & Irwin, E., 1992. PASP **104**, 441.

Persson, S.E., West, S.C., Carr, D.M., Sivaramakrishnan, A. & Murphy, D.C., 1992. PASP **104**, 204.

Wainscoat, R.J. & Cowie, L.L., 1992. A.J., **103**, 332.

Discussion

S. C. Russell: *What objection is there to stepping by several pixels in order to eliminate clumps of bad pixels?*

Glass: You could, but clumps of bad pixels occur on several scales, so to do a thorough job you might need to make, say, nine exposures.

The Avalanche Photodiode — A Promising Low Light Level Detector for Astronomical Photometry

G. Śzécésnyi-Nagy

Eötövs University of Budapest, Department of Astronomy, H-1083 Budapest, Ludovika tr 2. Hungary

Abstract

Astronomical photometry like other scientific methods is always forced to search for up-to-the-minute instruments and the most sophisticated devices. The photomultiplier tube (PMT) is the most widely used photon detector offering high sensitivity and gain with acceptable quantum efficiency and spectral range but it also has some disadvantages. During the past few decades all of these parameters have been improved and it seems likely that possibilities for their continued development are exhausted.

In the meantime, technical development — especially micro electronics and semiconductor technology — has produced a new detector, the avalanche photodiode (APD). The APD which is often referred to as a solid-state PMT is just at the start of its career and we may expect its rapid perfection. Although properties of APDs were first discussed almost twenty years ago their mass production did not start until 1990. Early silicon APDs had very limited dimensions compared to PMTs but unfortunately this also implied that their photosensitive area was less than 0.25 mm. One more year of development has been enough to increase the photon collecting surface by a factor of 1000 and APDs may soon become possible rivals of PMTs.

Following a brief description of the new device and discussion of its operation a comparison of photodetector performance will be given in order to draw attention to those parameters which are of great promise. Those sections of astronomical photometry which are ready to benefit from the suggested testing and use of this kind of photon detector will be highlighted.

1. Why do we need a solid-state photon detector with an internal gain mechanism?

At ultra-low light levels the electric current induced by incoming photons in the detector is definitely lower than the noise of any electronic amplifier connected to it. In order to be able to measure accurately the photon flux of very faint sources we need a detector which employs some kind of internal gain mechanism. Most generally this is internal multiplication or avalanching (multiplication of electrons produced by a photoemissive element or photocathode).

In spite of practical drawbacks, vacuum photomultiplier tubes (PMTs) have long been the most widely used photon detectors for high accuracy astronomical photometry. But nowadays the replacement of vacuum or low-pressure gas tube devices with solid-state detectors is a general trend which seems to be irresistible. The mass production of semiconductor chips and other photonic circuits and sensors will probably result soon in a performance/price ratio superior to that of PMTs which are very complicated and delicate products of state-of-the-art technology and craftsmanship. Although I am confident that the PMT contributed more than anything else to the spectacular development of high-precision photoelectric photometry, it is also evident that the good old tubes must have passed the culmination of their career and that astronomers specialized in photometry have to consider more modern and simple detectors for the photometers of the next century. The avalanche photodiode (APD) with its internal gain mechanism is already competitive with PMTs in many aspects and we can be optimistic about its future.

2. Principle of operation of APDs

In these devices which are usually made of germanium or silicon the incoming photons produce excess carriers in the active (PN or PIN) junction. It is crucial that *each* photon whose energy exceeds the bandgap of the semiconductor material has the ability to produce *one* electron-hole pair and consequently may affect the overall junction potential which can be measured.

In a typical photodiode the photocurrent is independent of the applied bias and its value gives immediate information on the incoming flux. Typical diodes are linear over eight decades of incident photon flux (Eccles *et al.* 1983) and this is their most promising characteristic. On the contrary, the magnitude of the applied bias voltage affects the speed of response of the device. Large reverse bias causes higher acceleration of the minority carriers across the junction, reduces their transit time and increases the response speed.

In a typical APD hundreds or thousands of volts is applied to the photo-diode as reverse bias. Consequently each electron generated in the P layer is accelerated by this enormous potential field to considerable energies as it passes the diode. Silicon has a bandgap of only 3.6 eV so in a SiAPD each photo-electron has enough kinetic energy to excite several hundred free electrons by collisions in an avalanche effect. Carrier production avalanches because any new secondary electrons are accelerated by the high field too. Thus *each* infalling photon causes 100-1000 charge carriers to flow across the junction.

It is evident even from this brief description that the APD is a theoretically complex device that combines photon detection with electrical amplification (gain) in a simple monolithic solid-state structure. Consequently APDs in principle are similar to PMTs while being much simpler (two wires only, no glass vacuum tube, no dynodes, no voltage divider etc.), smaller, lighter, more compact, more rugged and immune to magnetic fields. An extra advantage of an APD over a PMT is that the former uses a single high voltage which draws only the photocurrent whereas PMTs need 10–20 various voltage levels and they draw anode currents 20 times larger.

3. Recent developments in APD technology

While the APD has been recognized as a possible replacement for the PMT, until recently mass production has proved possible only for devices of small dimensions (diameters less than 1 mm). These small area APDs (SAAPDs) are not yet competitive with PMTs in low light level photon detection in astronomy.

One of the most serious problems with APDs has been the small size of their light sensitive area (Smith 1992). SAAPDs have a break down voltage of only some hundreds of volts which is insufficient to reach high enough gain for low light level photon flux determination. But a newcomer to the field, Advanced Photonix, has successfully developed large-area APDs (LAAPDs) that can satisfy the needs of many applications currently using PMTs (Noble 1991). The new LAAPDs combine the benefits of solid-state photodiodes with many of the performance characteristics of a typical PMT. They provide high internal gain as a consequence of the much higher reverse voltage which can be applied because the break down voltage of these devices is about 2.5 kV (Saloff & Madden 1992). Using transmutated doped silicon and a proprietary edge treatment this firm now offers LAAPDs with diameters larger than 0.5 inch, capable of operating at voltages in excess of 2 kV. It is important too that the typical dark current of the device at 2.4 kV reverse bias is only 200 nA (Advanced Photonix 1992).

LAAPDs with much larger active areas than traditional APDs and with a close functional similarities to PMTs and photodiodes offer a real alternative to these devices and require little or no system redesign while significantly improving the performance/cost ratio. Furthermore using APDs in a proper housing and electric environment offers overall reduction in system size, weight and complexity, improved ruggedness, reliability and longevity. In addition they have an enormous dynamic range with a perfectly linear response from a few hundred photons to 10 mW incident intensity. They are immune to damage by overload and do not have the long settling times typical of PMTs (up to 100 hours – THORN EMI Electron Tubes Ltd. 1986).

LAAPDs of Advanced Photonix typically use only 5% of the current than an equivalent PMT and operate from a single high voltage. Another benefit of these solid-state photon detectors is that they are immune to the effects of stray fields for which they require no shielding at all (Saloff & Madden 1992).

The maximum gain that has been reached by LAAPDs so far is considerably lower than that of a PMT used in astronomical photometers but it is not unduly optimistic to expect the appearance of SiAPDs 10–100 times more effective even in this decade. However, for ultra low light level applications like single photon counting, sophisticated PMTs remain the first-choice at least for five more years.

4. Spectral sensitivity and responsivity of the APDs

The spectral sensitivity of the photodiodes is affected by two phenomena. Reflection of the infalling photons from the surfaces of the semiconductor and its housing or window and absorption of the radiation by the material of the window fix the short-wavelength limit of the sensitivity range. On the other hand the bandgap of the raw material of the APD determines the long wavelength threshold. In the case of commercially available detectors UV-sensitivity is very low but may be extended theoretically to 200 nm and over practically the whole of the astronomically interesting (ground-based) UV-range to 290 nm, using proper window materials and/or phosphorescent coatings. For instance the *UV-to-visible* converter of Princeton Instruments is able to increase the quantum efficiency of silicon CCDs in the 250-400 nm range by a factor of 20 to 100 (Princeton Instruments Inc. 1991).

The IR-sensitivity of a photodiode is fixed by the band structure of its semiconductor material which depends on the temperature of the device. At room temperature silicon is capable of detecting photons with wavelengths below 1100 nm and germanium — the other popular raw material of photodiodes — is able to convert photons with wavelengths up to 1820 nm into carriers.

Earlier versions of SiAPDs had a spectral response range from 400 nm to 950 nm with a peak response at 800 nm. At this wavelength their typical peak responsivity is 150 A/W. (Responsivity R is the ratio of the photocurrent generated for every watt of incident light power with units of A/W or mA/mW. It is the preferred measure of a photodiode's response to light.) Normally responsivity ranges between 50 A/W and 150 A/W

The silicon photodiode response is usually linear within a few tenths of a percent from the minimum detectable incident light power up to several milliwatts and its linearity improves with increasing applied reverse bias and decreasing effective load resistance. Average SiAPDs normally have a dynamic range exceeding 10 million while the best devices give a linear response in a range at least ten times greater.

For photometrists who are more familiar with the quantum efficiency ($Q.E.$ or QE) of their photon detectors a similar feature, the external quantum efficiency (EQE) can be defined as the percentage of incident power which results in an electric current which flows when an external load is connected to the photodiode. It can be calculated with the following formula:

$$EQE = 123.96R/L$$

where EQE is the percentage of external quantum efficiency, R is the photodiode responsivity (measured in amperes/watt) and L is the wavelength of the incident radiation (in microns).

Typical values of quantum efficiency range from 50% to 95% depending on the wavelength of the incident light and the type of photodiode. EQE is always less than unity because of reflection losses at all wavelengths, surface loss mechanisms at VUV wavelengths and poor absorption of photons in the IR spectral range. Since at very low light levels the optimal exploitation of the incoming photons is an absolute necessity their QE is probably the most important feature of photodetectors. The table gives EQE and QE values for SiAPDs and PMTs respectively from 400 nm to 1100 nm (after MacGregor 1991 and THORN EMI Electron Tubes Ltd 1986). It is to be noted that typical PMT photocathodes have peak QE in the UV-range at about 380 nm while an SiAPD's EQE peaks in the near IR at about 850 nm. The fourth column gives the values of the ratio E in favour of SiAPDs.

Wavelength (nm)	SiAPD EQE (%)	PMT QE (%)	$E = EQE/QE$
400	25	27	0.926
400	25	27	0.926
500	80	22	3.6
600	85	10	8.5
700	90	5	18
800	>90	2	>45
900	60	<0.5	>120
1000	45	<0.3	>150
1100	25	<0.1	>250

Values in the column of PMT QE at any wavelength refer to the highest QE of the following three cathode types: Bi-alkali, S1 or S20.

It is evident from the table that SiAPDs should soon replace PMTs in the red-IR spectral range. In the yellow (visual) and orange colours the excess factor is at least an order of magnitude lower than that needed to compensate for the momentarily (and temporarily?!) much higher gain of photomultipliers. Nevertheless the quick and unprecedented development of the semiconductor technology and its products during the last decades predicts the possible and effective perfection of APDs. While the internal gain of these devices will be increased by a factor of 100, their overall signal will exceed that of the best PMTs over a much wider spectral range. And this is not a dream!

A team at EG&G by operating with low dark-count (small active area) APDs above the breakdown voltage has been able to achieve avalanche gains exceeding one million and producing current pulses that have been easily detected (MacGregor 1991 and Noble 1991). It means that APDs could be used even in photon-counting applications (at present only if they have a diameter $\leq$1 mm) provided that their bias and temperature are strictly controlled.

In view of the low ultraviolet quantum efficiency of APDs, the task of finding the best window materials and most effective *UV-to-visual* or *UV-to-NIR* converting layers is urgent. As it was mentioned, the efficiency of photo-electron production can be increased by a factor of up to 100 in this way.

It is time to mention some drawbacks of APDs. Dark current — an essential phenomenon which may limit the ultra low light applications of photodetectors — exceeds the acceptable level and it is higher in the case of LAAPDs which are otherwise more promising for astrophotometry. Another serious problem is the high temperature sensitivity of photodiodes. Until recently these devices have been used only at or near room temperature and even minute changes in the ambient temperature have caused measurable effects on the responsivity of SiAPDs. Heating the photodiode shifts its spectral response curve (including the peak) toward longer wavelengths whilst cooling shifts the curve towards shorter wavelengths (Advanced Optoelectronics, 1990). The following values are typical for the temperature dependence of responsivity:

at 1064 nm .... +0.75 to +0.90
at peak and somewhat below .... +0.1
from 500 nm to 700 nm .... practically 0
in the blue to UV .... -0.1 to -2.0

(Advanced Optoelectronics, 1990)

The stronger (and negative) temperature dependence of photodiode responsivity in the UV suggests the low-temperature application of this detectors which is permissible since the operating temperature of SiAPDs ranges from −40 to +45 deg C. Obviously in order to reach high reliability and photometric accuracy the temperature of avalanche photodiodes must be stabilized within 0.1 centigrade. Owing to the compactness of APDs, cooling and thermostatic control can be achieved with a simple Peltier thermoelectric cooler. According to Advanced Photonix Inc., LAAPDs with integral thermoelectric cooling are available on a customised basis.

Thermal stabilization will contribute also to gain uniformity and constancy since the breakdown voltage is also temperature dependent with a coefficient of +2 V/deg.

5. Prospects of further developments of SiAPDs

In order to get some basis for considerating the future of APDs let us first run through the brief history of PMT development!

1936 Discovery of the principle of the amplification of photoelectrons by means of secondary electron emission in an experimental PMT (Zworykin)

1937 Astronomical testing of an experimental RCA-tube (Kron)

1938 The start of 'mass' production of PMTs (Rajchman)

Complex cathode surfaces (caesium and antimony) $\gg$ fourfold sensitivity (Grlich). The QE of the new cathodes exceeded 20 times that of the best photoemulsions.

1939/45 WWII: military applications of the tubes, quick development of construction

1946 The introduction of the method of photon counting

1947 The start of the glory of the venerable 1P21 PMT (RCA). Small, compact design, S4 cathode in the centre of the tube, QE (at 400 nm) 22 %, sensitivity range 300 nm to 650 nm, 9 dynodes, amplification one million, dark current 1 nA.

This is the most extensively used PMT in astronomical photon detection.

1949/55 Observatoire de Paris program: 250 PMTs, up to 20 dynodes, UV, Visible and IR sensitivity, giant tubes and long contacting wires etc.(Lallemand)

1952 Photometric measurements with PMTs in the IR (Hardie)

1953 UBV-system of photoelectric photometry (Johnson & Morgan)

1957 extended sensitivity range (Ag-O-Cs cathode) up to 1100 nm (ITT)

1960/70 Wide offer of PMTs by various manufacturers (dimensions, number of dynodes, gain, photocathodes, window materials etc.)

1980- Thermoelectric coolers, commercially available photon counting systems, magnetic shielding

1984/85 NO SIGNIFICANT IMPROVEMENTS SINCE THEN ! The start of PMT-agony?

It is remarkable that astronomers of the thirties were so eager to benefit from the discovery of the PMT and tried to use it immediately. On the other hand it is obvious also that almost nothing happened in the past decade and next to nothing during the preceding two that could have revolutionized PM-theory and/or technology. This definitely suggests that the PMT-era is drawing to a close and in line with that semiconductor technology affects various areas of photodetector applications. My suggestion is to grasp the opportunities of solid-state optoelectronics.

6. Expectations of SiAPD development during the next 2-3 years

Having seen the sophistication of avalanche photodiodes explained briefly, the following essential improvements can be expected:

- extension of the spectral sensitivity to the blue-violet-UV range
- enhancement of overall sensitivity, detection of 10-photon and single-photon pulses
- lowering or compensation of the APD's dark current
- substantial increase or multiplication of the avalanche gain by operating the device above the breakdown voltage
- reduction of the LAAPD's price (this is only a desire of the author not of the manufacturers)

7. Those Fields of astronomy which could benefit soon from the use of APDs

The most important experimental advantages of APDs over PMTs can be summarized as follows:

- extremely high quantum efficiency in the visible-red-NIR spectral range
- unprecedented dynamic range
- compactness, ruggedness, immunity to magnetic fields, no settling time and very good tolerance of overload

Consequently every field of astronomical photometry could benefit from their use especially those which are aimed at the detection of red (low temperature) objects, R and I colours low-mass stars (dK, dM spectral type and brown dwarfs), eruptive variables (flare stars!), Hα emission objects and highly reddened objects. The use of APDs is very promising in space applications and also in campus observatories because they are astronomer- and student-proof. When the competition between manufacturers results a substantial reduction of their price, APDs should also be the preferred photon detectors for serious amateur photometrists.

Acknowledgements

This research was partly supported by grant (# 782/91) from the Hungarian Foundation for Higher Education and Research . I would like to acknowledge the kind help of Advanced Photonix Inc., Advanced Optoelectronics, Hamamatsu Photonics, Princeton Instruments Inc. and THORN EMI Electron Tubes Ltd.

References:

ADVANCED OPTOELECTRONICS, 1990, *Photosensor Product Catalog*, p.2.

Advanced Photonix Inc., 1992, *Large Area Avalanche Photodiodes*, p.2.

Eccles, M.J., Sim, M.E. & Tritton, K.P., 1983, *Low light level detectors in astronomy*, p.113.

MacGregor, A., 1991, Photonics Spectra, Vol.bf 25, p.139.

Noble, M., 1991, Lasers & Optronics, Sept. 1991, Elsevier.

Saloff, D. and Madden, M., 1992, Photonics Spectra, Vol. 26, p.111.

Smith, N., 1993, Poster Papers on Stellar Photometry, IAU Coll. 136, Dublin Inst. for Advanced Studies.

Princeton Instruments Inc., 1991, Charge Coupled Devices - CCD Area Array Detectors, p.2-3.

THORN EMI Electron Tubes Ltd., 1986, *Photomultipliers*, p.18.

Discussion

T.J. Kreidl: *Would you please comment on those manufacturers who currently supply the large-cathode APDs and indicate approximate prices? Also, have you taken any astronomical data with any APDs at this time?*

Ŝzécsényi-Nagy: Unfortunately I have no information even about approximate prices of LAAPDs. My data refer to information from Advanced Photonix, Inc., a Xinus Company, but other manufacturers, like Hamamatsu Photonics, offer LAAPDs too, (see the brochure *Product Guide* which was kindly placed by the organizers of this meeting into our portfolios).

D. Dravins: *At Lund Observatory, we have been testing an avalanche photodiode detector in the laboratory, for more than a year. The detectors seem very promising with regard to, for example, ease of use, and quantum efficiency. However, there is still a serious problem in obtaining acceptably low dark count rates with sufficiently large detectors. With our 0.15 mm diameter APD we get about 50 dark counts per second, which would be acceptable but an astronomical interesting detector would need to be one order of magnitude larger in diameter, and two orders of magnitude greater in area. Since the dark counts seem to increase in proportion to detector area, significant improvements are needed before such systems can be used for faint sources.*

Ŝzécsényi-Nagy: I agree with you. Amongst SAAPDs the newly developed SPCM-200-PQ has been reported with a dark count as low as 2 c/s (MacGregor. A., 1991), which is significantly lower than yours. For LAAPDs (0.5 in diameter) the dark current increases from 60nA to 200nA in a reverse bias range of 450 to 2250 volts. The gain of the device over the same range increases from practically 1 to 30 while with a dark current of 500nA the gain reaches about 250-300.

M. K. Tsvetkov: *I am so happy to hear about the future application of Avalanche Photodiode (APD) detectors in electrophotometry. APD's are based on the discovery of photoelectrets by the Bulgarian professor Georgi Nadjakov, in 1937. He worked at this time in Paris in the Laboratory of Professor Paul Langevin.*

Ŝzécsényi-Nagy: Thank you for your interesting comment.

N. Smith: *Using an RCA 30921 APD it is possible to operate in photon counting mode and with active quenching it is linear (to $\sim$1%) up to 100KHz. The major drawback is the small active area of approximately 0.25mm^2. Typical dark count rates are $<$ 200 c/s at -20°C. Using light cones it may be possible to increase the effective active area. Afterpulsing and problems with light emitted by the avalanche process in the APD can be overcome by introducing a deadtime of typically 40-100ns.*

R.M. Redfern: *There seems to be some confusion, in the questions, between APDs operating in the linear and photon-counting 'geiger' mode. Is there any suggestion that these large area disks can be operated in the 'geiger' mode?*

Ŝzécsényi Nagy: Yes, the manufacturer of the LAAPD mentioned here definitely suggests this kind of operation.

A Bidimensional Photon-counting Microchannel Plate Detector Using a Wedge and Strip Anode

R. Drazdys, J. Jukonis, A. Skrebutėnas, V. Vansevičius and G. Vilkaitis

Institute of Physics, Goštauto 12, Vilnius 2600, Lithuania

Abstract

A photon-counting detector with a 45 mm diameter active area has been developed using a microchannel plate (MCP) and the wedge and strip readout system. The detector uses a UV transparent glass input window proximity coupled to a stack of MCP. Quantum efficiency (QE) of the detector photocathode is $\sim 20\%$. The resolution of ~ 60 μm FWHM at a gain of $5 \cdot 10^7$ has been achieved.

1. Introduction

The paper describes the performance characteristics of a microchannel plate photon-counting detector developed at the Astrophysical Instrumentation Laboratory of the Institute of Physics, Vilnius, Lithuania.

2. Detector

A schematic view of the detector is shown in Fig. 1. Its first component is a photocathode where an incident photon may be transformed into electron. The electron crosses the 0.5 mm gap and is received by a stack of microchannel plates with the 15 μm channel diameter and the length to diameter ratio of 55:1. The MCP stack consists of two "V" configuration stacks separated by a 1 mm gap. MCPs in both "V" stacks are separated by a 15 μm electric contact ring.

As a result of traversing through the MCP stack, photoelectrons are multiplied by a factor of up to $\sim 10^8$, depending on the voltage applied across the stack. The electron cloud subsequently drifts through a region of uniform electric field to the wedge and strip anode. On this way the electron cloud grows to the size more than one repetition period of the wedge and strip anode pattern.

The wedge and strip anode (Martin et al., 1981) uses three electrods to divide the charge proportionally to X and Y coordinates of the event. The width of the wedges and strips varies linearly with Y and X, and the coordinates of the event may be deduced from the following simple algorithm

$$X = \frac{2Q_S}{Q_W + Q_S + Q_Z}, \quad Y = \frac{2Q_W}{Q_W + Q_S + Q_Z},$$

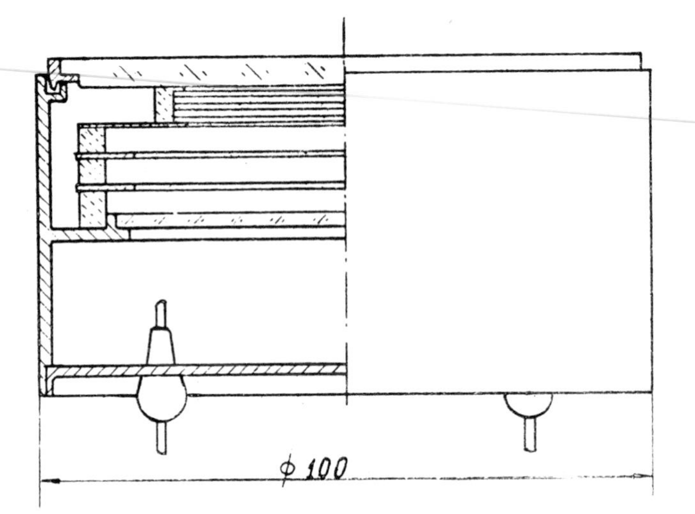

Figure 1. A cross view of the detector.

here Q_W, Q_S and Q_Z are the amounts of charge falling on the W, S and Z electrodes, respectively.

3. Electronics

The electric signals received at W, Z and S electrodes are amplified and shaped and then are passed to a signal processing unit which produces 10 bit addresses corresponding to X and Y coordinates of the event.

The data acquisition system is based on a PC/AT computer and uses a 1024×1024 video memory, allowing real time display of information. The video memory can store 2MB, and each memory element can register up to 65 000 events. Memory contents is displayed in real time on a monitor.

4. Test results

Quantum efficiency. The quantum efficiency (QE) of a bialkali cathode, combined with the UV transparent glass transmission, at the peak is approximately 20%. The detective quantum efficiency (DQE) of the detector is limited by the photoelectron detection efficiency of the unfilmed MCP, typically yielding $\sim 60\%$ (Clampin et al., 1988).

Position resolution. Principal factors affecting the spatial resolution of the detector are the electronic noise attributed to the amplifiers, the partition noise caused by the statistical deviation in the charge division among the three electrodes (Siegmund et al., 1983) and the lateral spread of the photoelectron cloud incident on the MCP (Clampin et al., 1988).

The experimental resolution was determined by imaging a resolution test chart.

The analysis of the bar pattern of one dimensional profiles yields the resolution of the order of 60 μm FWHM.

Image distortion and counting rate. The image linearity of the detector has been assessed by examination of a resolution test chart image. The most noticeable barrel-type distortion is observed in a few mm zone at the edge of the image. This distortion is attributable to variations in the field strength between the MCP and the anode.

The counting rate of the overall system, $5 \cdot 10^4$ counts per second, was limited by electronics.

5. Conclusions

The described detector is ready for tests in astronomical observations. We hope that this will be a highly competitive instrument for different low light level astronomical and other applications.

References:

Martin C., Jelinsky P., Lampton M., Malina P.F. and Anger H.O. 1981, Rev. Sci. Instrum., 52(7), 1067.

Clampin M., Crocker J., Paresce F., and Rafal M. 1988, Preprint of STScI, No. 276, 5.

Siegmund O.H.W, Clothier S., Thornton J., Lemen J., Harper R., Mason I.M., and Culhane J.L. 1983, IEEE Trans. Nucl. Sci., Vol. NS-30, No.1, 503.

Chakrabarti S., Siegmund O.H.W., and Hailey C. 1988, in Instrumentation for Ground-Based Optical Astronomy, ed. Robinson L.B., Springer - Verlag, 574.

The MEKASPEK project – a new step towards the utmost photometric accuracy

K. H. Mantel, H. Barwig, S. Kiesewetter

Universitaets–Sternwarte Muenchen, Fed. Rep. Germany

Abstract

We present the design of a new 4 channel, fiberoptic spectrophotometer which allows to measure the object, two comparison stars and sky background simultaneously throughout the whole optical wavelength range with high efficiency. The overall spectral resolution of about 50 in combination with oversampling allows to eliminate colour dependent extinction effects to a high degree and guarantees accurate transformations to other broad band photometric systems.

1. Introduction

Since the first eye estimated photometry by Ptolemaeus and Hipparchos, about 2000 years ago, photometric accuracy has increased from half a magnitude to some millimagnitudes. However, in spite of the enormous instrumental efforts that have been made after the adoption of photoelectric detectors in order to push forward the limits of accuracy, no further significant improvement could be achieved.

A careful reexamination and discussion of the whole measuring process and the reduction procedure by Young (1974, 1988) revealed that the low precission achieved so far, was caused by an insufficiant treatment of higher order terms within the theory of extinction correction and colour transformation. Young (1988) showed that even within the classical Stroemgren-King-Theory, second order terms can be treated correctly provided that the filtercurves fulfil the requirements of the sampling theorem (Shannon 1949).

Four years ago, the project MEKASPEK ('Multi Channel Spectral Photometer') was initiated at the Universitaets-Sternwarte Muenchen (Mantel et al. 1988). The major aim of this project was to develop a new 4-channel spectral photometer in accordance with the requirements of the sampling theorem which would be able to achieve the utmost accuracy reachable with today's technology. This paper discusses the main features of this instruments.

2. Instrumental Design

Within the MEKASPEK project a new photometer concept has been developed that may be characterized by the following features:

- high quantum efficiency throughout the whole optical wavelength range in order to keep the photon noise as small as possible
- simultaneous measurements of the object, two comparison stars and sky background for successful elimination of neutral atmospheric extinction effects
- spectral resolution in combination with oversampling allows to eliminate colour depended atmospheric effects and guarantees an accurate transformation to other broad band photometric systems
- optimized adaption of the entrance diaphragm's diameter to the actual seeing conditions
- high time resolution up to 10 ms in order to study rapid variable objects

The implementation of these features has been achieved by using four optical fibers which can be centered on the stars in the focal plane of the telescope.

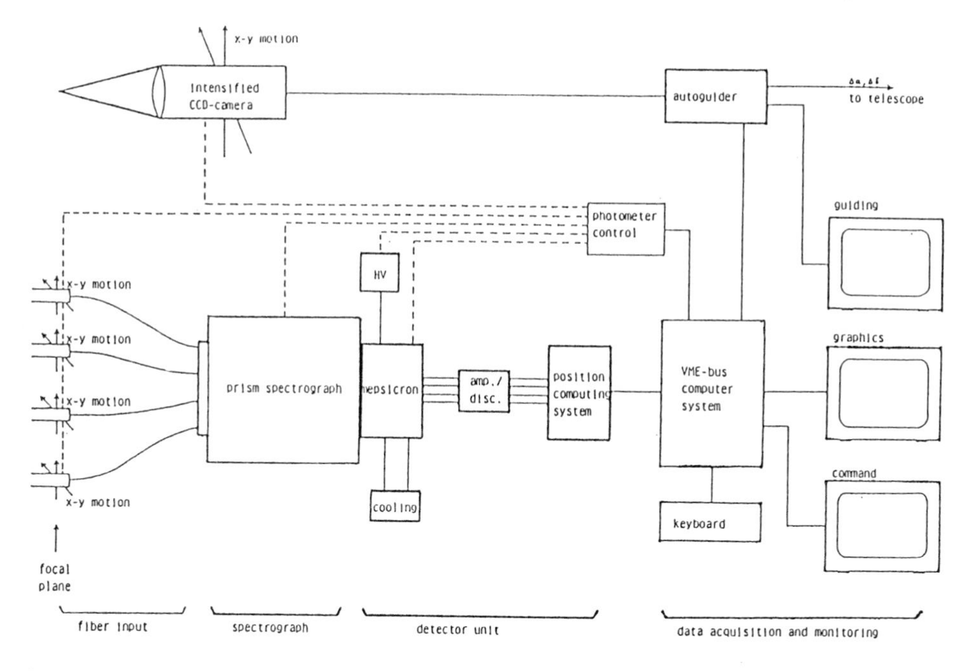

Figure 1: Block diagram of the 4-Channel-Spectral-Photometer.

The fiber bundles fed the slit of a two beam prism spectrograph which has been optimized to reach high transmission throughout the whole optical wavelength region. The four spectra are focussed onto the surface of a two dimensional photon counting detector (MEPSICRON). The images can be registered with a time resolution of up to 10 ms by a VME-bus based computer system. Figure 1 shows a block diagram of the instrument.

2.1 Optical fiber input channels

Light transfer from the telescope focus to the spectrograph is achieved by fiber bundles. Each fiber bundle has a diameter of 800 μm and consists of 130 single quartz fibers with a diameter of 70 μm each which transform the circular entrance diaphragm to a rectangular slit shape. A fabry lens with a focal length of 6 mm projects the entrance pupil of the telescope onto the entrance of the fiber bundle, thus transforming spatial motion to angular variations. The varying f-number at the entrance of the fiber bundle is smeared out by degradation effects yielding f-number variations of f/4.4 to f/2.5 at the entrance slit of the spectrograph. Each fiber bundle is equipped with an entrance diaphragm which may be exchanged under computer control and may vary between 0.6 mm and 2 mm in diameter.

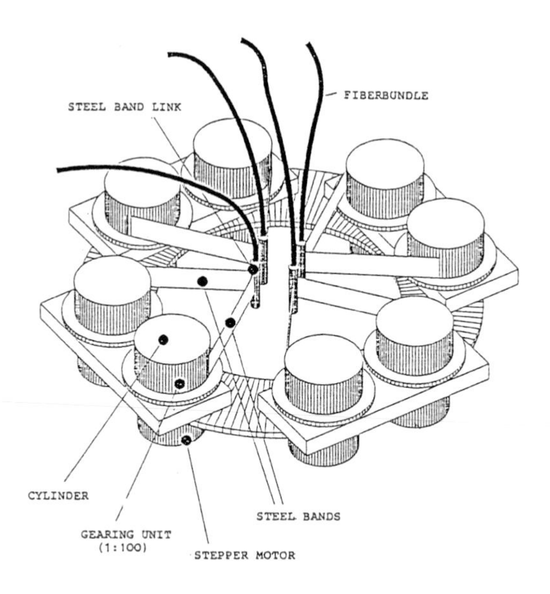

Figure 2: Fiber positioning unit.

Each fiber bundle is attached to two steel bands which are winded onto two steel drums (see Fig. 2). By rotating the drums via stepper motors, the length of the steel bands can be changed and thereby the positon of the fiber bundle in the focal plane. Test measurements have proved, that an positional accuracy of 10 μm can be achieved.

2.2 Spectrograph Unit and Detector

In order to get a uniform spectral resolution throughout the whole optical wavelenght range a double beam prism spectrograph is used. The light from the four fiber bundles is split into two optical wavelength paths by use of a dichroitic beamsplitter. The two wavelength paths cover the spectral regions of 3400-5000 Å and 5000-9000 Å . Each path is dispersed seperatly by use of a prism, yielding a spectral resolution of about 50 (see Fig. 3). The FWHM of a monochromatic image covers about 7 pixels on the detector surface. The overall transmission of the spectrograph is about 65 %.

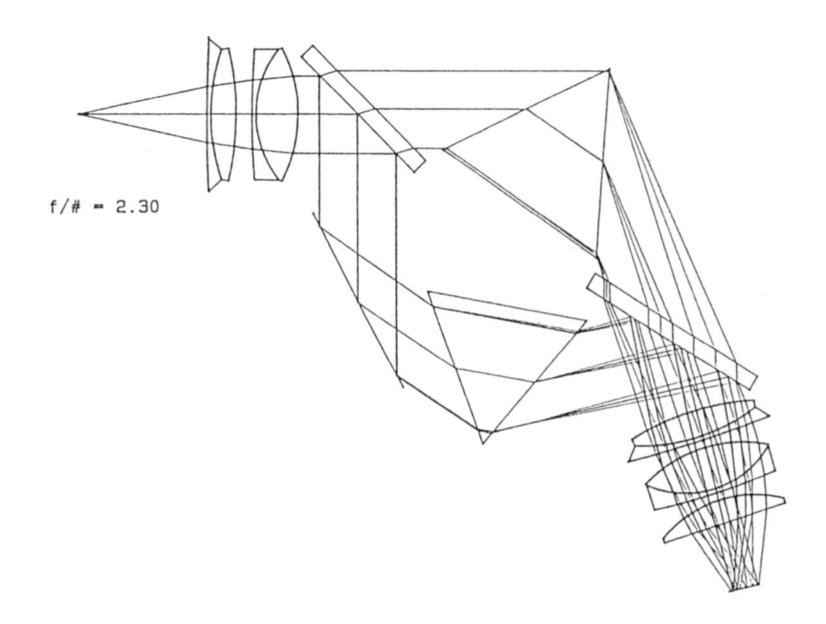

Figure 3: Double beam prism spectrograph.

The four spectra are recorded by a two dimensional photon counting resistive anode detector (MEPSICRON, Firmani et al. 1983). It has a sensitive area of 512 × 512 pixels and is equipped with a multialkali photocathode with a typical quantum efficiency fo 15-20% at 5000 Å. Five cascaded microchannel plates behind the photocathode accelarate the electrons to a resistive anode inducing a distinct charge distribution for each registered photon. The position of this distribution is determined by a positon computing system and send via an optical fiber transmitter to the host computer system. The detector is thermoelectrically cooled to $-25°$ C. In comparison to a CCD this new detector has the following advantages:

- no readout noise
- dark current less than 50cts/s or 1 count/pixel every 4 hours
- no overflow due to cosmic events
- time resolution of up to 1 ms
- high frequency flat field variations less than 1%

However care must be taken not to exceed the maximum count rate which is limited to 1×10^5 cts/s on the whole detector surface.

3. Measurement Process and Data Reduction

3.1 Online Reduction

For each object a blue and a red spectrum is generated which extends over about 10 pixel rows on the detector surface. In order to reduce the data stream, only the exposed rows of the detector are summed up during an online reduction process. However, due to imaging errors the spectral resolution may vary between different pixel rows. This variation is corrected by means of software.

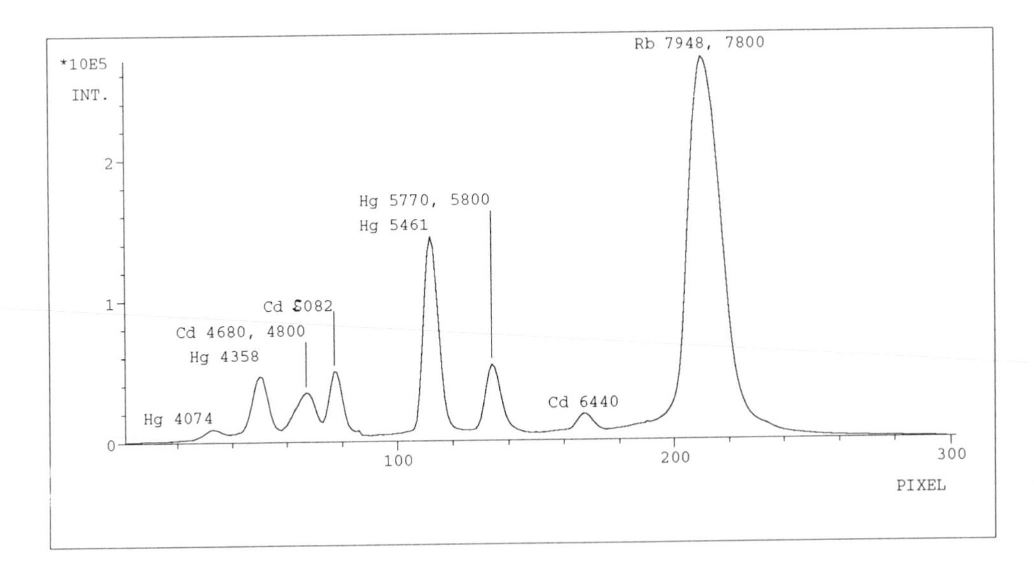

Figure 4: Hg-Cd-Rb calibration spectrum.

From a calibration spectrum of Hg-Cd-Rb lamps (see Fig. 4) the differential shift in wavelengths for each row and column on the detector is computed and stored in a transformation table. During the measurement process, this information is used to convert the individual spectrum to identical spectral resolution and dispersion throughout the 4 different input channels. This online reduction process diminishes the amount of data to be stored by about a factor of 200 and allows to achieve a maximum time resolution of about 10 ms.

3.2 Data Reduction

As a result of the online reduction of the measurement data, four one dimensional spectra covering 300 pixels are obtained.

From identification of single lines in the Hg-Cd-Rb calibration spectrum a transformation between pixel and wavelength can be determined and the pixel scale can be converted to a lambda scale. Figure 4 shows the results of the registration of a Hg-Cd-Rb calibration lamp.

As a monochromatic image covers 7 pixels on the detector surface, the data will be compressed by means of software to filtercurves with a FWHM of 7 pixels. The shift between succesive filtercurves is set up in order to obtain about 50 overlapping filtercurves which fulfil the sampling theorem (Shannon, 1949). The data will then be reduced to extraterrestic values by the standard second order Stroemgren-King Theorie as described by Young (1988). The overlapping filtercurves will guarantee a proper description of the second order moments required by the theory and will allow to reduce extinction effects properly. Furthermore the sampling theorem will also allow to transform the measured data with high accuracy to any broadband photometric system, which does not exceed the spectral resolution of the measured data.

This research program is partly supported by the Deutsche Forschungsgemeinschaft (grant BA 867/2-2).

References:

Firmani, C., Gutierrez, L., Ruiz, E., Bisiacchi, G.F., Salas, L., Paresce, F., 1983, in: *SPIE Instrumentation in Astronomy V*, 445, p. 192.

Mantel, K.H., Barwig, H., Kiesewetter, S., 1988, in: *Proc. of New Directions in Spectrophotometry*, Philip,A.G., Adelman,S.J., Hayes,D.S., (eds.), L. Davis Press.

Shannon, C.E., 1949: Proc. Inst. Radio Eng., 37, p. 10.

Young, A.T., 1974, in: *Methods of Experimental Physics*, Carleton,N., (ed.), Academic Press New York, Vol. 12, part A, p. 123.

Young, A.T., 1988, in: *Proc. of the second Workshop on Improvements to Photometry*, eds. W.J. Borucki, NASA Conference Pub. 10015, p. 215.

Discussion

R.M. Redfern: *I presume that the maximum count rate of 2.10 ^{5} c/s applies to a completely uniform illumination. Is there a problem with local deadtime?*

Mantel: The maximum count rate per pixel is about 3.10^4 c/s which is only a factor of ten below the integral maximum count rate. Therefore there is no problem with local deadtime.

J. Baruch: *Have you thought about the problem of modal drift in the optical fibres? Unless steps are taken to ensure no movement of the fibres, and no temperature changes, modal drift can produce a few percent change in the apparent light transmission for very small movement and a temperature change of 5°C. Movement of fibres in the focal plane will change the transmission and relative transmission of the bundles.*

Mantel: During observations, the fibres are fixed, so there will be no change in transmission due to bending variations of the fibres. Besides that, test measurements have shown that *f* numbers between *f*/30 and *f*/2.7 are smeared out by fibre degradation effects yielding variations of the light cone at the spectrograph entrance between *f*/4.4 and *f*/2.5. Therefore the spectrograph has been set up for an *f* number of *f*/2.3. So, even if temperature variations would change the light cone slightly, I do not expect any problems.

A. T. Young: *I was worried by your remark about adjusting the entrance diaphragm to match the seeing. In the Proceedings of the (first) Workshop on Improvements to Photometry, Charles Ken Knight had an interesting paper on the wings of the star image, which contains a wavelength-dependent and large fraction of the stellar energy. Changing the diaphragm changes your photometric system.*

Mantel: I'm aware of the problems that might be caused by the large wings of the stellar image. Nevertheless it is important to adapt the size of the entrance diaphragm to the focal-scale of the telescope and to the seeing condition in order to optimize the signal to noise ratio.

E. O'Mongain: *At UCD we were responsible for the Hipparchos photometric calibration plan development, but now are involved in a similar transition in Remote Sensing from discrete band radiometry to low resolution spectral radiometry. Rather than reducing your data to discrete bands, have you considered using singular value decomposition methods to determine star spectral type and brightness directly? In this case extinction effects will have their own spectral characteristics and can be discriminated against.*

Mantel: We are interested in accurate broad band photometry, therefore the data will be binned to widely overlapping filter curves. But depending on the scientific problem, it is also possible to use the whole spectrum, e.g. for determination of the spectral types.

J. Tinbergen: *You seem to depend for passband wavelength stability on the stability of the spectrograph and detector. Do you monitor the wavelength stability, and if so, how?*

K.H. Mantel: Prior to and after each observing run we do extensive calibration measurements with HgCdRb-line lamps. From these measurements the transformation between pixel and wavelengths, as well as the wavelength stability, can be determined.

A Four Star Photometer

J.C. Valtier, J.M. Le Contel, P. Antonelli, P. Michel, J.P. Sareyan

O.C.A., Departement Fresnel, U.R.A. 1361, B.P. 229, 06304 Nice cedex 4, France

Abstract

A new photometer is presently being developed at the O.C.A. Observatory. It consists of four arms and a CCD camera situated in the focal plane of the telescope. Each arm can move in both directions and support a diaphragm and a liquid optic guide that directs the light to a photomultiplier. The simultaneous acquisition of the four signals enables to obtain magnitude differences between the objects in real time. A typical use of this photometer is to observe at the same time one or two variables, comparison stars and the sky background.

Introduction

Differential photometry is usually performed on bright stars using a single telescope which is pointed sequentially at the variable, the sky and at one or two comparison stars. This operating mode is time-consuming and assumes that the sky does not vary during the movement of the telescope : that point is one of the major sources of errors in precise differential photometry (there are a lot of other ones !). A first solution is to observe in good photometric sites where atmosphere is very stable. In Europe such sites are few, although many telescopes are operating in different countries.

So, how could we improve the efficiency of the photometers in order to reach the millimagnitude level of precision in poor sites?

It is evident that, for every one, if we can measure simultaneously the program star, two comparison stars and the sky background, we should obtain data from which the atmospheric fluctuations could be easily removed. Different solutions have been proposed to solve this problem at least partially : use of two telescopes to point simultaneously at two different objects (Querci et al, 1990), two channel photometer (Piccioni et al 1979), three or four channel photometer (Barwig et al, 1987, Mantel et al, this conference).

As part of a project to install a photometric automatic telescope at the south pole, we have been studying a four channel comparative photometer in collaboration with N. Walker. The South Pole project was stopped because of funding problems but we could develop and build the photometer. We present here a description of that prototype and the results of preliminary observational tests.

Description

We can divide the four star photometer into three distinct parts : a "focal plane" part mounted on the telescope, for collecting the light, a "driving" part, essentially electronics and software, these two parts are now complete, finally an "acquisition" part, with filters, cooling box, optical switching, etc, which is currently in design phase.

The basic principle of the instrument is the following : Four arms, driven by step by step motors move in the focal plane of the telescope. On each arm a liquid optic guide coupled with a diaphragm collects and transmits the light to four photomultipliers. In addition an intensified CCD camera on its own arm is used to recognize the observed field and if necessary for telescope guiding. The photometer is completely automatic and fully controlled by a micro-computer.

The observing schedule is the following :

The observer first chooses the field by entering the corresponding coordinates and the characteristics of the telescope in the computer. The computer then displays the field (SAO catalogue is included in the software) and allows the observer to identify stars to be observed and to choose the best mechanical positions of the arms and of the CCD camera in the focal plane. Then the telescope is pointed and the real field controlled by means of the CCD camera. The arms are sent to the positions of the stars. Automatic centering is then performed for each of them. After this, the data acquisition program is run and the different signals are registered and displayed in real time on the computer screen.

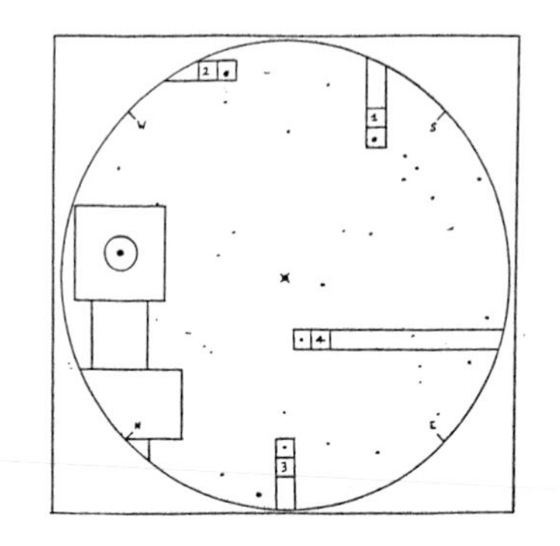

Figure 1 : Picture of the focal plane part of the photometer.

Technical Characteristics

The linear dimension of the field is 200 mm, the precision on each arm position is 10 microns, each arm can move in two directions in half of the field, the minimal distance between arms being 4 mm.

The diaphragms in front of the light guides have a 0.6mm diameter, the fluid optic guides are 2 meters long and 3 mm large (useful diameter of 2 mm) with a numerical aperture of 0.47. The CCD camera is a light-amplified one (figure 1).

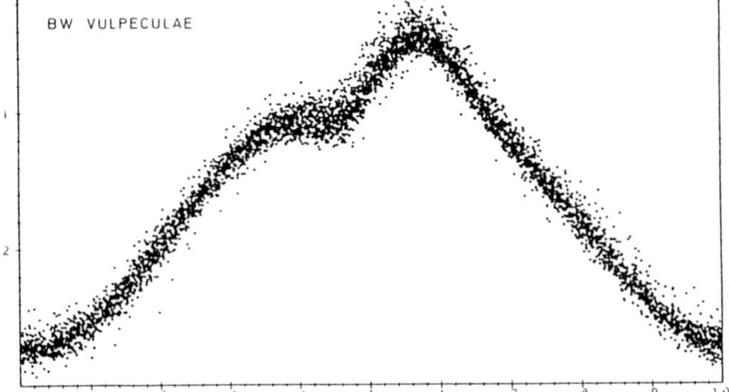

Figure 2 : Mean light curve of BW Vul obtained from international campaign.

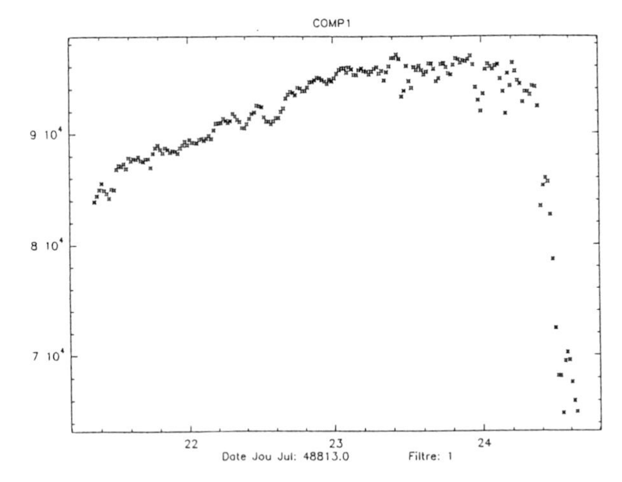

Figure 3 : Raw signal of HD 198820 on July 9.

Initial Tests

These were performed in July 92 at Nice observatory on an old 40cm astrograph with a two meter focal distance. That corresponds to a total field of 5° and 40' and diaphragm diameters of 1'. Four I.P.21 uncooled photomultipliers mounted in a single copper box were used with Johnson B filters.

The external temperature was $25°C$. Due to that high temperature and sea proximity, the sky was foggy and the absorption highly variable. Due to the light pollution of the city and moonlight during some nights, the sky background was almost as bright as the second comparison star (a 7th mag star).

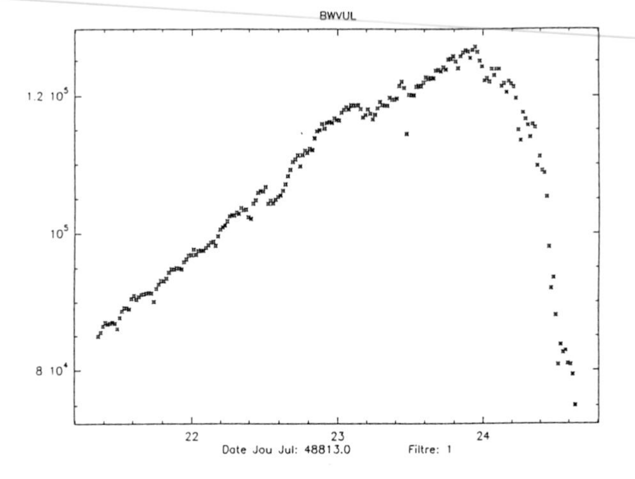

Figure 4 : Raw signal of BW Vul on July 9.

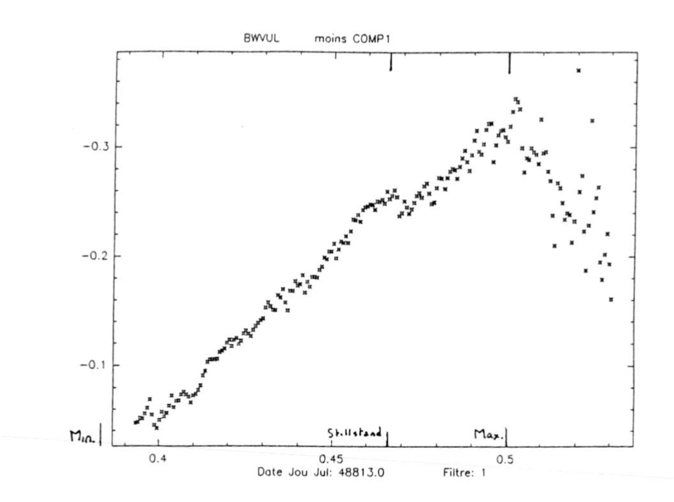

Figure 5 : Magnitude differences BW Vul - HD 198820 on July 9.

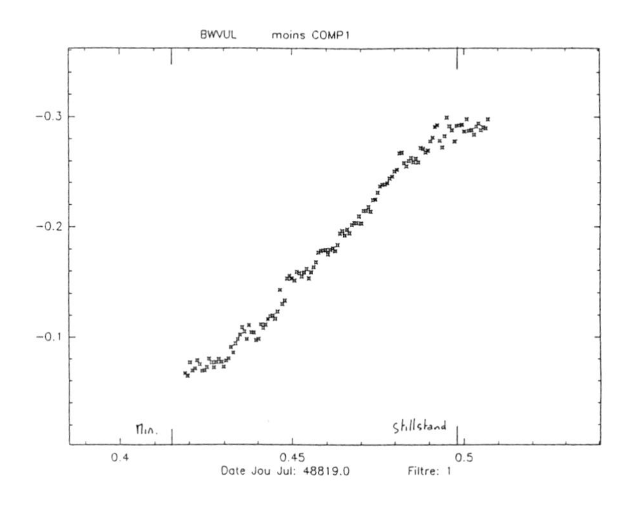

Figure 6 : Magnitude differences BW Vul - HD 198820 on July 15.

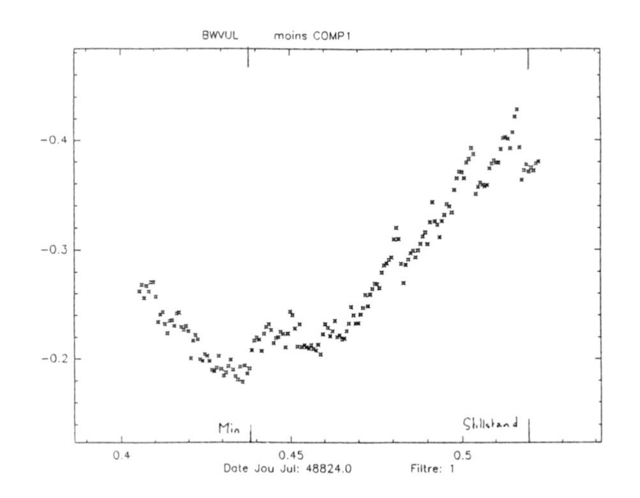

Figure 7 : Magnitude differences BW Vul - HD 198820 on July 20.

In brief, the observing conditions were worse than anything that an Astronomer could imagine in his worst nightmares.

We chose to observe the well-known Beta Cephei star BWVul (B2 III, V = 6.44), with the comparison stars C1 = HD 198820 (B3 III, V = 6.34) and C2 = HD 199102 (B9, V = 7). A very well defined mean light curve for this variable had been previously derived from an international campaign (Sterken et al, 1986) (figure 2).

It is important to note that the C1 star is situated 4° 20' south of the variable. The chosen integration time was 10 sec.

Figures 3, 4 and 5 show respectively the raw signals of C1 and BWVul and the final magnitude differences with the theoretical corresponding ephemeris (Chapellier and Garrido, 1990) obtained on July 9. Figures 6 and 7 show the same as figure 5 for the 15th and 20th of July. One can note that on night 9 the bump (so called "stillstand") in the lightcurve is clearly seen.

Discussion and Conclusion

These figures clearly show that the four star photometer enables us to obtain interesting results on variable objects even in extremely poor conditions. In addition, the possibility of moving arms independently enables us to use this instrument to carry out photometry on extended or moving objects (asteroids for instance).

We now need to perform complete tests on this prototype, particularly on fiber stability and response with defocusing, changing temperature, etc.

We also plan to finish the acquisition system quickly that should have eight cooled photomultipliers, at least two filters and an optic switching system in order to have a self-calibration of the instrument. We also plan to check new receptors, such as AP diodes for instance, that could be mounted directly on the arms in the focal plane.

References:

Barwig, H., Schoembs, R. 1987, The Messenger **48**, 29.

Chapellier, E., Garrido., R. 1990, A and A, **230**, 304.

Piccioni, A., Bartolini, C., Guarnieri, A., Giovannelli, F. 1979, Acta Astron., **29**, 463.

Querci, F.R., Querci, M., Fontaine, B., Klotz, A. 1990, Variable Star Research: An international perspective, edit. J.R. Percy, J.A. Mattei and C. Sterken, Cambridge University Press, 221.

Sterken, C. et al. 1986, A and A Supp. **66**, 11.

Discussion

C. Sterken: *Tinbergen, this morning, said that in the past all money was put into the telescope, and none into the photometer to be attached to it. You seem to do it the other way round. You put a lot of effort into the design and construction of the photometer, but where is the telescope on which the photometer will be mounted. It also seems to me that the photometer is of a complex and fragile design, and that it most likely will need to be mounted permanently on a dedicated instrument.*

J. C. Valtier: The photometer now is not so fragile and can be changed from one telescope to another without great difficulty. It should be installed on a one meter telescope in the south of Spain.

G. Szecsenyi-Nagy: *As you told us the four star photometer just has been tested on a 40 cm refractor. Was it an astrograph of the type made by Carl Zeiss Jena? If you definitely intend to use the device to measure stars with angular distances of 4-5 degrees you may* *evaluate the possible use of Schmidt type cameras. Do the weight and dimensions of the photometer allow it to be mounted at the prime focus of telescopes or Schmidt cameras?*

J. C. Valtier: No, the photometer is to be installed at the cassegrain focus of a telescope.

K.H. Mantel: *What is the transmission of your liquid optical fibre?*

J. C. Valtier: The transmission given by the maker between 3500Å and 6000Å varies from 60% to 80%.

How do you make sure you are observing in the same passband with each of the four channels?

J. C. Valtier: It is just the problem of choosing the same filters - but anyway it is a problem. It is the reason planning to switch between the P.M.

W. Tobin: *Why are you using eight PMTs for four channels?*

J. C. Valtier: Because in the 'switching' configuration we need one P.M. per filter and we plan to observe with two filters.

Gordon Herries Davies and Patrick Wayman at the Banquet

Session 4

Automatic Photoelectric Telescopes and Extinction

Robotic Observatories: Past, Present and Future

Russell M. Genet and David R. Genet

AutoScope Corporation, PO Box 2560, Mesa, Arizona 85214, USA

Abstract

We briefly describe the history of robotic observatories, give details on an example of the current state-of-the-art in robotic observatories, and suggest several key areas for future development.

1. Past

The first fully automatic electrical measurements of starlight were made at the University of Wisconsin (McNall, et al, 1968) and at Kitt Peak National Observatory (Meinel and Meinel 1990, and Maran 1969) about 25 years ago. These pioneering, fully automatic photoelectric telescopes (APTs) were mainly technological demonstrations, and were little used for scientific research. The Wisconsin group went on to develop automated space telescopes, which were very useful scientifically although outside the scope of this Earth-bound review. KPNO quickly reverted to manual operation, however, never to return to automatic operation. As Aden B. Meinel, KPNO's innovative first Director, mentioned to me, "Astronomers just weren't ready for automation yet." It also seems likely that robotic telescopes were not really ready for astronomers yet either. The Wisconsin system only had 4K of RAM, punched paper tape as an input, and a printer as an output, while the computer for the KPNO system was not very reliable.

About 15 years were to transpire before automatic photoelectric measurements were to be made on a regular basis. The next fully automatic system was able to take advantage of the much more capable (and less expensive) microcomputers that had been developed in the interim. This first really useful APT was developed by Louis J. Boyd, with some assistance from one of us (RMG). This system, the "Phoenix-10", saw first operation in October of 1983, and has been producing useful science ever since. One of us (RMG) developed a somewhat similar system, the "Fairborn-10" APT, which began operation in September of 1984, eight months after the Phoenix-10 APT began its operation. The Fairborn-10 APT was transferred from Fairborn, Ohio, to Mt. Hopkins in southern Arizona in 1985, and has been operating there ever since. The Phoenix-

10 and Fairborn-10 APTs have been described by Boyd and Genet (1984), and in more detail by Trueblood and Genet (1985). The history of all four pioneering systems mentioned above, as well as that of a semiautomatic system developed by David Skillman, have been discussed in detail by Genet (1986) and by Genet and Hayes (1989). Genet and Hayes (1989) also discuss the development of the Automatic Photoelectric Telescope Service on Mt. Hopkins, jointly run by the Fairborn Observatory and the Smithsonian Institution. With some five APTs in operation (and two more about to begin operation), the APT Service now serves some four dozen institutions world wide, and is a primary source of photoelectric measurements of variable stars.

There have been a number of mainly independent developments of APTs elsewhere. An outstanding example is the 0.5-meter "Danish APT" at the European Southern Observatory, described elsewhere in these proceedings by its developer, Ralph Florintien-Neilson. Russell Robb developed a nearly automatic system that has been used to observe faint X-ray binaries at the University of Victoria. The Bulgarians recently brought a system on line that is being used to observe flare stars. R. Kent Honeycutt, and his associates at Indiana University, developed a fully automated APT that utilizes a CCD camera as the detector and is able to make automatic photometric observations of very faint stars. Edwin Budding and his colleagues in New Zealand are using a modified C-14 to make automated observations. There are also many other APTs under development but not yet operational, such as those by James O'Mara in Australia, David Killkenny in South Africa, by an Irish amateur astronomer, and by a group in India, etc.

Five years ago, one of us (DRG) founded AutoScope (a contraction of *auto*matic tele*scope*) to commercially produce robotic observatories. Both of us thought that the astronomical community should now be ready for robotic telescopes and, more importantly, thanks to technological advances across a wide front, that robotic telescopes should, at last, be ready for astronomers. A number of complete observatories or telescopes have now been purchased from AutoScope by various institutions. The Jet Propulsion Laboratory purchased three complete observatories that are being used in a network in the southwestern US. The Lawrence Berkeley Laboratory purchased a 30-inch aperture telescope for use in their supernova search. Tennessee State University and the Smithsonian Institution have teamed up to place a 32-inch system on Mt. Wilson in southern California. The University of California at Berkeley is placing a 32-inch aperture system at Lick Observatory for faint object CCD photometry. Catania Observatory placed a 32-inch system on Mt. Etna. Its primary use is the observation of flare stars. The Fairborn Observatory operates a system in Mesa, Arizona. Buhl Science Center, in Pittsburgh, operates a system with a video wall display in their public science center of the live output from a CCD camera. A 40-inch system is nearing completion for the Korean Astronomical Observatory, and is described in some detail below. Systems are also being built for the University of California at Irvine, NASA Ames Research Center, and New Mexico State University. The latter is a 40-inch alt-az system with an autoguider.

2. Present

We thought that the best way to provide the reader with a feel for the current state-of-the-art in robotic observatories was to describe a recently-produced system in some detail. We have chosen, as the example state-of-the-art system, the 1-meter (40-inch) aperture telescope, instruments, and observatory manufactured for the Korean Astronomical Observatory (KAO) by AutoScope. Unlike some earlier AutoScope systems, whose intended use was strictly just for aperture photometry, the "KAO-40" was intended to also be used to obtain long-exposure CCD images and, perhaps later on, for fully automated spectroscopy. The KAO-40, as well as all currently built AutoScope systems, are now general-purpose automatic systems with highly precise optics and very accurate pointing and tracking, and are *not* just automatic photoelectric telescopes (APTs). As these systems usually include, besides an automatic telescope, automatic instruments and an automatic enclosure (with weather station, etc.), we often refer the them as *robotic observatories*.

AutoScope systems have made a sharp break with past tradition in terms of their size and especially their weight. Research telescopes have, in the past, been rather gargantuan in proportions, and have been generously fabricated from steel with weights measured in the tens or hundreds of tons. Their design philosophy was somewhat akin to that of battleships and bridges. Our telescopes, however, are unusually compact, and they primarily utilize aluminum alloy in their construction. They are philosophically similar to aircraft or spacecraft in conception (we are both airplane pilots, so such thinking comes naturally), and are very lightweight. Lightweight telescopes, other things being equal, have higher natural resonant frequencies and are thus less effected by wind buffeting than heavier telescopes. Furthermore, their moments of inertia are an order of magnitude less than conventional telescopes, allowing very rapid acceleration, deceleration, and settling. Such "crispness" is well suited to the inhumanly rapid and precise robotic nature of these systems.

Lightweight automatic telescopes demand lightweight mirrors, as well as lightweight and compact (or off-telescope) automatic, instruments. The KAO-40 and all other AutoScope systems now use primary mirror blanks manufactured for us by HexTek. For apertures up to 1 meter, these HexTek blanks use a gas fusion process that sandwiches vertically stacked Pyrex tubes between fairly thin horizontal Pyrex front and back plates. The entire assembly is then brought to fusion temperature, slumped over a mold and, at the right moment, gas is forced, under pressure, through holes in the bottom plate into each of the Pyrex tubes. This causes all the tubes to simultaneously expand until they are in contact along their entire surfaces, forming hexagonal patterns (hence the company's name, HexTek).

Not only are these blanks light in weight, but they are very stiff. Also, with only 20% of the mass of solid mirrors of the same thickness, HexTek blanks have a much lower thermal mass, allowing them to more closely track changes in the temperature of the night air, thus helping to preserve the excellent seeing available at many mountaintop sites. Secondary blanks are also light in weight, and are typically made from a low-expansion material. The secondary for the KAO-40, for instance, was made from Cervit.

As mentioned above, the KAO-40 (and other AutoScope) mounts are made of aluminum alloy to increase their resonant frequencies and greatly reduce their moments of inertia. Aluminum is also highly corrosion resistant, especially when protected with an electrostatically attracted, baked-on, powder-coat enamel. Light weight also eases handling during assembly, disassembly, and shipping. Two persons can, for instance, easily handle the major subassemblies of the KAO-40 telescope without any assistance from cranes, etc.

The KAO-40 mount is shown in Fig. 1. As can be seen, it is an equatorial horseshoe mount somewhat similar in layout to the 4-meter telescopes at KPNO and CTIO and a number of other large telescopes. The secondary mirror is mounted on a swivel that is supported by a precision shaft passing through two linear bearings, thus totally eliminating all side play. Mirror movement, either focus or tilt, is provided by three fine-pitch linear stepper motors. The square top of the open-frame optical support structure, which is more efficient than the more conventional round top (which tends to "bow" out when compressed by the spiders), was modeled after the 3.5-meter telescope at Apache Point Observatory. We are grateful to the designers of the 3.5-meter telescope, Walter Sigmond, Charles Hall, and Edward Mannery, for this helpful suggestion.

Figure 1 The KAO-40 mount uses aluminum alloy in its construction and a lightweight HexTek primary mirror.

The primary mirror cell, which is constructed of aluminum alloy, as is all of the telescope mount, provides support for the primary mirror via a central radial support and a rear 27-point flotation system. Two overlapping dust cover sets on the front of the mirror cell provide protection when the system is not observing. The dust covers are under automatic computer control. A fan (also under computer control) is used to draw filtered air through the rear of the fully enclosed mirror cell.

Very rapid response, very high torque, DC servo systems (Dynaserv DC servos made by Compumotor) are used to drive the telescope in both RA and Dec. These servo motors, mainly used by others for strong, high-speed industrial robots, are of such high precision, resolution, and torque, that only a single stage of friction drive reduction is needed. A small roller on each motor shaft directly drives, respectively, the large RA and Dec disks shown in Fig. 1. The acceleration and top speeds of the telescope have to be limited by software for safety's sake (and an audio chime sounds to warn humans during movement). There is nothing sedate about the movement of this telescope, it is extremely rapid.

The telescope's initial position is sensed by Sony "Magnaswitches" which provide about 1 micron of precision or, in an angular sense, a couple of arc seconds. Limit switches prevent accidental motion outside of the allowable observing window which is preset in the software.

All of the computers and electronic control equipment for the telescope, its instruments, and the observatory itself are contained within the single cabinet shown in Fig. 2. Rather than trying to run the telescope, all of its instruments, and the observatory itself with a single, expensive, multi-tasking computer, a hierarchial network of lower-cost computers is used to improve modularity and to ease the later addition of new automatic instruments. At the highest level is a single *Master Computer* (an industrial rack-mounted -386 PC). This Master Computer communicates to the outside world via modem or computer network, and to second-tier computers (also industrial, rack-mounted PCs) via Ethernet and well-defined ASCII messages to RAM disks. Thus a new instrument can be added at any time via a simple Ethernet connection. The computer used to control this new instrument can be of essentially any type (and can use any operating system and programing language). All that is required is the Ethernet connection. The third level of computers in the hierarchy consists of specialized control computers on cards within the PCs. These are used, for instance, to execute telescope motion commands, change filters, etc.

The KAO-40 uses five custom AutoScope controllers, each of which is described briefly below. The controllers are mounted in the left side of the equipment rack shown in Fig 2.

The Telescope Controller controls 2 or 3 axes telescopes (either equatorial or alt-az), with the third axes being an instrument rotator. Optical encoders on the main telescope axes provide positional feedback information. Besides control of telescope motion, sensing limits, sounding alarms, etc., the Telescope Controller positions the secondary mirror and also controls a four-port instrument selector that switches, under automatic control, the optical beam to any one of four permanently mounted instruments. A WWV or GPS clock provides precise time.

The Observatory Controller uses weather and environmental sensors to determine if it is safe to open the observatory, when it needs to be closed, when unsafe conditions are encountered, etc. This information is used by the Observatory Controller to open or close dome shutters or roll-off roofs. A "watchdog" timer is frequently reset by the computer, and if it fails to do this the timer will time out, control will be taken away from the computer, and the dome shutter or roll-off roof will be forced closed by the Roof Controller (which contains its own battery power). A Power and Environmental Controller manages the power and temperature environments for the equipment, executes emergency stop commands, etc.

Figure 2 Computers and Controllers for the KAO-40 automatic telescope, instruments, and observatory.

The Photometer Controller manages the ultra-precision photometer, including very tight control of filter and PMT temperatures, humidity, the positions of two filter wheels, a diaphragm wheel, and a photometer / CCD camera flip mirror, etc.

The KAO-40 and other AutoScope systems can be controlled in real time, either locally or remotely, or can be run automatically. Automatic operation adheres to the Automatic Telescope Instruction Set (ATIS) standard that specifies the format and content of observational requests and results.

3. Future

We feel robotic observatories have a bright future, perhaps eventually dominating research astronomy. Robotic observatories nicely solve two old problems: (1) the usually considerable physical distance between astronomers (and students) and the best observatory locations; and (2) the difficulty in staying up at nights to observe and teaching or attending classes during the day. Furthermore, the smart, automatic scheduling used by robotic observatories is much more efficient than the "block time" scheduling used by almost all non-robotic observatories. Finally, robotic observatories are very cost-effective. Large amounts of very high quality data are obtained at low cost. Robotic observatories are mainly "run" by their remote users, with only occasional on-site technical support being needed. Observers and large staffs are expensive to transport to and support on remote mountain tops, and robotic observatories do away with all of this. We briefly mention, below, some of the areas we are currently working on and hope to bring to fruition in the near future.

We plan to extend what full automation is currently doing for photometry (and imaging) to spectroscopy. We were very pleased to receive a Small Business Innovation Research (SBIR) award from the National Science Foundation for the development of a fully automated spectrograph. With R. Kent Honeycutt (Indiana University) and with the kind advice of many leading spectroscopists, we have completed Phase I of the SBIR grant -- the design of the spectrograph. During Phase II we plan to build and operate the prototype spectrograph. The spectrograph is a fiber-fed Echelle that covers 3900 - 9000 Angstroms with a single, fixed grating setting and output format. The spectrograph is housed in an off-telescope enclosure whose environment is rigidly controlled.

We also plan to extend the capabilities of our current automatic scheduling system, and to implement automated diagnostics. We were very pleased to receive a Small Business Innovation Research (SBIR) award from the National Aeronautics and Space Administration for the development of these advanced capabilities which should not only greatly benefit the operation of robotic telescopes here on Earth, but could lead, in a natural evolutionary process, to robotic telescopes at the Lunar Outpost operated in a similar fashion (i.e. operated directly by the users themselves without any significant permanent staff).

Working with Mark Drummond and his associates at NASA Ames, and with current users of robotic observatories such as Gregory Henry, we are extending the capabilities of current automatic scheduling system to include capabilities for automatically filling in phase (light or radial velocity) curves, scheduling observations during eclipses, switching between observational instruments (such as a photometer and spectrograph) depending on photometric and seeing conditions, and many other advanced features. We are also working on international standards for networking robotic observatories together. The advantages of such networking are discussed elsewhere in this volume by David L. Crawford.

Working with Ann Patterson-Hine and her associates, also at NASA Ames, we are developing extensive capabilities for automated equipment monitoring and diagnostics. Each of our current controllers has all of its key monitoring points wired to a "test jack", and these test jacks are now being connected to a switching high-speed analog-to-digital converter that allows the monitoring and diagnostics computer direct access to over 100 test points in the controllers. Besides allowing self checks at start up (or whenever desired), keeping track of possible adverse trends (health monitoring), and allowing remote human-assisted or on-site automatic diagnostics, we are planning on having a capability for automatically switching to on-line "spares". By having essentially two of everything, any failure would only cause the system to be down for an instant while the fault was automatically diagnosed and the backup unit automatically switched in. The system would then keep right on operating, and humans would be notified to replace the unit at their convenience. We feel that such an approach would be useful for systems at very remote locations, such as the South Pole or Lunar Outpost, or for systems which were critical, large, or expensive, where essentially no system downtime could be tolerated.

With the assistance of Butler Hine (NASA Ames) and Jack Burns (New Mexico State University), we are giving serious consideration to the problems of operating robotic observatories at such very remote sites as the South Pole and the Lunar Outpost. We plan to build a system with all of the advanced features mentioned above and eventually place it at the South Pole.

Finally, we are working on the design of an "advanced technology" 2.5-meter production robotic observatory. This would be the largest aperture telescope, instruments, and enclosure every produced commercially in quantity -- robotic or otherwise. We expect that it will be unusually compact and light weight -- even more so than our already lightweight systems -- perhaps employing carbon composite fibers in a few critical areas. We expect that the system will incorporate adaptive optics and, on nights of excellent seeing, will provide images of fractional arc second quality.

References:

Boyd, L. J., and Genet, R. M. 1984, in *Microcomputers In Astronomy II*, Ed. by K. A. and R. A. Genet, Fairborn Press, Fairborn.

Genet, R. M. 1986, in The Study of Variable Stars Using Small Telescopes, Ed. J. R. Percy, Cambridge University Press, Cambridge.

Genet, R. M. and Hayes, D. S. 1989, *Robotic Observatories*, AutoScope, Mesa.

Maran, 1969 in *Stellar Astronomy*, Ed. H. Y. Chiu, J. Remo, and R. Warasila, Gorden and Breach, New York.

McNall, J. F., Miedaner, T. L., and Code, A. D. 1968, *Astronomical Journal*, **73**, 756.

Meinel, A. and Meinel, M 1990, in *Robotic Observatories: Present and Future*, Eds. S. L. Baliunas and J. S. Richard, Fairborn Press, Mesa.

Trueblood, M, and Genet, R. M. 1985, *Microcomputers In Astronomy*, Willmann-Bell, Richmond.

Discussion

W. Tobin: *What will a 2.5-metre APT cost?*

Genet: We expect that the cost for a complete telescope, control system, observatory, and instruments would be about US$2.5m to US$3.5m .

I. S. Glass: *What happened to the amateur space telescope mooted by Rochester?*

Genet: I wrote to them about their project, but never received a reply, so I do not know the status of their system. I have never met any of them at meetings.

E. F. Milone: *Do you envisage the large Autoscope telescopes to be single-instrument telescopes? There may be advantages to single instrument telescopes; maybe relialibity and precision can be better maintained. If so, it makes sense for a consortium of like-minded users rather than institute/university consortia.*

Genet: I expect that typically there will be four instruments permanently installed, such as a high-resolution Echelle spectrograph for bright objects, a low-resolution spectrograph for faint objects, a CCD camera with filters for imaging and red photometry, and perhaps an aperture photometer. It may be necessary that IR telescopes be different from optical telescopes.

J. B. Hearnshaw: *For your proposed automated spectroscopic telescope, there are many different spectrograph parameters one could select (such as resolution, wavelength region etc.) Will you select a fixed set of parameters or have the ability to change these from observation to observation? Ideally the science you want to accomplish should dictate the instrumental design of any spectrograph. What scientific program do you envisage for an AST?*

Genet: To begin with we plan to have only a fixed configuration, without changes except minor adjustments. Spectral coverage will be from 3900Å to 9000Å or so, with 1 pixel covering 0.5Å or slightly less. We are aiming for a fairly simple but fully automatic spectrograph that would appeal to many, but not all, users.

M.S. Bessell: *To answer John Hearnshaw's question: with a large format 2048^2 CCD, a 31.6 g mm^{-1} echelle and prism cross dispersion one can get full wavelength coverage without any wavelength adjustment at a resolution suitable for almost all stellar spectroscopic work using a camera with a focal length of about 80 cm.*

Genet: It is correct that a fibre-fed Echelle spectrograph allows the entire spectrum to be displayed simultaneously. This helps to minimize adjustments. Our goal is to have no user adjustments that must be regularly made.

J. Tinbergen: *Whatever type of millimagnitude photometry your users are going to do, with a Nasmyth arrangement you will have to do something about polarization-induced photometric errors. Depolarizers unfortunately are small, so an intermediate focus is needed before the Nasmyth flat. Do look at the LEST designs for alternatives.*

Genet: This is an important consideration. We have not fully dealt with it yet and will be looking into it after the meeting. You are certainly correct that it must be considered!

J. Tinbergen: *There is a world of difference in observing five extinction stars per night or doing so every half hour. In the interests of disentangling extinction gradients from local or global time variations, could you tell us what your slewing speed is and how many seconds it takes to come to a dead stop, on the star, from full slewing? (take a 1-metre telescope, for example).*

Genet: For safety reasons we limit slew speeds to 1.0 degrees/second, although faster speeds would be possible. The system which weighs only a tenth as much as conventional telescopes, (a hundredth of the moment of inertia) can stop in one second or less from full speed. If centering on the star is required, this may take a few seconds more.

J. Tinbergen: *That's fantastic news for extinction fighting! You mention having an offset guider. Have you considered using those CCD frames for monitoring extinction variations during a long photometric exposure?*

Genet: This is a good idea. There is no reason why it could not be done as the data is there in the computer.

Determination of Data Quality and Results from two Mount Hopkins Robotic Telescopes

Diane M. Pyper[1], S.J. Adelman[2], R.J. Dukes, Jr.[3], G.P. McCook[4], M.A. Seeds[5]

[1]*University of Nevada-Las Vegas, USA,* [2]*The Citadel, USA,* [3]*College of Charlestown, USA,* [4]*Villanova University, USA,* [5]*Franklin and Marshall College, USA*

Abstract

We present various methods for determining night quality from observations obtained with totally unattended automatic photoelectric telescopes at the Fairborn Observatory site on Mt. Hopkins, AZ. Telescopes are the Phoenix 10-inch (P10) and the Four College 75-cm (CAPT). Filter systems used are Johnson UBV, Kron-Cousins RI and Strömgren uvby. As a preliminary data filter for the P10 "rent-a-star" data, all observations with standard errors of the mean (SEM) above 20 mmag are discarded; a summary of nightly SEM's, number of aborts and total observing time is sent to each user. The CAPT data are evaluated by standard deviation of magnitudes or counts for photometric groups and for an entire night. The Geneva photometric statistics are computed for these latter data and have proved useful as a preliminary screen for bad data. Various criteria for good and bad groups and nights are discussed. Scientific results are presented for several magnetic CP stars that were observed using both telescopes. Differences in precision of the data and strategies for observing are discussed.

1. Introduction

Russ Genet has just given a very thorough description of the Mt. Hopkins robotic telescopes (this volume). I would like to describe the data taking process, evaluation and some results from two of these telescopes, the Phoenix 10-inch (P10) and the Four College 75-cm (CAPT).

The P10, for which Mike Seeds is the Principal Astronomer (PA), carries out a basic program of Johnson UBV differential photometry for a large number of users on a "rent-a-star" basis. Each group, consisting of variable, comparison and check star, is observed once per night; check stars are observed twice and comparison stars four times per group. Reductions are made using average extinction and transformation coefficients determined on several standard star nights per observing season.

The CAPT is able to carry out a much more versatile program, as it uses the Automatic Telescope Instruction Set (ATIS) (Genet and Hayes 1989), which provides many more options for observing. At present the Four College users are four of the authors; Pyper, Dukes (PA), Adelman and McCook; plus Ed Guinan at Villanova. We are observing a wide variety of variable stars using Johnson UBV, Kron-Cousins RI, Strömgren four-color, β, and Hα systems. A group can be repeated many times a night and all-sky observations of standards and program stars can also be made. For every differential photometry group, at least two check star and four comparison star measurements are made per group. At present, we also use average coefficients for reductions.

The basic problems in evaluating these data are that there is no on-site evaluation of atmospheric conditions and instrument performance and no real-time interaction while the observations are being carried out. Therefore, the quality of the data must be evaluated after the observations are completed.

2. Data Evaluation - P10

As is the case for many photometric observers, we find the scatter and standard error of the mean (SEM) of individual observations of comparison and check stars within a group to be a useful criterion for the evaluation of a night. For the P10, SEM's are calculated for all stars, variables, comparison and check stars and the data for any star in a given filter are discarded if the SEM 20 mmag. Each quarter, the user receives a data file containing his/her data for that quarter, and a night report file for the P10. The night report file contains information about the quality of individual nights, including the length of time the roof was open in hours, the total number observations deleted by the 0.02 mag. filter, total number of observations made during the night, and the average and median SEM for the night. There is also a mountain log file which contains information about roof closings (clouds, rain) and data for the P10 concerning the number of starts, stops and number of groups aborted. The Rent-A-Star philosophy holds that users must be given all potentially useful data. The user is the only person who can decide which data are most useful in a given case. In judging the quality of the night, the user examines the night report, looking at all the information provided. A good night will be many hours long; will include many observations with only a few rejected by the 20 mmag filter; and the average and median SEM will be low. A bad night will be short, perhaps only a few hours, and may include only a few observations that were not rejected; the SEMs would probably be quite high. Some nights may be many hours long, but will contain few observations, a signal that the telescope spent a large part of the night closed, probably because of clouds. A long night with many observations made but few accepted by the 20 mmag filter was probably a poor night with scattered haze or clouds; the SEMs are typically high. Most troublesome are long nights with many observations and SEMs slightly higher than normal, which are proabably clear nights which were not quite photometric. The mountain log files give further clues. If the roof closed due to rain or thick clouds, this is stated in the log but the number of stops and starts and group aborts for the P10 should also be examined. On a good night, the telescope will have only one start at the beginning of the night and one stop at the end. Multiple starts and stops signal a night broken by clouds not quite thick enough to cause the roof to close. A high number of aborts is also a warning that the night may have been poor. An abort commonly occurs when the telescope is unable to locate the next star in the sequence, and such aborts can again be caused by clouds.

Ultimately the evaluation of P10 data is left to the user. The 20 mmag filter is designed to reject only the most serious errors such as those caused by twilight. All other data are sent to the user. Users observing large amplitude stars may need little further filtering, but users watching low amplitude stars will need to examine their data more carefully. Some users reject any data from incomplete groups. Because U is so sensitive to sky conditions, it often has SEMs over 20 mmag and the U data is missing from a group. On less than photometric nights, groups may be missing data from other filters as well. By rejecting all such groups, the P10 users report precision of 5 mmags or better. While this certainly throws out some good data with the bad, it assures the highest quality. For example, the P10 data for CU Vir discussed below in Section 4 was pre-filtered to eliminate all groups with missing data. These data were further filtered to remove all observations with SEMs 10 mmag, leaving 626 observations in three observing seasons. This example further illustrates one of the big advantages of robotic telescopes; an abundance of data. With observing programs of variable stars on conventional telescopes, the amount of data is ususally sparse and takes many years to collect, thus the observer is tempted to retain any data that possibly may be useful and probably keeps much marginal data that should be discarded. Even though old habits are hard to break, it is a lot easier to discard questionable data when one knows there is a large body of better data available.

3. Data Evaluation - CAPT

With the CAPT data, ATIS permits further options for the evaluation of night quality. It is possible for each user to examine the SEM and residuals of each comparison and check star observed during a night (either as magnitudes or as a percentage of the mean photon count). McCook (1991) has written software that does this. An initial look at a plot of residuals vs time or SEMs vs time gives a good idea of what the night was like and examination of such plots can immediately eliminate terrible nights. ATIS also provides the user with detailed information concerning the reasons for group aborts, such as "star not found", "too near moon", "outside of observing window", etc. As with the P10, a check for nights with many "star not found" aborts, small numbers of groups observed, etc. can flag poor nights. A further screening process is made possible by the fact that each integration interval consists of a number of 100 ms sub-intervals in order to generate the Geneva Statistics (Bartholdi et al. 1984). The telescope output for each observation thus includes the total count, $N = N_i$, where N_i is the count in a given sub-interval, plus the Geneva statistics Q, R, and G. Within the total integration interval, Q is sensitive to scintillation, R to oscillation and drift and G to sharp spikes and drops. We have found (McCook 1992) that the R statistic correlates fairly well with sky conditions, but simply removing readings with bad statistics will not insure high quality data. However, we have found the Geneva statistics to be useful as the next step in the filtering process, which is to remove all data with $Q > 10$, $R > 1.5$ or < 0.5, and $G > 0.006$ or < -0.006 (G seems to be only sensitive to instrumental problems).

The SEMs are clearly the best indicator of sky and data quality and can be used with confidence to identify and eliminate questionable data from robotic telescopes. Each of the CAPT users has his/her own favorite method of deciding which data to finally keep.

One way of evaluating the data (used by DMP) is to calculate SEMs on the photon counts for each constant star and to express these as percentages of the mean counts. If any constant star in a group has an SEM > 2%, it is called a bad group. The number of bad groups in a night's observation is then counted and that number is expressed as a percentage of the total number of groups (=%BG) successfully observed that night (i.e., groups that have survived the previous filtering processes). Any night with %BG > 90% can be eliminated immediately; those with %BG between 70% and 90% are usually no good, but individual groups are examined anyway. On all the other nights, all of the bad groups are examined; the decision to keep or discard the data is based on the program and the individual variable star (most of such data are usually discarded).

4. Some Scientific Results for Upper Main Sequence Chemically Peculiar Stars

As mentioned previously, robotic telescopes present great advantages for the observer of variable stars. A much greater amount of data can be collected compared to conventional telescopes, as the observer's presence is not required; time-sharing in a given night by a number of different observers is also advantageous as this results in greater phase coverage for a given variable. On the CAPT, short-period variables and eclipsing binaries can be observed in long runs over one night or several nights as well. The observer is also able to collect data over a period of several years, giving him/her a homogeneous data set for year-to-year comparisons.

Because of these advantages, two of the authors (DMP and SJA) have begun studies to search for changes in the shapes of the light curves of the two magnetic chemically peculiar (MCP) stars with the shortest known periods. CU Vir and 56 Ari have periods of about 0.52 days and 0.73 days, respectively. Light, spectrum and magnetic variations in MCP stars are believed to be due to rotation and inhomogeneous atmospheres with large magnetic fields. Shore and Adelman (1976) suggested that MCP stars that were not in binary systems could experience free body precession due to a distortion in the shape of the star by the magnetic field. Since the precession period would be many times that of the rotation period of the star, the shortest period variables are the most likely candidates to display changes in their light curves due to precession. For stars with periods less than one day, the precession periods would be expected to be about 5 to 10 years. Adelman and Fried (1992) have found changes in the shape of the UBV light curves of 56 Ari compared with earlier published data. Two years of CAPT uvby data obtained by Adelman may also show some changes in the shapes of the u and v light curves, although the results are preliminary .

Three years of P10 UBV data for CU Vir were studied by DMP and compared with UBV data of Hardie (1958). An improved period of 0.5206800 days (Adelman et al. 1992) was used. There is some evidence for changes in shape between the Hardie and P10 data, although this is not compelling due to phase gaps in the Hardie data. The three years of the P10 data were separately compared. There appear to be subtle differences from year to year, the principal of which are a slight change in the time of minimum for the 1989 data with respect to the 1987 and 1988 data; and systematic changes in average brightness and perhaps amplitude for the U data from year to year (Figure 1). It is clear that higher precision is necessary in order to further pursue such studies. Some CAPT uvby data are available for CU Vir in 1991-92 but are not sufficient to come to any conclusions about yearly changes in the light curves; the star is on the program for the next few years including some long nightly runs to check on cycle-to-cycle variations. Additionally, there is difficulty in fitting all the available data for CU Vir to a single period, a problem which will be further investigated in the future.

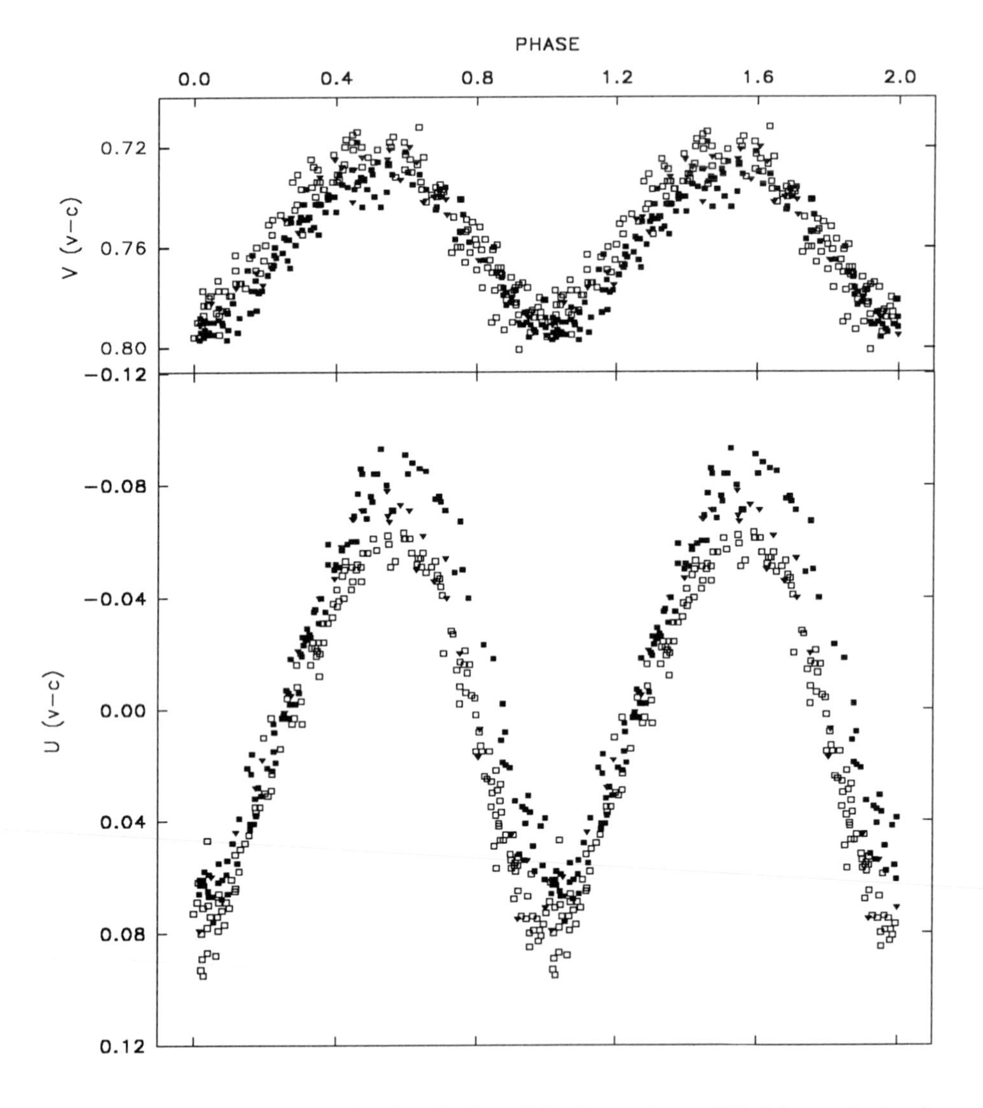

Figure 1 V and U photometry for CU Vir from P10 observations. Filled inverted triangles, open squares and filled squares are 1987, 1988 and 1989 data, respectively.

5. Future Improvements

The CAPT group will carry out two major projects in the next year or so to improve both the quality of our data and our observing efficiency. The first of these is to incorporate observations of standard stars into the nightly observations rather than relying on standard star nights and average extinction and transformation coefficients. We also plan to use a scheme to incorporate all the constant stars into the characterization of a night or a run of nights, similar to that described by Manfroid and Heck (1983).

Secondly, as mentioned above, we schedule on a time-share basis, rather than granting blocks of time to a single observer. Because of this, we need a much more sophisticated way to plan observations than we have at present. Even with our five regular observers plus students and collegues who need occasional observations, the scheduling procedure becomes very complicated for the PA, especially since he must insure that each school gets its fair share of telescope time. Additionally, we are now receiving an increasing number of requests for observing time by observers organizing international campaigns and those who have groups of stars they would like observed. Mark Drummond and others at NASA/Ames is writing scheduling software based on artificial intelligence for robotic telescopes with an eye toward future networks of such telescopes and possible lunar-based networks. The CAPT group hopes to be an early "guinea pig" for this software; we feel that a system such as this will be essential for efficient operation of multi-user robotic telescopes.

References:

Adelman, S.J. and Fried, R., 1992, in preparation.

Adelman, S.J., Dukes, Jr., R.J., Pyper, D.M., 1992, Astron. J., Astron. J., **104**, in press.

Bartholdi, P., Burnet, M., Rufener, F., 1984, Astron. Astrophys. **134**, 290.

Genet, R.M. and Hayes, D.S., 1989, *Robotic Observatories*, AutoScope Corp., Mesa, AZ, p. 28.

Hardie, R., 1958, Astrophys. J. **127**, 620.

Manfroid, J. and Heck, A., 1983, Astron. Astrophys. **120**, 302.

McCook, G.P., 1991, in: *Advances in Robotic Telescopes*, ed. M. Seeds, Fairborn Press, Mesa, AZ, p. 263.

McCook, G.P., 1992, in: *Robotic Telescopes in the 1990s*, ed. A. Filippenko, Astr. Soc. of Pacific Conf. Series, in press.

Shore, S.N. and Adelman, S.J., 1976, Astrophys. J. **120**,816.

Discussion

T.J. Kreidl: *To what accuracy is the time actually recorded for each data point?*

Pyper: We get fractional JD to six decimal points, or about 0.1 sec.

E. F. Milone: *I noticed that the check and comparison stars are observed at different times in the 'group'. It would be better to have an extra comparison star observation at each end of the sequence to get the same mean time for it as the check star observations.*

Pyper: The P10 sequence is "set in stone" but ATIS enables us to vary our sequences, so we now mostly do follow your suggested sequence.

An Intimate Relation with Two Automatic Telescopes for Almost Nine Years

Douglas S. Hall[1], Gregory W. Henry[2]

[1]*Dyer Observatory, Vanderbilt University Nashville, Tennessee 37235, U.S.A.*
[2]*Center of Excellence in Information Systems, Tennessee State University, Nashville, Tennessee 37203, U.S.A.*

Abstract

From Oct. 1983 through Dec. 1987 most of the observing capacity of the Boyd–Genet prototype 10-inch APT was devoted to a program of 92 variable stars the output of which was sent to us for analysis. From Nov. 1987 through the present the entire observing capacity of the VU-TSU 16-inch APT has been devoted to a program of 136 mostly magnetically active variables. We review the technical performance of both APTs in terms of malfunctions, down-time, resulting photometric accuracy, phase coverage, and scientific results to date.

1. Introduction.

We want to give a frank account of our experience with two automatic photoelectric telescopes over a time interval of nearly nine years, for the benefit of others just embarking on or contemplating involvement with these remarkable instruments.

We were the principal scientific users of the prototype automatic photoelectric telescope, the so-called Phoenix 10-inch, at the invitation of its developers Louis J. Boyd and Russell M. Genet, from its first night of observation in October 1983 until the last night of calendar year 1987. During these 4.2 years we received differential photometry of a grand total of 92 program stars, 72 of them known or suspected variables of the chromospherically active type.

During that time a 16-inch automatic telescope was acquired with the help of an N.S.F. grant to Vanderbilt University and its continued operation has been funded by N.A.S.A. and N.S.F. grants to Tennessee State University. This VU-TSU 16-inch saw first light in November 1987, has observed variable stars for the 4.7 years between then and the date of this Colloquium, and is still working as we write this paper. At this moment 136 program stars are on its menu, along with 20 UBV standards.

Specifics detailing the operation of these two automatic telescopes have been published elsewhere, references in Table 1, and will not be repeated here. Moreover, the paper by Henry and Hall (1992), presented at the Workshop on Robotic Telescopes in Kilkenny just before this Colloquium, is useful as a companion to this paper. Let us explain, however, that these telescopes do obtain differential magnitudes between a program star and a nearby comparison star and, at the same time, differential magnitudes between a check star and that same comparison star. The sequence of 10-second

Table 1. References pertaining to the 10-inch and the 16-inch.

I.A.P.P.P. Communications	P.A.S.P.
1984 - No. 12, p. 20	1986 - Vol. 98, p. 618
1985 - No. 19, p. 41	1987 - Vol. 99, p. 660
1985 - No. 21, p. 59	1991 - Vol. 103, p. 221
1985 - No. 22, p. 47	
1986 - No. 25, p. 32	
1986 - No. 25, p. 43	Ap. J. Supplement
1988 - No. 33, p. 10	1988 - Vol. 67, p. 439
1990 - No. 42, p. 44	1988 - Vol. 67, p. 453
1990 - No. 42, p. 54	1989 - Vol. 69, p. 141
1991 - No. 45, p. 11	1990 - Vol. 74, p. 225

Table 2. New Variables Discovered with the 10-inch and 16-inch

omi Dra = omi Dra [A]	HD 28591 = V492 Per [A]
eps Hya = eps Hya [A]	HD 31738 = V1198 Ori [A]
33 Psc = BC Psc [B]	HD 43930 = ? [A]
xi UMa = xi UMa [A]	HD 71071 = LU Hya [A]
	HD 80715 = BF Lyn [A]
HR 454 = OP And [A]	HD 90385 = ? [A]
HR 1362 = EK Eri [A]	HD 116204 = BM CVn [A]
HR 1970 = V1197 Ori [B]	HD 136901 = UV CrB [B]
HR 3337 = LO Hya [C]	HD 144515 = ? [A]
HR 4430 = EE UMa [A,B]	HD 152718 = ? [A]
HR 6469 = V819 Her [A,B,C]	HD 155989 = ? [A]
HR 6626 = V826 Her [B]	HD 160952 = ? [A]
HR 6902 = [C]	HD 163621 = ? [A]
HR 6950 = [B]	HD 181219 = ? [C]
HR 7428 = V1817 Cyg [A,B]	HD 181943 = ? [A]
HR 7578 = [A]	HD 191011 = ? [A]
HR 9024 = OU And [A]	HD 191262 = ? [A]
	HD 193891 = ? [A]
HD 1405 = ? [A]	HD 209943 = ? [A]
HD 6286 = ? [A]	HD 212280 = ? [A]
HD 9313 = ? [A]	HD 217188 = AZ Psc [A]
HD 12545 = ? [A]	HD 218153 = KU Peg [A]
HD 19485 = ? [A]	HD 219989 = OT And [C]
HD 19942 = ? [A]	HD 222317 = KT Peg [A]
HD 25893 = V491 Per [A]	

A = spots, B = ellipticity, C = eclipses

integrations which results in a 'group observation' is K-S-C-V-C-V-C-V-C-S-K, where K = check star, S = sky, C = comparison star, and V = variable or program star. A group observation is accomplished in about six minutes, of which about 75% is spent counting photons, the remaining 25% spent deciding which star to observe next, moving to that star, finding it, verifying its identity, centering it in the diaphragm, and switching filters. Henry and Hall (1992) explain the 'first-to-set-in-the west' rule which the telescope control program uses to assure that the maximum number of program stars on the master menu are observed once each night, this being the optimum observing frequency for these stars most of which vary on time scales generally longer than one day. Moreover, Henry and Hall (1992) explain the function of the 'cloud filter', the procedure by which group observations having a mean differential magnitude uncertain by more than $0^{m}.02$ are not archived, i.e., eliminated. We see later that the cloud filter was recently tightened from $0^{m}.02$ to $0^{m}.01$.

2. Accomplishments

One measure of the performance of these two automatic telescopes is a simple listing of the new variable stars they have discovered. As shown in Table 2, the total to date is 47, of which 16 are of naked-eye brightness, *i.e.*, in the *Yale Bright Star Catalogue.* The official variable star designation is given except for those discovered too recently to have been named. It happens that all 47 vary in brightness by one or more of the same three physical mechanisms: starspots (rotation), the ellipticity effect (tidally distorted shapes), or eclipses. Note that all three mechanisms contribute in HR 6469 = V819 Her.

That is, though, just one measure. Both telescopes have been programmed to observe a large number of stars approximately once each night on as many nights throughout a year as is possible and continuously for as many years as is possible. Most of the program stars are chromospherically active and hence heavily spotted. As a result, their variability is on a variety of time scales (Hall 1992) and the multiple periodicities are a challenge to sort out. Rotation periods are days, weeks, or months; starspots or active regions live for weeks, months, or years; and magnetic cycles of years or decades modulate the mean brightness. These time scales all are long enough that continuous photometry throughout one night is not warranted, and so these automatic telescopes in their one-point-each-night year-after-year mode are proving ideal for the challenge. Examples of spotted stars with photometric coverage approaching or exceeding a decade, much of it with these two telescopes, are V1764 Cyg (Lines *et al.* 1987), σ Gem (Strassmeier *et al.* 1988), V478 Lyr (Hall, Henry, Sowell 1990), HR 1362 (Strassmeier, Hall, Barksdale, Jusick 1990), V1817 Cyg (Hall, Gessner, Lines, Lines 1990), HD 181943 (Hooten and Hall 1990), HK Lac (Oláh, Hall, Henry 1991), EI Eri (Strassmeier 1990), V711 Tau (Henry and Hall 1991), τ Per (Hall *et al.* 1991a), λ And (Hall *et al.* 1991b), V1149 Ori (Hall, Fekel, Henry, Barksdale 1991), BM Cam (in preparation), and DK Dra (in preparation).

Table 3. Problems causing data to be lost or accuracy to be diminished.

problem	effect on data	stars affected	APT	time interval affected
leap year	lost	all	10	1 nt
computer	lost	half	10	Δ t = 180 nts
electronics	$< 0\overset{m}{.}05$	bright	10	2 nts
floppies	lost	all	10	9 nts
stuck filter	lost	all B,U	10	3 hrs
dead time	$< 0\overset{m}{.}05$	bright	10	Δ t = 485 nts
filter fell out	salvageable	all V	10	Δ t = 28 nts
power supply	$\sigma_{ext} = 0\overset{m}{.}02$	all	10	Δ t = 150 nts
worm gear	$\sigma_{ext} = 0\overset{m}{.}02$	all	16	gradually worse up through 2Q 90
centering	σ_{ext} large	all	16	Δ t = 12 nts
secondary mirror	lost	20%, high k	16	2 quarters
lady bug	lost	all	16	3 nts
$\epsilon = f(T)$	$< 0\overset{m}{.}02$	large Δ(B-V)	both	through 2Q 91
aborts	lost	a few	both	throughout

Table 4. Important dates in the history of the two automatic telescopes.

10-inch

12-13 Oct. 1983 was first night of data with the 10-inch.
No data were taken on 29 Feb. 1984, the first leap year.
On 7 Feb. 1985 the 'dead time' problem was corrected.
On 1 July 1985 the yellow filter fell out.
On 1 Aug. 1985 the yellow filter was cemented back into place
~1 Dec. 1985 symptoms of power supply malfunction were first apparent.
On 1 May 1986 the power supply was repaired.
During July, Aug., and Sept. of 1986, the 10-inch was being moved from downtown Phoenix to the top of Mt. Hopkins.
30-31 Dec. 1987 was our last night of data with the 10-inch.

16-inch

12-13 Nov. 1987 was first night of data with the 16-inch
Through 1Q 1988, UBVRI photometry with a GaAs photomultiplier.
BV photometry with a new photomultiplier began with 2Q 1988.
During 3Q 1990, worn worm gear drive replaced with new belt drive system.
Operation with ATIS began with 4Q 1990.
In mid October 1990, transient bug in centering algorithm corrected.
During 2Q 1991, first noticed symptoms of misaligned secondary mirror.
From July 1991 through Feb. 1992, precision photometer installed.
Beginning in March 1992, master menu includes standards, for nightly determination of extinction and transformation.
In mid-June 1992, secondary mirror misalignment corrected.
Operation with $0\overset{m}{.}01$ cloud filter began with 2Q 1992.
Lady bug problem in June 1992.

3. Problems

Table 3 lists all of the problems, other than clouds or daylight or time off for equipment upgrade, which have caused data to be lost or accuracy to be diminished. The second column indicates whether data were lost altogether, or the external error σ_{ext} became larger, or a systematic error was introduced. The third column indicates how many program stars were affected, for example, only the brightest stars, or only stars with a large color difference between variable and comparison, or mostly stars at high declination. The last column indicates the time interval affected. In some cases we give an exact number of nights on which data could have been taken but were not, or were actually taken and then lost, or were actually taken but with diminished accuracy. In other cases we give a time interval Δt during which a problem persisted, noting that many of the nights within Δt were unusable for other reasons.

A detailed account of the problems which affected the 10-inch can be found in Hall, Kirkpatrick, Seufert (1986) and Boyd, Genet, Hall, Busby, Henry (1990). Of the remaining problems, which have affected the 16-inch, Henry and Hall (1992) have described in detail the worn worm gear problem, the centering problem, and the temperature-dependent transformation coefficient problem, but not the secondary mirror problem or the lady bug problem. The first of these was simple but nasty. A loose screw caused the secondary mirror to slip, ruin the telescope alignment, and foul up the acquisition and centering process. The telescope failed to locate a number of stars, mostly at higher declinations, which should have been no problem. The second of these affected three nights in June 1992 when a lady bug (or maybe three different lady bugs) sat on the infrared LED which serves as a limit switch to signal that the telescope has reached the horizontal 'home' position. This made the telescope control computer think the night was over and turn off power to all instruments. The last problem in the list, the aborts, has affected both telescopes intermittently throughout the nine years. An abort occurs when the telescope executes its outward spiral search for the next star to be observed and cannot find it within 15 arcminutes of the starting point. The problem, other than a cloud in front of the star, usually proves to be an incorrect right ascension or declination, a nearby star of comparable brightness which we had not recognized, or a 'hunt magnitude' (the nominal brightness used to verify a star's identity during the acquisition stage) set too bright.

The number of group observations which were actually made and then lost or which could have been made but were not, when compared to the total number made successfully, amounts to a loss of only 10% or so. This is minimal compared to the three months of down time every summer, the monsoon season in southern Arizona, during which time we also perform most of the necessary repair, maintenance, and upgrade work.

Table 4 gives a list of dates and time intervals during which telescope operation began or ended, problems developed and ended, equipment upgrades were effected, observing or data reduction procedures were changed, etc.

Table 5. History of accuracy with the two automatic telescopes

APT	σ_{int}	σ_{ext}	situation
10	$0^m_.005$	$0^m_.010$	before power supply problem
10	–	0.016	during power supply problem
10	–	0.008	after power supply problem, on Mt. Hopkins
16	–	0.007	first year of operation
16	0.010	0.016	just before replacing worm gear drive
16	0.005	0.008	after installing belt drive
16	0.003	0.005	after precision photometer, nightly k, constant ϵ
16	0.003	0.003	best photometric nights only

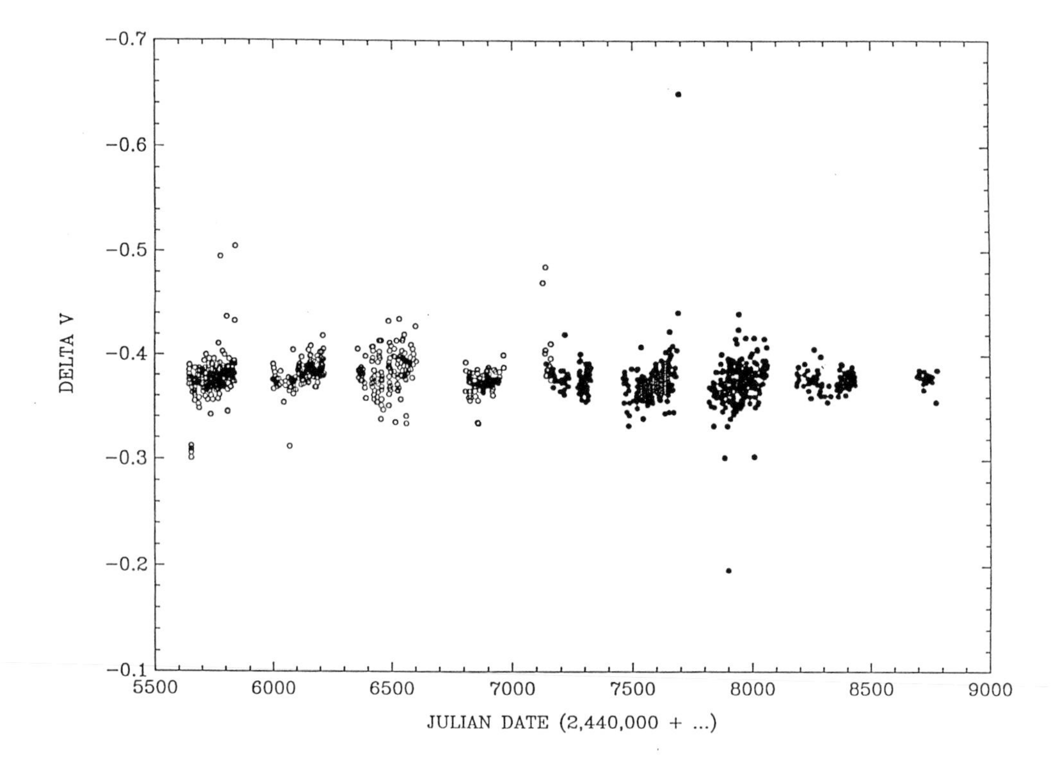

Figure 1. Each point is a group mean differential magnitude of the star pair 27 and 28 LMi, open circles are from the 10-inch, filled circles from the 16-inch. Note the 1-month overlap in late 1987. The rms deviation from the 9-year mean is σ_{ext}. A periodicity close to the tropical year shows up in these data, a consequence of the temperature-dependent transformation coefficient, with the large color difference between 27 and 28 LMi making this star pair particularly vulnerable. The last bunch of points as shown has not yet been passed through the 'second filter' but, after filtering, would show $\sigma_{ext} = 0^m_.003$.

4. History of the photometric accuracy

Table 5 is a summary of the evolution of the photometric accuracy achieved by both the 10-inch and the 16-inch. As explained by Hall, Kirkpatrick, and Seufert (1986), σ_{int} is simply the standard error of the mean of the three differential magnitudes within each group observation, calculated from the standard deviation of those three from the mean. On the other hand, σ_{ext} is the standard deviation of all group means from an average over some long interval of time, in the case of a program star which is constant, or from a calculated light curve fit in the case of a periodic variable.

Note that σ_{ext} on the 10-inch began at $0\overset{m}{.}010$, worsened during the power supply malfunction, and recovered after its repair. Similarly, σ_{ext} on the 16-inch began at $0\overset{m}{.}007$, worsened as the worm gear wore, recovered after its replacement, was improved to an impressive $0\overset{m}{.}005$ by the precision photometer and nightly observation of standards, and improved more still — to the theoretical limit — when a 'second filter' was used to discard data from all but the first-class photometric nights, as judged by the photometry of the standards in the all-sky mode.

Figure 1 illustrates the content of Table 5 in graphical form. Each point is a group mean of the star pair 27 and 28 LMi, with the latter considered the variable. Open circles are from the 10-inch, filled circles from the 16-inch. Note the 1-month overlap in late 1987. The rms deviation from the 9-year mean is σ_{ext}. A periodicity close to the tropical year (376 $\pm$ 2 days) shows up in these data. As Henry and Hall (1992) show, this is surely a consequence of the temperature-dependent transformation coefficient, with the large color difference between 27 and 28 LMi making this star pair particularly vulnerable. The last bunch of points as shown in Figure 1 has not yet been passed through the 'second filter' described in the paragraph above but, after filtering, would show $\sigma_{ext} = 0\overset{m}{.}003$.

Acknowledgements

The 16-inch telescope was obtained with an NSF grant to Vanderbilt (AST 84-14594). Its continued operation and data analysis are supported by grants from NASA (NAG 8-111) and NSF (HRD-9104484) to Tennessee State University. D.S.H. thanks Aer Lingus and the Local Organising Committee of the Colloquium for support.

References:

Boyd, L.J., Genet, R.M., Hall, D.S., Busby, M.R., Henry, G.W. 1990, I.A.P.P.P. Comm. No. 42, 44.

Hall, D.S. 1992, in *Robotic Observatories in the 1990's*, edited by A. Filippenko (Provo: A.S.P.), in press.

Hall, D.S., Fekel, F.C., Henry, G.W., Barksdale, W.S. 1991, A.J. **102**, 1808.

Hall, D.S., Gessner, S.E., Lines, R.D., Lines, H.C. 1990, A.J. **100**, 2017.

Hall, D.S., Henry, G.W., Sowell, J.R. 1990, A.J. **99**, 396.

Hall, D.S., Kirkpatrick, J.D., Seufert, E.R. 1986, I.A.P.P.P. Comm. No. 25, 32.

Hall, D.S. et al. [21 authors] 1991a, A.J. **101**, 1821.

Hall, D.S. et al. [31 authors] 1991b, J.A.A. **12**, 281.

Henry, G.W. and Hall, D.S. 1991, in Robotic Observatories, edited by S.L. Baliunas (Mesa: Fairborn Press), p. 307.

Henry, G.W. and Hall, D.S. 1993, in *Workshop on Robotic Observatories, Kilkenny, July 1992*, eds. B.P. Hines and M.F. Bode, Ellis Horwood.

Hooten, J.T. and Hall, D.S. 1990, Ap. J. Suppl. **74**, 225.

Lines, R.D., Lines, H.C., Kirkpatrick, J.D., Hall, D.S. 1987, A. J. **93**, 430.

Olàh, K., Hall, D.S., Henry, G.W. 1991, A.& A. **251**, 531.

Strassmeier, K.G. 1990, Ap.J. **348**, 682.

Strassmeier, K.G., Hall, D.S., Barksdale, W.S., Jusick, A.T. 1990, Ap.J. **350**, 367.

Strassmeier *et al.* [24 authors] 1988, A.& A. **192**, 135.

Discussion

W. Lockwood: *The APT and manual telescopes have the same duty cycle ~ 75%. But the APT centres better and moves faster than our manually-operated 21-inch telescope. Yet we get the same precision, ~ 0.003 mag rms. Evidently, therefore, time-dependent extinction and imprecise centering are not the causes of low precision, to within ~0.003 mag.*

Hall: It seems as if we both are at our 'theoretical limit', i.e., the precision limited by the combination of photon noise and scintillation.

W. Tobin: *Are the discovery times for problems that you showed really such that you need to be apologetic about them? I suspect it can take just as long and in many cases longer for such discoveries to be made with manual operations.*

Hall: You are probably correct, and I thank you for your kind perspective. Let me add something. In the beginning, when we received our data in batches on a quarterly basis, we were necessarily limited to discovery on a time scale of months. Now we can receive data on a morning-after basis and normally do examine it on the spot, making for discovery on a time scale of days, at least for the more obvious problems. Let me add, in conclusion, that our APT operation probably encounters fewer problems than would have been so with manual operation. That is because an APT, the ultimate dedicated telescope, experiences the absolute minimum number of equipment changes, modifications, human interference, etc.

R.L. Hawkins: *Can the APT system get data points for the variable spaced by two minutes or less?*

Hall: Yes, the system can be programmed to do so, but since I work on long period variables, I have not done so.

The Strömgren Automatic Telescope

Ralph Florentin-Nielsen

Copenhagen University Observatory

Abstract

Copenhagen University Observatory operates a 52cm Cassegrain telescope at ESO, La Silla (Chile) for photoelectric photometry in a completely automatic mode. Alternatively, the telescope may be operated, remotely, from Europe. We report on the operational experience from more than four years of successful use and on the decision by ESO to establish a similar system on their photometric telescope.

1. The telescope and Instrumentation

The Strömgren Automatic Telescope (SAT) is a 52-cm Cassegrain telescope. The optics are a Dahl-Kirkham design with an ellipsoidal primary and a spherical secondary. The telescope is operated by Copenhagen University Observatory at the European Southern Observatory, La Silla, Chile. It is used solely for stellar photoelectric photometry. A special purpose spectrophotometer (R. Florentin-Nielsen, 1985 and Helt, Franco and Florentin-Nielsen, 1987) for Strömgren uvby, H photometry is permanently mounted at the SAT. This spectrophotometer uses a reflection grating in Littrow configuration. The uvby spectral bands are defined by interference filters in conjunction with the output slots. The filters have very high transmission mainly due to the fact that the spectrograph slots removes the need for blocking for side bands.

The telescope is used mainly for the observation of eclipsing binaries, Andersen & Clausen, 1989 and for large scale classification work, Gray and Olsen, 1991 and references therein. Simultaneous photometry and radial velocities measurements done with the Coravel instrument at the Danish 1.54-m telescope at La Silla have yielded absolute dimensions of eclipsing binaries. A review paper by Andersen, 1991, shows that about 60 % of all 45 systems with radii and masses known with an accuracy to 1 - 2 percent have been determined from the Danish observations.

The telescope was equipped with a computer controlled six channel photon counting system, Klougart, 1984a and a digital telescope drive system, Klougart, 1984b. However, due to the very large amount of photometry carried out with that telescope it was decided to make the observations completely automatic. The telescope drive and control system already computes the apparent places for the stars to be observed and applies a pointing model correction (look-up table in declination and hour angle) before setting the telescope. Also the photon counting systems have built-in routines to test the quality of the photometry by comparing the actual spread of multiple 1 second integrations with the spread expected from the photon shot noise alone. Finally, the photon counting system may be set to either use specified integration times, to integrate until a required photometric accuracy is obtained or a combination of the two methods, Claudius and Florentin-Nielsen, 1981.

What remained to be established to allow an automatic observing mode was a method to center the telescope at the stars and a program to let the telescope control system and the photon counting system communicate with each other and to arrange for the proper sequence of the observations.

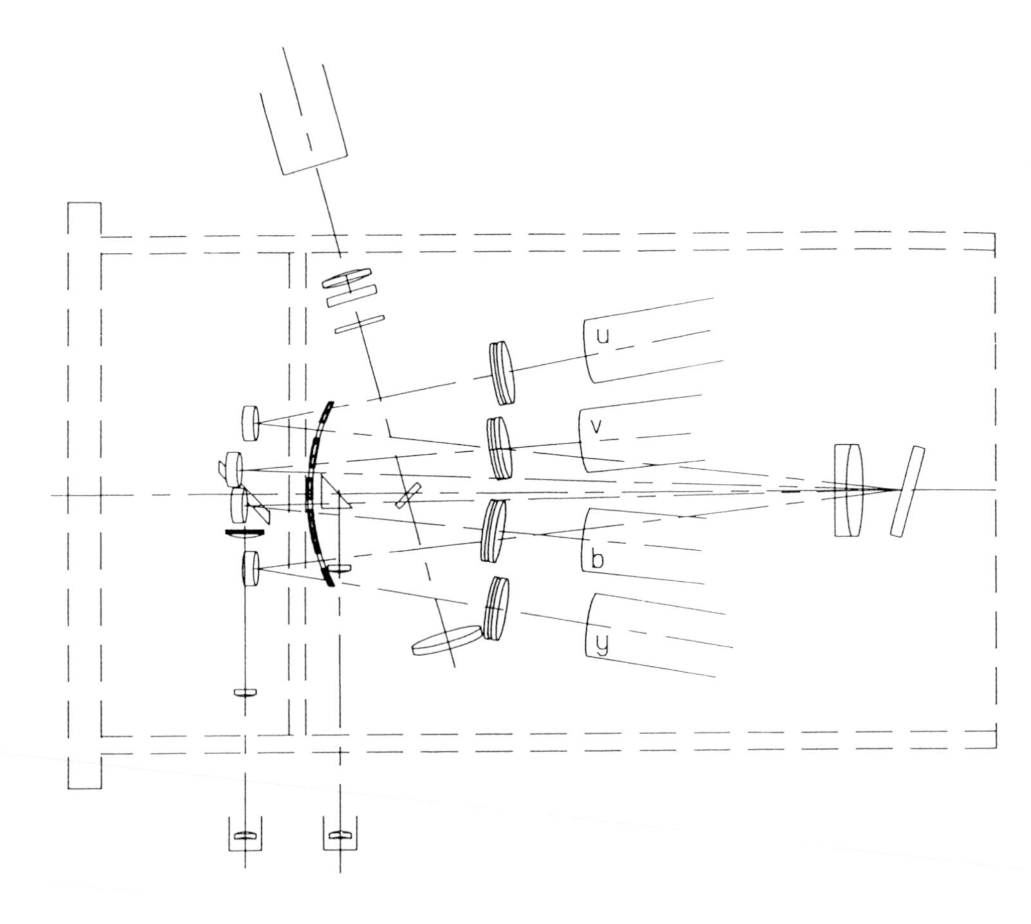

Figure 1a. The optical layout of the 6-channel uvby, Hβ spectrophotometer of the Strömgren Automatic Telescope.

2. The Auto-Centering Device

We have chosen not to incorporate any additional two dimensional detector (CCD) to the existing spectrophotometer. The setting error of the telescope is determined by scanning the star across a V-shaped slit plate in the focal plane. As can be seen from figure 3 the V-shaped slits are formed directly in one of the positions of the focal plane aperture wheel. The transits of the star over the slits are detected as the sum of the signals from the photomultipliers for all four channels, u, v, b and y.

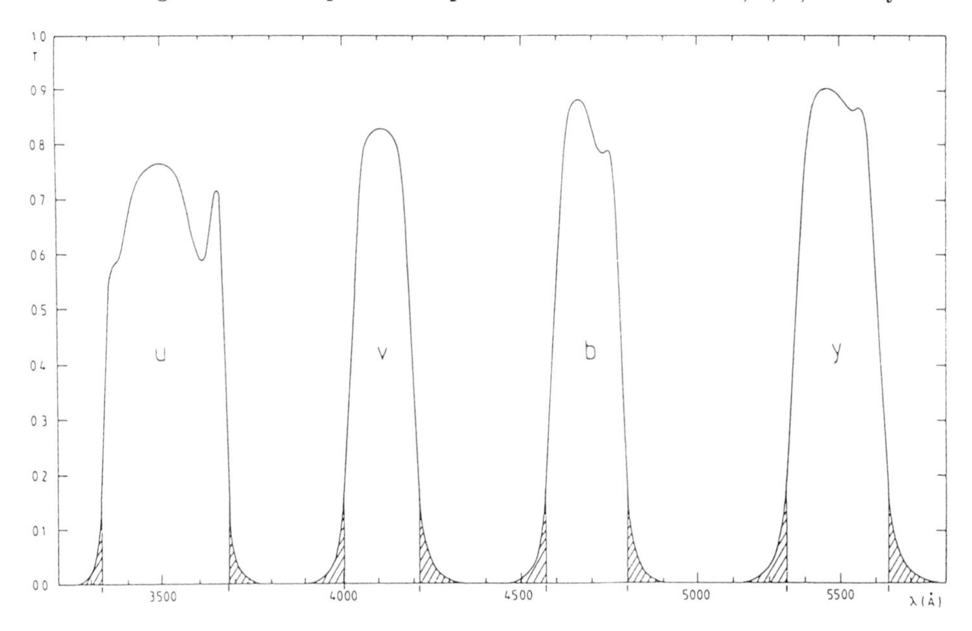

Figure 1b. The optical transmission of the u,v,b and y filters. Shaded areas of the passbands are blocked by the output slot section of the spectrograph.

A preset to a new star to be observed is then performed by setting the telescope to a position which is offset by 2 arcminutes in declination, while at the same time turning the aperture wheel to the V-slit position, immediately followed by a linear scan in delta. The first 60 arcsec is utilized to settle the servos to a constant scanning speed of 40 arcsec/second. During the subsequent 60 arcsec the sum of the photon counts from the 4 photomultipliers are sampled every 50 millisec. The double-peaked intensity profile obtained over this 60 arcsec (1.5 seconds) scan is analysed to calculate the setting errors in RA and Dec, $\delta\alpha$ and $\delta\delta$. The times for the median of the peaks from the transit of the upper and lower part of the V-slit are found by numerical convolution of the scan data with a line spread function, and are denoted t_1 and t_2. The start time of the 60 arcsec scan is denoted t_0,. Refer figure 3. It can be seen that the distance in time between the two transits t_1 and t_2 is a measure of the setting error in RA, whereas t_1-t_0 and t_2-t_0 defines the setting error in declination.

$$\delta\alpha = \text{-v(scan)}/2 * [(t_2\text{-}t_1) - (t'_2\text{-}t'_1)]$$
$$\delta\delta = \text{v(scan)} * [(t_1+t_2)/2 \text{ -}t_0 - (t'_1+t'_2)/2 \text{ -}t'_0]$$

where the scanning speed, v(scan) is 40 arcsec/sec, and t'_0, t'_1 and t'_2 are the times for the scan start and the two transits for a star which is perfectly centered ($\delta\alpha = \delta\delta = 0$). For the scan start time, t_0 defined as 0 sec the nominal values of t'_1 and t'_2 are 375 and 1125 msec respectively.

Immediately following the search scan the position errors are calculated and the telescope sets to the corrected position, and the aperture wheel is turned to the desired circular focal plane diaphragm. The total autocentering procedure takes slightly less than 5 seconds. The autocentering device works reliably down to a limiting magnitude of 11m in uncrowded fields. In cases, where positions of the program stars are not known very accurately the autocentering may perform multiple, longer scans parallel to each other. This, however, is quite wasteful in terms of time, and

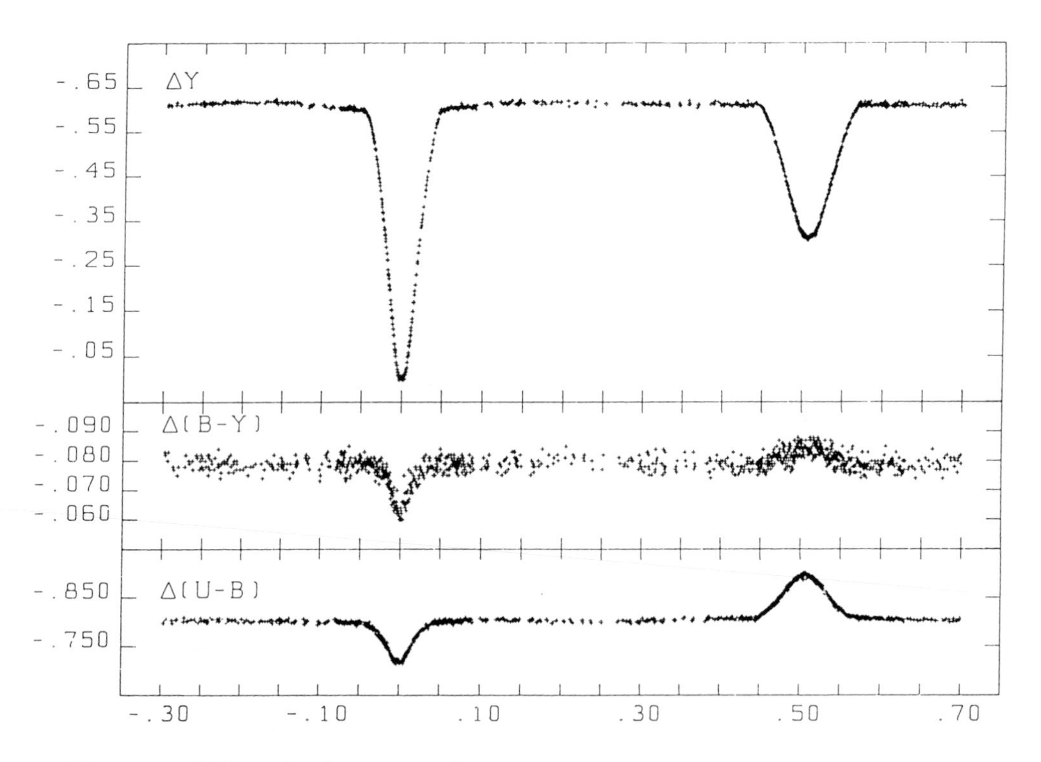

Figure 2. Light and colour curves for GG Lup observed at SAT by Anderson, Clausen and Giminéz (not yet published). Magnitude differences are accurate to 0.004 mag rms.

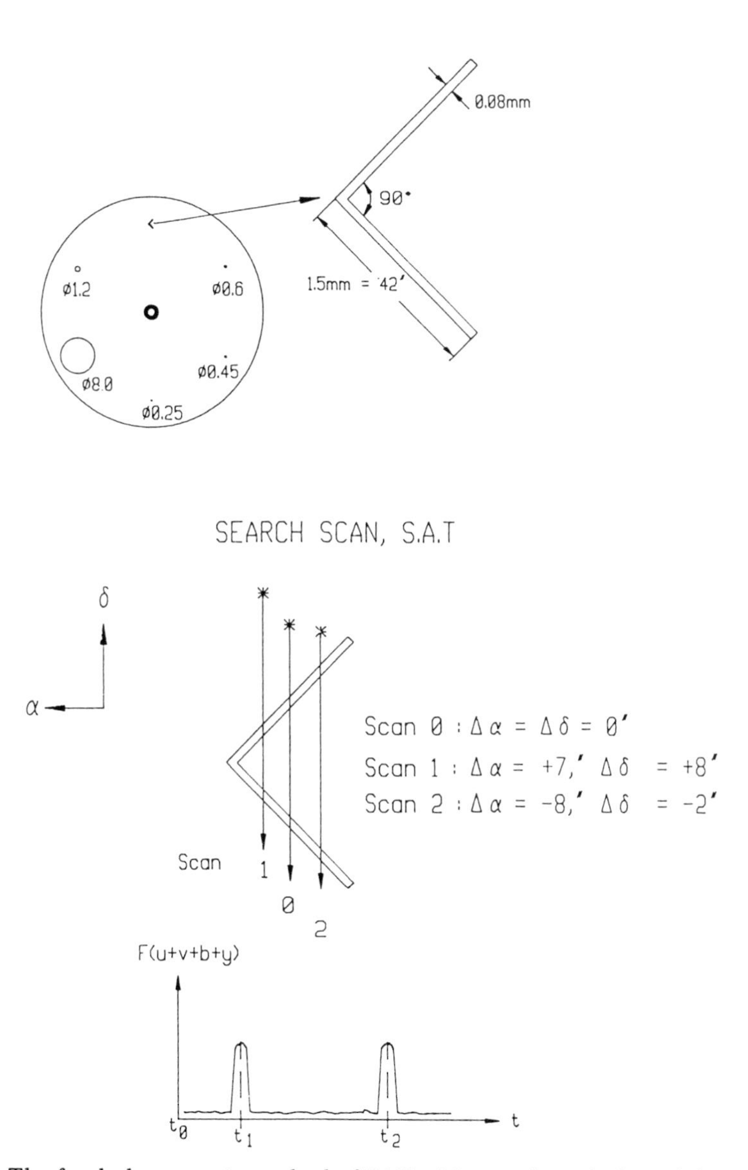

Figure 3. The focal plane aperture wheel of SAT with an enlarged view of the V-shaped slits. The star to be centred is scanned across the slits. The lower half of this figure shows three search scans. Scan 0 is for a star which was perfectly centred. Scans 1 and 2 have setting errors in RA and Dec of +7 and +8 arcsec (scan 1) and -8 and -2 arcsec (scan 2). Setting errors of the telescope are calculated from the profiles of the sum of the u,v,b and y channels.

hence to maintain telescope efficiency observers are always encouraged to prepare good positions, including proper motions if available. The observing rate for stars brighter than 9th magnitude is typically 30 stars per hour.

If a search field is crowded by other stars, or if the scan profiles indicate that the star is double, then the telescope will make no attempt to observe that star, but leave a mark in the observing list indicating that this star need to be inspected and set at manually. If the separation of a double star is less than about two arcsec (the limit depends somewhat on the difference in magnitude of the two components), then the telescope will set on the center of gravity of the flux from the two components.

3. Automatic Sequencing of the Observations

To operate automatically the system needs at least three files stored on hard disk. They are: 1) An observing list or star catalogue with the positions, equinoctium and proper motions. 2) An output file to store results, error messages and comments by the observer. 3) A command file containing a string of instructions on how the observations should be carried out. The most important instruction is OBS, which will preset the telescope, start a search scan, autocenter the telescope, select the diaphragm size, select the uvby or β section and initiate an integration sequence. Following the OBS instruction a record number specifies the star from the star catalogue. Alternatively OBS N sets on the next star in the catalogue and observes it. A jump (JMP) instruction can be used to repeat the observations of a number of stars (e.g. variable, comparison and check star), for example until a specified sidereal time.

All functions of controlling the telescope, dome, auto-centering, and data acquisition from the 6 channel photon counter, observing sequencer and data logging on the SAT is governed by one VME-10 computer. Nörregaard, 1988.

The system has been continuously in operation since 1987 (Florentin-Nielsen, Nörregaard, Olsen, 1987). Besides occasional problems with a disk drive and dome rotation the system has been remarkably reliable. Over the last four years no observing night has been lost due to technical reasons.

4. Operation in the immediate future

Transport of observing lists, command files and observational data is done on VME diskettes. As several user institutes have no access to read or write VME diskettes, it has been decided to interface a PC AT to the VME computer. (European astronomers generally tend to prefer PCs for most of their work). Not only can data then be transported on PC diskettes, but also the observers may add extra PC software to do on-line data reductions or display light curves of variables, show extinction plots or any other program that the observer may find useful. Some such data reduction PC software already exists, but it is the general policy that while individual observers may not modify the VME software, they may add as much PC software as they like.

In practical use observers may attend the observations from the office downstairs to watch astronomically relevant data being created or displayed, or just occasionally verify that the quality of the photometry is good. Several Danish observers have had telescope time at the SAT and at the Danish 1.54-m telescope at La Silla at the same time, thereby saving travelling expenses.

ESO has this year decided to have their 50-cm photometric telescope rebuilt, such that it will become a close copy of the SAT. The ESO 50-cm has an ordinary one channel filter photometer. This upgrade is presently been made (mechanics at ESO, electronics and software in Denmark), and is planned to go in operation in February 1993. After that time ESO will not generally offer travel grants for observers to go to the SAT or ESO 50-cm telescope. One ESO night assistant will load the observing list into the VME and start the observing procedure in the beginning of night. All data transport between Europe and Chile will be by PC diskettes or electronic mail.

References:

Andersen, J.: 1991, The Astron. Astrophys. Rev. 3, 91.

Andersen, J., Clausen, J.V.: 1989, Astron. Astrophys. 213, 183.

Claudius M., Florentin-Nielsen, R.: 1981, Astron. Astrophys. 100, 186.

Florentin-Nielsen, R.: 1985, A 6-channel uvby-beta Spectrophotometer for the Danish 50 cm Telescope, Copenhagen Univ. Obs. Internal Report 8, 1.

Florentin-Nielsen, R., Nörregaard, P., Olsen, E.H.: 1987, First Fully Automatic Telescope at La Silla, ESO Messenger, No. 50, 45.

Gray, R.O., Olsen, E.H.: 1991, Astron. Astrophys. Suppl. Ser. 87, 541.

Helt, B.E., Franco, G.A.P., Florentin-Nielsen R.: 1987, Proc. ESO Workshop on SN 1987a, 89.

Klougart, J.: 1984a, Photon Counting System (Photsys) Danish 50 cm, Users manual, 1.

Klougart, J.: 1984b, Telescope Control System, Users manual, 1.

Nörregaard, P.: 1988, Control System on Strömgren Automatic Telescope, Version 3.0, Users manual, 1.

Discussion

R.R. Shobbrook: *How are sky positions chosen? Highest accuracy would be attained by choosing a position from a deep photograph.*

Florentin-Nielsen It is certainly true that observers working in clusters or crowded fields have been selecting their sky background positions carefully from maps. However, much of the work on bright stars in uncrowded fields has been done using a nominal position offset, and then verifying that this position was alright.

C.L. Sterken: *We did long-term monitoring during the manual lifetime of the telescope, and also after automation. We indeed found that we could gain more than a millimagnitude precision by replacing the sky offset selected previously by visual inspection, by sky positions selected from sky photographs.*

When observing in NGC3293, applying a three-star and sky sequence during 5 or 6 hours, I experienced more than once that the system would centre all my stars during several hours, but it would happen once in a while that the system would fail, and search endlessly for the star it had previously been able to find without problem. This happens about once in 3-4 nights. Do you have any clue as to what may be at the origin of this problem?

Florentin-Nielsen: This has probably been a software bug. It is a large control programme, and several corrections and additions have been made to the software.

A. T. Young: *How accurately does your pointing model find stars?*

Florentin-Nielsen: The pointing accuracy is 5-10 arc sec over the sky.

D.L. Crawford: *Can you comment on the stability of the photometric system over the years (as the hardware has been very stable over a long period)?*

Florentin-Nielsen: The instrumental system has been remarkably stable over the years, so much so, in fact, that variable star observers have had no need to transform to the standard system for stability reasons.

Instantaneous determination of a variable extinction coefficient in photoelectric photometry

E. Poretti[1], F. Zerbi[2]

[1]*Osservatorio Astronomico di Brera, Merate, Italy,* [2]*Dipartimento di Fisica Nucleare e Teorica, Università di Pavia, Italy*

Abstract

Observational evidence was found that even in good photometric sites the extinction coefficient can display large variations during a night. The authors propose a method to determine instantaneous values of the extinction coefficient: it is based on the knowledge of the extra-atmosphere instrumental magnitude m_o for a reference star. An analytical procedure is proposed to determine a reliable value for m_o and results are very satisfactory.

1. Introduction

At the effective wavelength λ, the ground observed instrumental magnitude m is related to the extra-atmosphere instrumental magnitude m_o by Bouguer's law

$$m \;=\; m_o + k'_\lambda \sec z \tag{1}$$

where k'_λ is the extinction coefficient (measured in mag/airmass) and $\sec z$ is a good approximation for the airmass value. When expressing the magnitude difference between two stars we have

$$\Delta m \;=\; \Delta m_o + k'_\lambda \Delta \sec z + k''_\lambda \Delta(B - V)\, \overline{\sec z} \tag{2}$$

where k'_λ is the first-order coefficient and k''_λ the second-order coefficient; in this paper we shall deal with UBV photometry and hence we use $B - V$ as colour index. Very accurate differential measurements can be performed in not high-quality photometric conditions by choosing the comparison stars among the stars not only located close to the program star but also having a very similar colour index; the only request is a regular variability of the sky transparency over an observing cycle. However, when searching for small amplitude variations, the contribution of the product $k'_\lambda \Delta \sec z$ can rarely be considered to be negligible and a precise determination of k'_λ is necessary.

In the data reduction, the extinction coefficient is supposed to be a constant quantity. A posteriori, this constraint is generally justified by the goodness of the results, but its verification is rarely undertaken.

2. Deviations from Bouguer's linear law

In our observing programme of δ Sct stars (Poretti and Mantegazza, 1992) we collected a large sample of extinction coefficient behaviours. We used in our analysis the simultaneous least–squares fitting technique proposed by Vaníček (1971) and the power spectra of the magnitude differences between two comparison stars always show an increasing number of peaks at the lowest frequencies: if the peaks are observed at integer values of c/d we ascribe them to meteorological effects. With a large number of measurements, these effects become more evident because in the power spectrum the white noise level decreases and also periodicities (spurious or real) with a very small amplitude can be detected. By analyzing very rich datasets, we noticed that a spurious periodicity can be introduced in the data if, in presence of sky transparency variations, a constant value of k'_λ was used (see below, Fig. 4).

In our observing experience at La Silla we generally found Bouguer's lines with a correlation coefficient very close to 1, but in September and October 1991 only on 6 nights (out of the 16 with a sufficient time baseline) we observed a constant k'_B coefficient. Figure 1 shows an example of the behaviour of the instrumental magnitude of HD 225086, a comparison star, in function of sec z: an interpolation by means of two lines cannot be considered a satisfactory solution because there is no physical reason for an abrupt change in the k'_B coefficient in correspondance with the meridian crossing. Actually, k'_B changes continuously during the night.

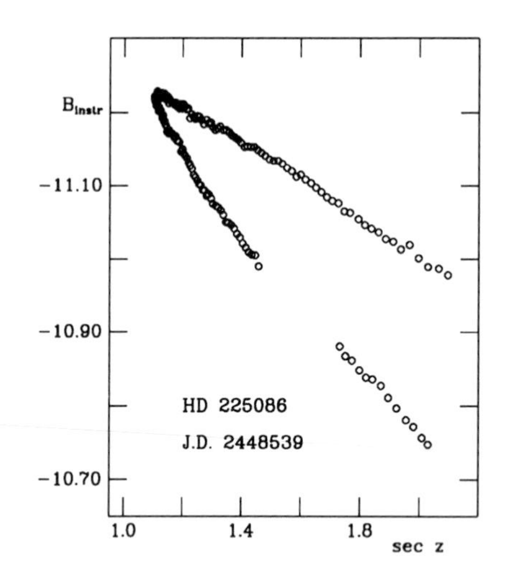

Figure 1. Behaviour of the instrumental magnitude showing the effect of a variable extinction coefficient (ESO measurements).

The non–linear behaviour of the extinction law observed not only at Merate Observatory, but also at such a good photometric site as La Silla gave us the warning sign that a method not requiring a constant value of k'_λ should be developed.

3. Instantaneous determination of the first–order term k'

The starting point for the correction of the extinction in presence of a rapid variability of the k'_λ coefficient can be found in the reliability of a photon–counting instrumentation that works in conditions which can be maintained constant over a long time baseline. The constant response to a standard signal (e.g. a constant artificial light source) guarantees that the above conditions are fulfilled.

If the stability is maintained over the whole observing run, the extra–atmosphere instrumental magnitude m_o will be a constant quantity. If we can calculate m_o for one of the two comparison stars (used as reference star), the extinction coefficient can be easily determined at the time t of the measurement $m(t)$ by means of

$$k'(t) = \frac{m(t) - m_o}{\sec z} \tag{3}$$

The main difficulty in the application of this equation is to have a reliable value for m_o. A good opportunity is offered by the analysis of the errors resulting from a wrong choice of the value for m_o. If we can neglect the colour effect, for each measurement of the magnitude difference $\Delta m(t)$ between the two stars we can calculate the extra–atmosphere magnitude difference Δm_o by means of

$$\Delta m_o = \Delta m(t) - \frac{m(t) - m_o}{\sec z} \Delta \sec z \tag{4}$$

where $m(t)$ and m_o are the magnitudes of the reference star inside and outside atmosphere. If we are considering an m_o value which differs by δ from the true value m_o^*, we will observe a linear trend when plotting Δm_o (or the residuals $\Delta m_o - \overline{\Delta m_o}$) against $\Delta \sec z / \sec z$; the angular coefficient yields exactly δ:

$$\Delta m_o = \Delta m(t) - \frac{m(t) - (m_o^* + \delta)}{\sec z} \Delta \sec z = \Delta m(t) - \frac{m(t) - m_o^*}{\sec z} \Delta \sec z + \delta \frac{\Delta \sec z}{\sec z} \tag{5}$$

Hence, we can determine the m_o value which does not yield any trend in the residuals or, similarly, for the m_o value which minimizes the height of the peak at integer values of c/d in the power spectrum. Figure 2 shows the satisfactory results obtained by applying this procedure to the 2715 measurements between HD 18768 and HD 19279: for different values of m_o, the plot of the residuals (left side) displays a progressive change in slope, while the power spectra of the magnitude differences (right side) show the corresponding flattening of the noise at the lowest frequencies. We can conclude that m_o^*=–11.75. It should be noticed that the iterative procedure is not necessary, but we recommend it in order to avoid unreliable determinations.

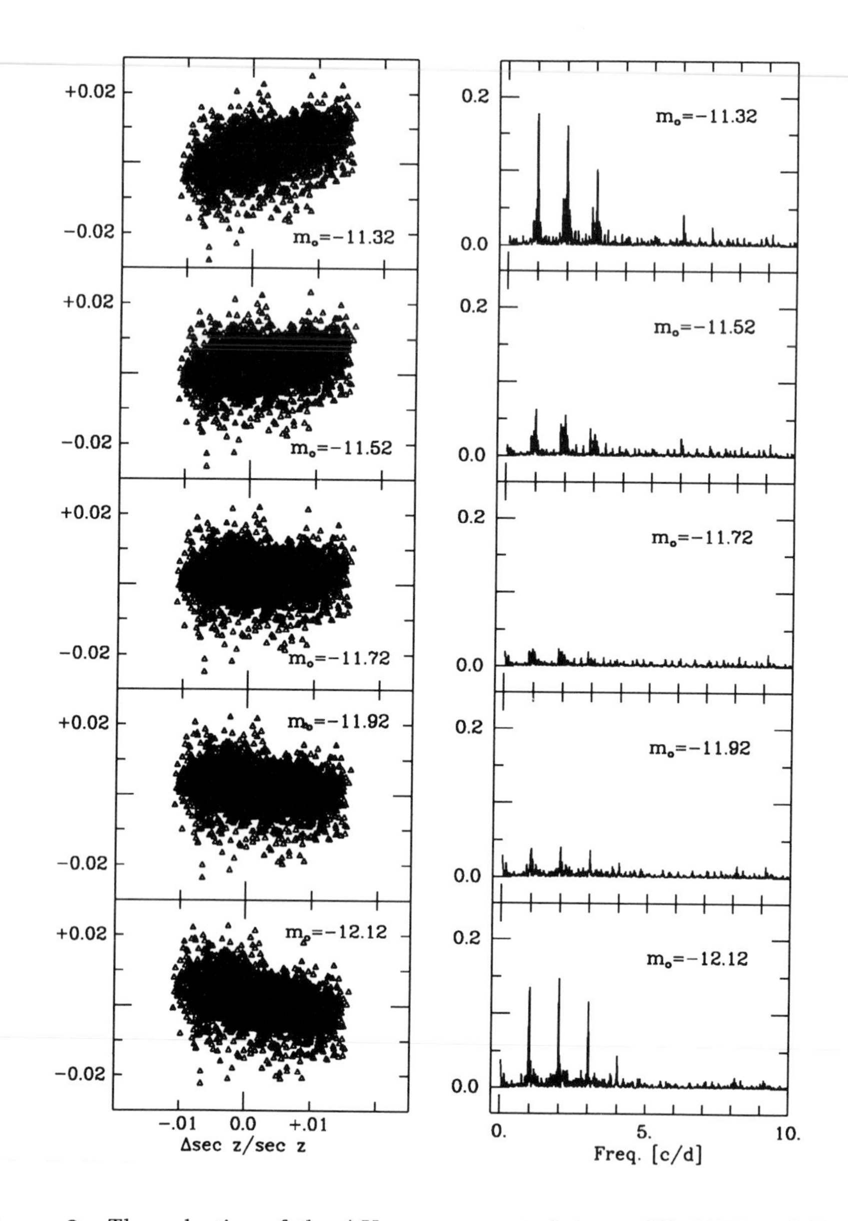

Figure 2. The reduction of the ΔV measurements between HD 18768 and HD 19278 assuming different values for m_o. Left panels: the linear trends have an angular coefficient equal to $m_o^* - m_o$. Right panels: the power spectra evidence an increasing number of peaks at the lowest frequencies for increasing values of $\mid m_o^* - m_o \mid$.

This procedure allows us to reconstruct the behaviour of the k'_λ coefficient. Figure 3 shows the behaviour of the k'_B and k'_V coefficients at La Silla: slow, but sometimes large variations are present. k'_B values are ranging from 0.26 to 0.58 mag/airmass (the maximum amplitude for a night is 0.12 mag/airmass, see J.D. 2448539 and compare the behaviour of the k'_B coefficient with that of the instrumental magnitude reported in Fig. 1), while k'_V values are ranging from 0.20 to 0.28 mag/airmass. We notice that these values are slightly higher than the normal ones. Figure 3 also shows that the k'_λ coefficients change slowly, without any well-defined periodicity or behaviour: for this reason any attempt to fit their variability by means of a periodic term cannot solve the problem.

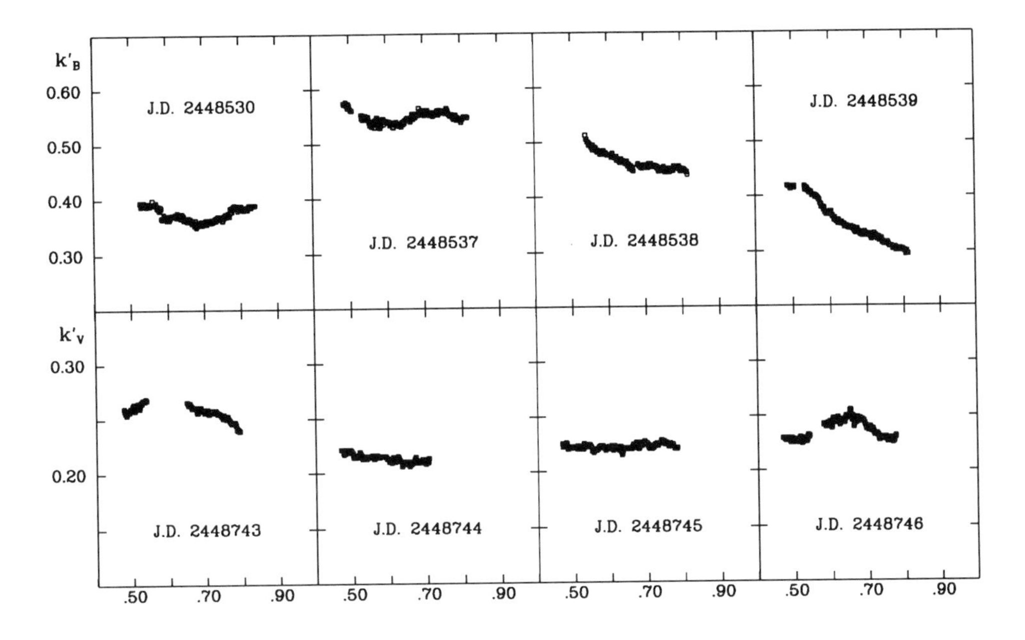

Figure 3. Behaviour of the extinction coefficients k'_B and k'_V on some nights at La Silla. It was determined by applying our method to the measurements of HD 225086 (*B* light) and HD 105654 (*V* light).

4. Methodological conclusions

The method described here provides a satisfactory correction of the extinction effect; its application to differential photometry does not require any additional measurement and allows us to clearly recognize spurious periodicities, due to the variability of the k'_λ coefficient during the night. Figure 4 shows the power spectra obtained from the ΔV measurements between HD 18768 and HD 19279 calculated by assuming a constant k'_V value throughout each night (left panel) and by applying the described procedure (right panel). The height of the spurious peak at 1.001 c/d, well above the noise level in the former case, is strongly reduced in the latter. On the other hand the peak at 14.454 c/d is present in both spectra and it can be ascribed to the physical variability of HD 19279.

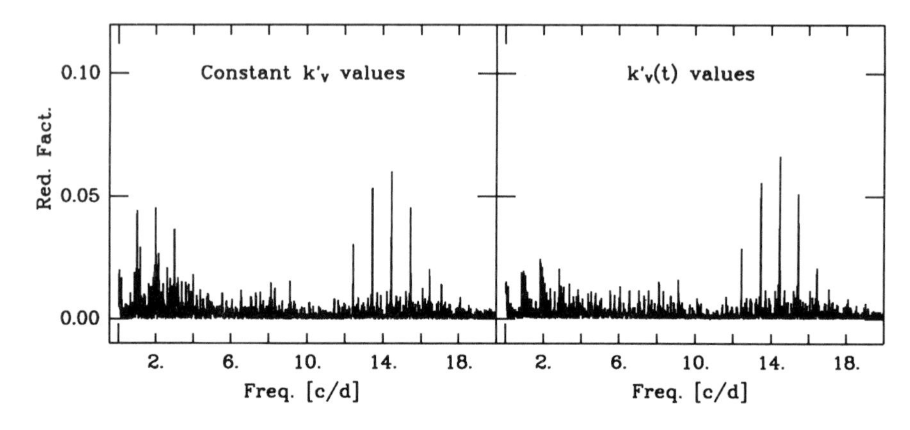

Figure 4. The noise at the lowest frequencies is strongly reduced when using the instantaneous values of $k_{V'}$ (right panel) instead of the constant values (left panel).

As regards the colour effect, it is a well known fact that the analysis of the residuals allows us to correct it . It is sufficient to plot the residuals $\Delta m_o - \overline{\Delta m_o}$ versus $\overline{\sec z}$; if a colour term is present, the angular coefficient of the least-squares line will now give $k''\Delta(B-V)$. The two k''_B values obtained from the ESO measurements are in good agreement with each other: k''_B=–0.034±0.005 (September and October 1991) and k''_B=–0.031±0.002 (April 1989).

The extension of the method to standard photometry, where rapid variability of the extinction coefficient can generate erroneous results, can be ensured by the frequent observation of one standard star, which will provide the requested k'_λ behaviour during the night. A particular care should be used when calculating both transformation and extinction coefficients in presence of a variable extinction coefficient since least-squares routines can converge to not-physical values.

The observed evidence of a variable k'_λ coefficient strongly suggests the need of a more rigorous monitoring of the extinction behaviour; volcanic eruptions are the most probable causes for this effect (the effects observed at La Silla are surely related to the eruption of Pinatubo), but the lifetime of dust and aerosol clouds in the Earth's atmosphere does not allow us to consider them as time-limited phenomena.

References:

Vaníček P., 1971, Astrophys. Space Sci. **12**, 10

Poretti E., Mantegazza L., 1992, The Messenger **68**, 33

Discussion

R. Shobbrook: *How certain are you that your extinction change from east to west is not in fact a drift in gain (due to electronics or telescope orientation)? East to West differences in extinction are not likely at many sites. The absorbing part of the atmosphere in the light path lies only 3 to 5 km at the most from the observatory so unless the site is on the edge of a major geographical change (e.g. land/water) no effect is likely.*

Poretti: My answer to this question is reported in the answer to A. Young's comments.

A. T. Young: *In response to Bob Shobbrook's question, the examples shown are very typical for an extinction coefficient that changes linearly with time. A zero-point drift produces a rounded corner, not a sharp angle, in the Bouguer plot. The two slopes, of the rising and setting branches contain aliasing effects due to the correlation of airmass and time, neither slope is a 'real' extinction coefficient.*

Poretti: Daily tests should be performed to check instrumental drifts as we did. Moreover, instrumental drifts would be regular. In our case, we observed changing behaviour (and constant ones, too) when observing the same comparison star on different nights; this suggests the atmosphere was responsible for the variability of the instrumental magnitude.

W.Z. Wisniewski: *Can the different behaviour of extinction when it is rising and setting be explained by asymmetry between east and west?*

Poretti: A continuous drift of the k′ coefficient seems to be more plausible than an east-west dicotomy; see also Bob Shobbrook's comment. Moreover, our method does not show this dicotomy, which, if real, should be observed.

C.L. Sterken: *A question to observers at good and less good sites: have you ever experienced any periodicity or quasi-periodicity in extinction coefficients on time scales of about 10 days to a fortnight?*

Poretti: Seasonal variability of k′ extinction coefficient was recorded in many observatories. We performed frequency analysis of k′(t) values, but we failed to find evidence of well defined periodicities or quasi-periodicities.

Correlations between Atmospheric Extinction and Meteorological Conditions

H.-G. Reimann[1] V. Ossenkopf[2]

[1]*Astrophysikalisches Institut und Universitäts-Sternwarte, Jena, Germany*
[2]*Max-Planck-Gesellschaft, Arbeitsgruppe "Staub in Sternentstehungsgebieten" Jena, Germany*

Abstract

A retrospective investigation covering about twenty years of wide-band and intermediate-band photometry at Jena University observing station allows an analysis of variations of atmospheric extinction at different time scales. A comparison with an earlier investigation by Wempe (1947) permits to draw conclusions about the evolution of air pollution by aerosols over nearly half a century. Complementary weather observations obtained at the meteorological observing station of the Jena University allow to find correlations between the variation of extinction parameters and the changing weather situations. Parallel Mie calculations allow to interpret the variations of the measured extinction data in terms of size variations of water-rich aerosols.

1. The observations

Since 1968 atmospheric extinction observations have been performed at Großschwabhausen observing station of Jena University Observatory. Photoelectric photometry in three colour UBV, four colour $uvby$, and six colour standard IHW comet filters have been carried out at the 90-cm telescope. The atmospheric extinction coefficients and the zero points for each photometric night were determined from a sample of about 15 standard stars. The aerosol extinction coefficient k_{D} was determined by the correction for standard Rayleigh scattering and ozone absorption.

In general, an extinction law for aerosols of the form

$$k_{\mathrm{D}}(\lambda) = 2.5 \lg e^{-\beta\lambda^{-\alpha}} \; ; \; [\lambda/\mu\mathrm{m}] , \qquad (1)$$

is assumed (α – wavelength exponent, β – turbidity factor).

The figure 1 contains the α-$k_{\mathrm{D}}(V)$-diagram for all observations. We see that most of the observations follow one general law: α decreases with increasing $k_{\mathrm{D}}(V)$. All α-values from nearly pure Rayleigh scattering up to completely grey light scattering occur. The dashed line represents the best fit for the data given by the exponential function $\alpha = 4 \exp(-2.18 k_{\mathrm{D}}(V))$. The observations at different time intervals are marked by different symbols. Full triangles: observations before 1970, open circles: observations 1975 – 1984, full squares: observations since 1985.

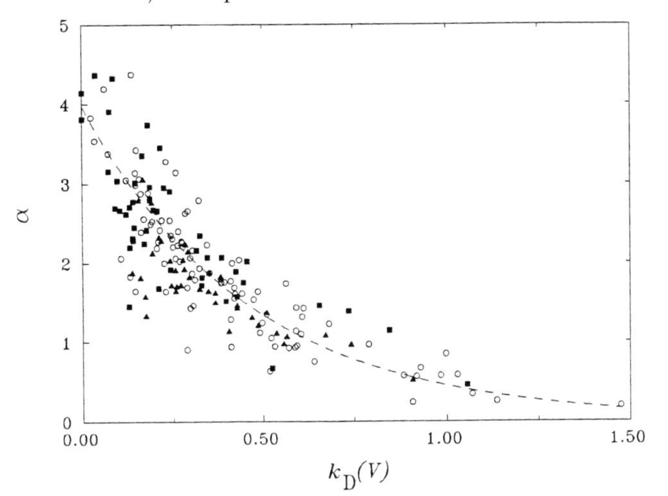

Figure 1 The α versus $k_D(V)$ diagram.

2. Simulation by Mie theory

By means of Mie calculations we have tried to simulate the observed aerosol extinction. To characterize the models of typical aerosol distributions we used standard parameters as given by Koepke and Hess (1988) and optical constants from Twitty and Weinmann(1971) and Volz (1973). We have found that we can reproduce the full range of α-values only by assuming that most of the extinction is produced by a water-soluble aerosol component characterized by a water-like refractive index and low absorption. Then, the different α-values may be produced by the same material but different average aerosol sizes.

In figure 2 we demonstrate the effect of particle growth. We have plotted the evolution of a distribution of water-soluble particles within the $\alpha - k_D(V)$ diagram. Two limiting cases are considered: number conservation according to a process of aerosol growth by condensation from the gas and mass conservation representative for a growth driven by aggregation of the aerosols. The lines mark the way of growth connecting the points of particles with neighbouring mean radii.

A comparison of this diagram with the $\alpha - k_D(V)$ plot of the observations shows that the number conserving process reproduces the set of observations very well. Therefore, the different points of the observations may be caused by aerosols which have started with a similar value of the number density but have grown to different sizes due to their different history.

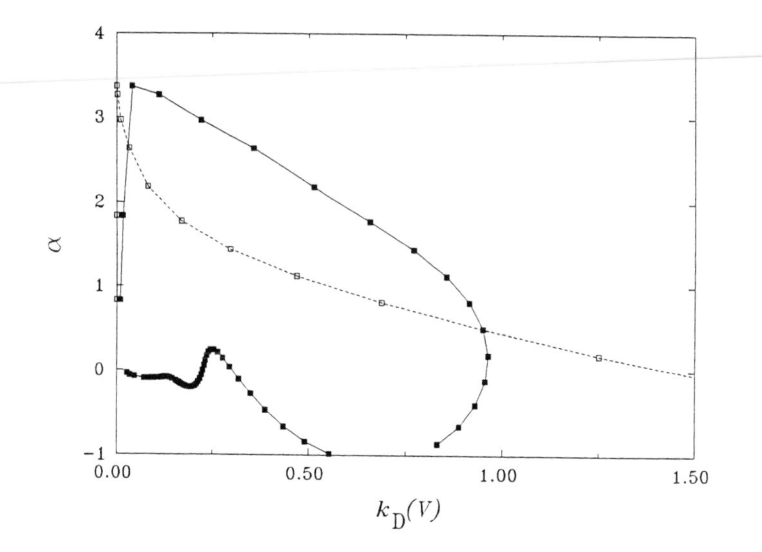

Figure 2 Wavelength exponent α versus absolute aerosol extinction $k_{\rm D}(V)$. Evolution of water-soluble aerosols under mass conservation (filled squares) and under number conservation (open squares).

3. Correlation with meteorological conditions

Parallel weather observations at the meteorological station of the Jena University allowed us to find correlations between the variation of extinction parameters and the weather conditions.

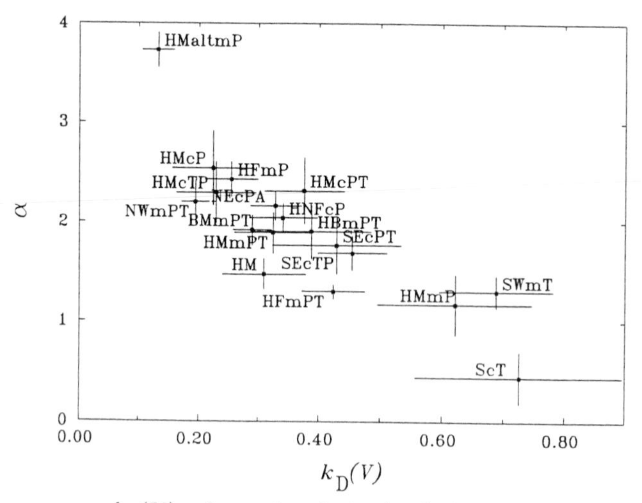

Figure 3 Mean α versus $k_{\rm D}(V)$ values related via circulation types and air masses. Error bars are given.

In the figure we present the mean α versus $k_{\rm D}(V)$ values for given meterological conditions. The meteorological conditions were classified by the atmospheric circulation type (see Heyer 1988) and the air-mass (see Critchfield 1960). The first capitalized letters of the abbreviations characterize the circulation type and the following letters denote the air-mass.

It is significant that there exist at least three different groups in the arrangement of the data points following the same tendency of decreasing α with increasing $k_{\rm D}(V)$. Although it is not possible to find a simple relation between extinction and meteorological conditions we found two general tendencies: Air-masses which come directly from higher latitudes (letter symbol on second position "P") contain smaller particles then those directly transported from lower latitudes (letter symbol on second position "T"). Best observing conditions occur in air-masses carrying low air-moisture (cTP, cP, altmP) while more humid maritime air-masses produce high extinction and a low wavelength exponent α.

4. Nightly variations of the extinction

Nightly time dependent extinction changes have been known for some times and are one of the main sources of errors in all-sky photometry (Angione 1984, Stickland et al. 1987).

In case of intermediate band and narrow band photometry where the deviations from a monochromatic behaviour are small and under the assumption that the extinction varies slowly and isotropically in the whole azimuth angle considered we replace the classical Bouguer line by

$$m_{\rm s} - m_{\rm obs} = L(\lambda_\circ) - k(\lambda_\circ)X + D(\lambda_\circ)t\,. \tag{2}$$

$m_{\rm s}$ means the standard magnitude of the star, $m_{\rm obs}$ the ground-based measurement at zenith distance z, $L(\lambda_\circ)$ the instrumental zero point at some photometric passband characterized by its mean wavelength $\lambda_\circ$ and $D(\lambda_\circ)$ represents a drift coefficient of extinction with time t (Reimann et al. 1989). A statistical analyse shows that the visual drift coefficient can reach relatively high values whereas the drift coefficients in the colour indices remain generally low.

To get a plausible explanation for the observed drift coefficients in agreement with the results from the Mie calculations we related $D(y)$ to the absolute humidity of the air at the observing site in terms of the meteorological conditions (see figure 4). It seems significant that there exist at least two groups of different mechanism causing variable extinction.

Within the first group exhibiting characteristic low $D(y)$ values there exists a clear dependence that the drift coefficient grows with increasing moisture. This indicates that in this group the main process which is responsible for the nightly drift is the shrinking or swelling of the aerosols due to the water vapour content of the atmosphere at the observing site.

Separated from the first there exist a second group of meteorological conditions connected with high $D(y)$ values which does not seem to show any trend with moisture. In these cases the high values of the drift coefficient are caused by the exchange of different air-masses in higher layers over the observing site in a relatively short time interval.

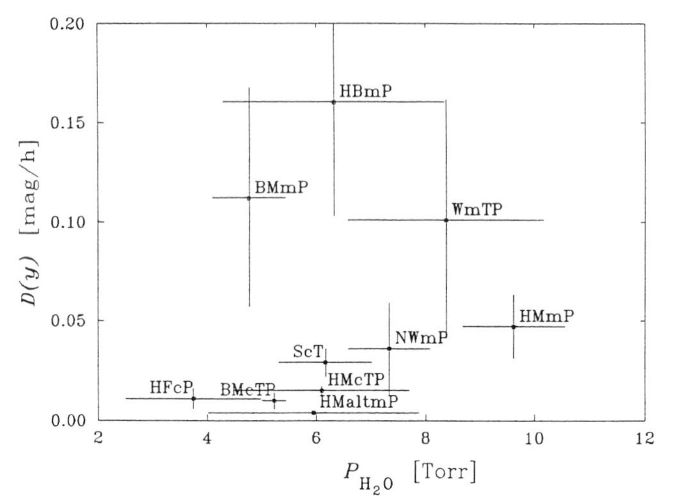

Figure 4 The drift coefficient versus water-vapour pressure diagram. Each point represents the mean value for the specified meteorological condition. Error bars are given.

5. Longtime variations of extinction

There is no doubt about the fact that the state of the atmosphere as well as its changes is influenced by anthropogeny activity contaminating the air with strange kinds of gaseous or dusty aerosols. But also natural catastrophes as large volcanic eruptions could induce long-time perturbations of the normal distribution of stratospheric aerosols. We found in our data a considerable increase of the visual extinction after the eruption of El Chichon in 1982 with the maximum in 1983. Taking the time interval since 1989 with definite lower extinction values as baseline interval characteristic for an unperturbed stratosphere we derived a maximum extinction excess caused by the El Chichon eruption of about 0.1 mag. This is in good agreement with estimated excess values for the sites of Gornergrat and Jungfraujoch between 0.06 and 0.1 mag (Rufener 1986). Other important volcanic eruptions as e.g. the Mount St. Helens event are not traceable in our data.

Beside the normalization of the stratospheric aerosol content since the eruption of the El Chichon there seems to be a second long-time trend present in our data. We found that the mean values of the measurements from the earliest to the latest observations has continuously shifted towards higher α-values (see figure 1). This small growth of the mean α values since the classic measurements by Wempe 1947 is probably a result of a slow change of the general chemical constituents of the anthropogenic induced aerosols.

6. Conclusions

The analysis of the visual drift coefficient $D(y)$ shows that the main part of variable nights is caused by swelling or shrinking aerosols due to the changing water vapour content in connection with certain meteorological conditions. As we have shown this is not only restricted to the lower tropospheric layers and, therefore, also conclusive for mountain observing stations. Even during clear weather conditions qualified as photometric we highly recommend frequent observations of standard stars to detect nightly variable extinction using Eq. (2).

An improved relation between meteorological conditions and extinction parameters as we have found could have predictive value for a better organization of different types of photometric observations and the reverse approach – the extinction correction of remote sensing images.

A theoretical explanation of the observed extinction data is possible when one assumes that the main part of the aerosol extinction is contributed by an aerosol with optical constants typical for the water-soluble type. Almost the full range of the observed points in the $\alpha - k_{\rm D}(V)$ diagram can be reproduced with nearly the same number density of scatterers but different sizes. This indicates condensation as the main process of aerosol growth.

Several analyses of volcanic produced aerosols (e.g. Cardelli & Ackerman 1983) show that dust and ash aerosols are considerably larger than aerosols associated with gas-to-particle conversion. In analogy, we would interpret the long-time increase in the α-values as a historical shift of the composition of the antropogeny atmospheric pollutants from the more dust/soot-like types to the gaseous types producing aerosols by gas-to-particle conversion. This requires, of course, further confirmation by direct aerosol measurements.

References:

Angione R.J., 1984, in: Proceedings of the workshop on improvements to photometry, NASA Conference Publication 2350, 1

Cardelli J.A., Ackerman T.P., 1983, PASP 95, 451

Critchfield H.J., 1960, General Climatology, Prentice-Hall, Inc., Englewood Cliffs, N.J.

Heyer E., 1988, Witterung und Klima: eine allgemeine Klimatologie, Teubner Verlagsgesellschaft, Leipzig

Koepke P., Hess M., 1988, Applied Optics 27, 2422

Reimann H.-G., Böhm M., Pfau W., 1989, Astron. Nachr. 310, 41

Rufener F., 1986, A&A 165, 275

Stickland D.J., Lloyd C., Pike C.D., Walker N.E., 1987, The Observatory 107, 74

Twitty J.J., Weinman J.A., 1971, J. Appl. Meteor. 10, 725

Volz F., 1973, Appl. Optics 12, 564

Wempe J., 1947, Astron. Nachr. 275, 1

Discussion

A. T. Young: *The relation between humidity and aerosol optical properties has been very extensively investigated in the geophysics literature (see the review of G. Hamel, Adv. in Geophys.* **19**, *73 1976). The physical mechanism is understood: above 70% relative humidity, salt particles dissolve and form large droplets, between 30% and 70% R.H., the biosols swell and shrink; there is little effect below 30%. It is the relative rather than the absolute humidity that is important. This model has been used for 20 years or more in the AFGL atmosphere models LOWTRAN, MODTRAN, etc.*

Reimann: The relative humidity of air-masses differs strongly between summer and winter months and therefore a general relation between drift coefficient and meteorological conditions via the relative humidity cannot be easily found.

B.Nicolet: *A simple method to monitor the extinction coefficient* k_λ *has been used for* $\sim$*30 years, (Rufener). Two extinction stars, one ascending and one descending, are observed several times a night to deduce extra-atmospheric magnitudes and* k_λ*. Even during the best photometric nights, variations in* k_λ *are observed; they mimic a phenomenon incorrectly interpreted as anisotropic extinction.*

Reimann: We followed another method to detect time dependant extinction. We frequently observed a greater number of standard stars, carefully selected according to colour and air mass, in a solid azimuth angle. We have done this under the assumption that the extinction is isotropic.

On Improving IR Photometric Passbands

A.T. Young[1], E.F. Milone[2], C.R. Stagg[2]

[1]*ESO, Garching, Germany*
[2]*RAO, U. of Calgary, Canada*

Abstract

The passbands of the Johnson JHKL broadband photometric system used at a number of major observatories have been compared to the atmospheric window transmissions calculated by MODTRAN, and a family of solar-composition model stellar fluxes from Kurucz (1991 private communication) have been used as input to model the atmospheric extinction under different water vapor content, altitude, and airmass conditions. A figure of merit related to the slope of the extinction curve at zero airmass describes the sensitivity of the response function to atmospheric extinction. We have compared passbands used at several observatories, and have designed an improved set of passbands.

1. Introduction

This paper is a progress report on work undertaken by a Working Group on infrared extinction and standardization of IAU Commission 25, to carry out the recommendations from a two-session meeting of Commissions 25 and 9 at the 1988 Baltimore General Assembly. The work has been carried out primarily at the University of Calgary for the past year, beginning in Fall 1991.

Although most astronomers are difficult to classify, essentially two groups of people participated in that meeting: photometrists who have been wary of working or at least publishing in the infrared because of the difficulties, and a bolder group, who forged ahead and did the best that could be done.

Part of the problem is that there is no generally accepted definition of broadband infrared photometric systems. Although Johnson's (1965) system survives in name, no observer that we know of is actually using anything like his passbands, for the following reasons. The extinction line curves upward in the range $0 < M < 1$, i.e., there is a strong Forbes (1842) effect. Those spectral components that have the highest extinction coefficients are absorbed at very low air masses, leaving only the weaker absorptions to appear at higher airmasses. This is much more severe in the infrared than in the visible part of the spectrum (Milone et al. 1993; Young et al. 1993). Hence, there are systematic variations with water vapor content that cause loss of precision and accuracy, which has led to different attempts to improve matters at each observatory where these effects are noticed. Consequently, there has been a proliferation of passbands, and no single, accepted system.

Bessell & Brett (1989), Glass, and Carter, among others, have by systematic and careful attention to details, been able to wring what was possible from the current systems. The general strategy in the past has been to observe at the same dry sites, to reduce the data to standards taken at the same airmass, and to redefine the passbands to better match the transmission windows at particular sites. But this approach has produced site-dependent results that cannot be accurately duplicated elsewhere, particularly at the wetter locations of most telescopes. Here we discuss a new system which we hope will overcome these difficulties.

2. Proposed Global Solution

At the IAU General Assembly joint sessions of Commission 25 and 9 held in Baltimore in 1988, suggestions were put forward which appeared to have strong support among both those present and others who subsequently saw the proceedings (Milone 1989):

1. The passbands should be redesigned to be better centered in the atmospheric windows, and made narrower so as not to be defined by the water vapor absorption bands;

2. The atmosphere should be modeled in real time to better track the extinction variations, thereby partly eliminating the previously unknown variability in the observations due to water vapor variability.

In the simulations carried out to date, we have concentrated on recommendation 1.

In the simulations, we have used 10 stellar atmosphere models from Kurucz (1991) as stellar flux sources, ranging from a 3500 K supergiant to a 35,000 K dwarf, as well as solar and Vega models. For the atmospheric transmission spectra, we have used MODTRAN, a moderate-resolution (1 cm^{-1}) version of LOWTRAN. We have concentrated on the MODTRAN mid-latitude summer and tropical models, which are the wettest ones likely to be met in IR observing, at 1 and 4.2 km (the altitude of Mauna Kea) above sea level. The mid-latitude summer model has about 1.8 cm of precipitable water above 1 km, and the tropical model has about 0.4 cm above 4.2 km. Our extinction-curve model (Young 1989) is

$$m = (a + bM + cM^2)/(1 + dM)$$

where a is the magnitude at $M = 0$, d is effectively $1/M$ at the "corner" of the curve, and b and c may be color-dependent. At $M = 0$, the slope is $b - ad$; at large air masses, m approaches the line $(b/d) + (c/d)M$, plotted as dashed lines in the figures.

Figs. 1 – 5 show simulated extinction curves for two source fluxes. Figs. 1 and 2 are for the original Johnson (1965) K passband. The two atmosphere models differ by a factor of 4.5 in water content, less than the time variations at a single site. If one linearly extrapolates the observable part of the curves outside the atmosphere, the inferred stellar magnitudes vary by more than 0.2 mag, because the water absorption between 0 and 1 airmass is included in the effective passband. Hence, if the water varies during a night, the photometer appears to have an unstable zero-point.

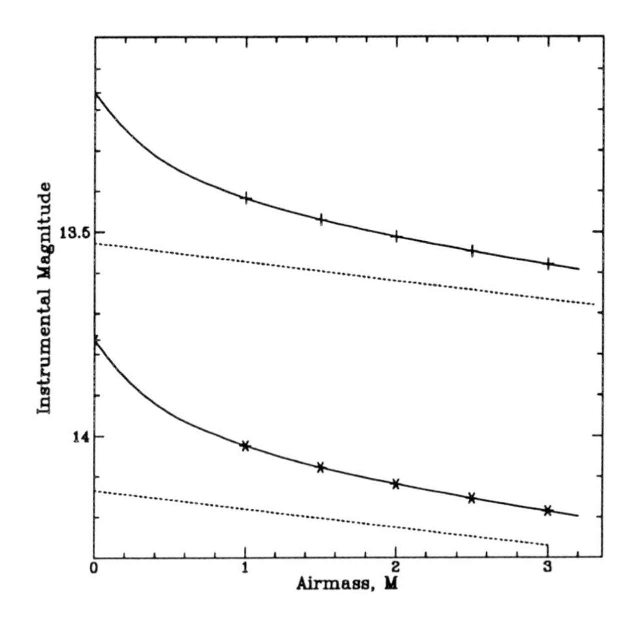

Figure 1 Extinction curves for Johnson's K band at Mauna Kea. The lower curve is for the model atmosphere with 3500 K and log $g = 0$; the upper curve is for T = 5250 K.

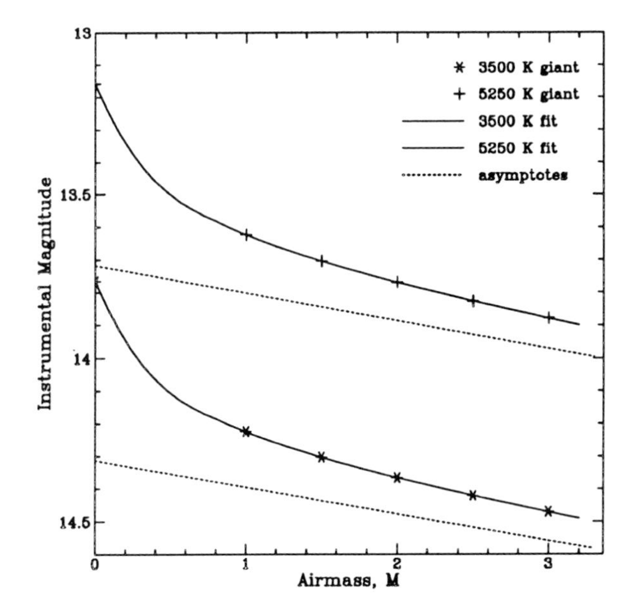

Figure 2 Extinction curves for Johnson's K band at 1 km for midlatitude summer atmosphere.

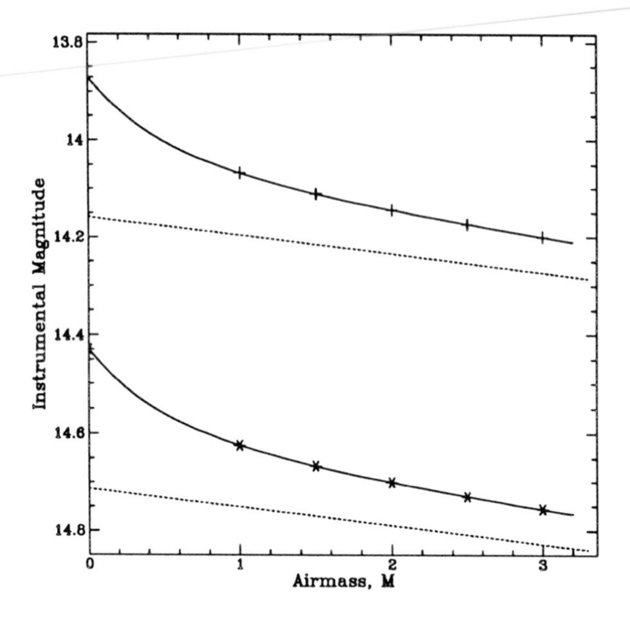

Figure 3 Extinction curves for Wainscoat-Cowie K′ band at Mauna Kea.

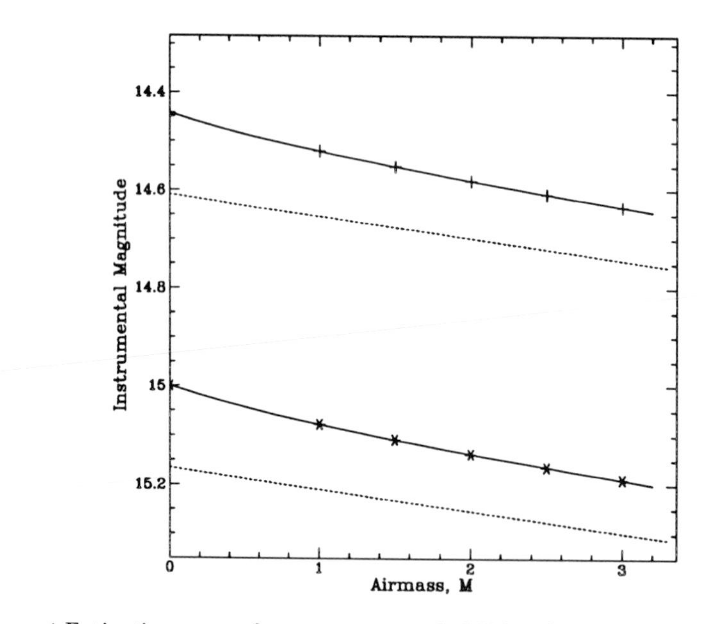

Figure 4 Extinction curves for our recommended K band at 1 km for midlatitude summer atmosphere.

Fig. 3 shows a similar plot for the K′ filter proposed by Wainscoat & Cowie (1992). Again, the extinction line is strongly curved, and similar instabilities in the measurements result from variable water vapor.

Figs. 4 and 5 are extinction plots for our proposed K replacement for the same two atmospheres as Figs. 1 and 2. The extinction curves for the new filters are nearly linear for both the high- and low-altitude sites. The long-wavelength cutoff of our K passband is similar to that of the Wainscoat-Cowie filter, so we should enjoy a similar freedom from thermal noise. These results promise excellent reproducibility and may well augur a new era of milli-magnitude precision for near-IR passbands in the near future.

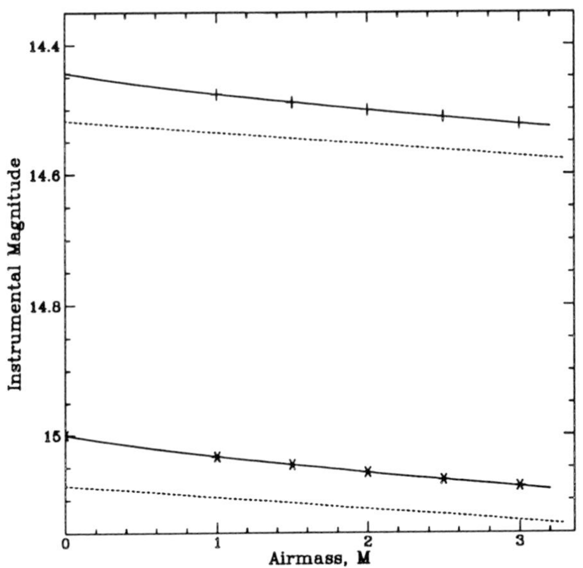

Figure 5 Extinction curves for our recommended K band at Mauna Kea.

The figure of merit that we used to evaluate passbands is the angle θ, by which the stellar spectral irradiance function, regarded as a vector in Hilbert space (Young 1993; Young et al. 1993), is rotated as a consequence of passage through the terrestrial atmosphere. This approach has proven very useful. The θ vs. FWHM relation starts out slowly and then rises rapidly. We try to approach that turning point, within manufacturing tolerances. While decreasing the sky noise, we have kept the filters as broad as possible to maximize flux. Nevertheless, some loss of throughput is inevitable, especially at the higher altitude sites. The benefits are improved transformability and, especially at lower or wetter sites, much improved reproducibility.

We cannot exactly predict expected gains in S/N because of variability in the sky background. We expect substantial improvement from the elimination of portions of older passbands that block source flux and contribute unwanted thermal flux to the detector; but there are non-thermal background sources, such as OH emission from airglow and auroral emission. The airglow may be chopped out in single detector work, but it is not clear what can be done about aurorae, which have spatial and temporal variability on many scales. The airglow is a problem for stare-mode detectors, but by eliminating much of the thermal sky background, we can use longer integrations, so that clear gains in S/N are possible.

The next stage is to test these filters in practice. We expect to obtain a set of filters made to our new specifications in the near future, and should have the first observations early in 1993.

Acknowledgements

We thank Bob Kurucz for providing his model-atmosphere fluxes; Jim Chetwynd at AFGL for recommending MODTRAN instead of LOWTRAN 7; P. Zvengrowski and K. Salkauskas, of the Mathematics Department of the University of Calgary, for helpful discussions with ATY of Hilbert space and functional analysis; Ted Ziajka of the Academic Computer Services of the University of Calgary, for providing ATY with an account on the IBM RS 6000; Barbara McArthur, of the University of Texas, for much assistance with GaussFit; A.W. Harrison and L. Cogger of the Physics and Astronomy Department of the University of Calgary for helpful discussions with EFM on atmospheric emissions; members of the Working Group on Infrared Extinction and Standardization, particularly Ian McLean for commissioning the WG; and M.S. Bessell, T.A. Clark, G. Rieke, and Mike Skrutskie for useful discussions. Roger Heatley of Barr Associates and Mike Larro of OCLI provided much helpful information about interference filters. Sun Kwok assisted with the loan of computer hardware. This project was funded by grants from the University of Calgary Research Grants Committee and the Natural Sciences and Research Council of Canada to EFM, for which we express gratitude.

References:

Bessell, M.S., Brett, J.M., 1989, in: *Infrared Extinction and Standardization*, (Lecture Notes in Physics, Vol. 341) ed. E.F. Milone, Springer, Berlin, p. 61.

Forbes, J.D., 1842, Phil. Trans. **132**, 225.

Johnson H.L., 1965, Astrophys. J. **141**, 923.

Milone E.F. (ed.), 1989, *Infrared Extinction and Standardization*, (Lecture Notes in Physics, Vol. 341) ed. E.F. Milone, Springer-Verlag, Berlin.

Milone E.F., Stagg C.R., Young A.T., 1993, in: *Workshop on Robotic Observatories, Kilkenny, July 1992*, eds. B.P. Hine, M.F. Bode, Ellis Horwood.

Wainscoat, R.J., and Cowie, L.L., 1992, Astr. J. **103**, 332.

Young A.T., 1989, *Infrared Extinction and Standardization*, (Lecture Notes in Physics, Vol. 341) ed. E.F. Milone, Springer-Verlag, Berlin, p. 6.

Young A.T., 1993, in these proceedings.

Young A.T., Milone E.F., Stagg C.R., 1993, in preparation.

Discussion

V. Straizys: *Let me congratulate your working group for making such good order with the infrared photometric systems. It would be nice if a similar order could be made in the visible and the near ultraviolet spectrum.*

Milone: Thank you for the encouragement. There is, of course, less tradition in the IR, and we are most fortunate in getting strong support from a number of people who have done a bit of standardization work with existing systems. In the V-band, a reformation is more difficult to accomplish, but Chris Sterken is working on pairs of filters in the old UBVRI passbands, following suggestions by Andy Young.

R.F. Wing: *you spoke about the filters, but not the detectors. If the filters are wide, the detector plays a role in defining the system. I am worried about the proposed z filter because most detectors which can be used in the one-micron region have a response that either increases or decreases steeply with wavelength. If the z filter is as wide as you propose, observers with different types of detector will be working on different systems. Why not minimize this problem by defining a narrower filter from the outset?*

Milone: Your comment is well taken. There are detectors which are sensitive in this window but probably are not flat here. Most IR people want maximum throughput so one of our constraints has been to maximize this while not sacrificing our primary purpose. We will need to check the effects on transformations of convolutions of our passband with various detectors to see what we can tolerate. Narrowing the filter is certainly one way of responding to it, should the problem appear to be very serious. Keep in touch!

Ian Elliott congratulates Russ Genet on his birthday

Session 5

Global Networks

GNAT -- A Global Network of Automatic Telescopes

D.L. Crawford

Kitt Peak National Observatory, PO Box 26732, Tucson AZ 85726 U.S.A.

Abstract

All of us "have-nots" need more telescope time, for ourselves and for our students. There are also many programs where a global linkage is needed to accomplish the objectives. In addition, the world needs more and better science education; astronomy can be a leader if it has adequate facilities to do so. A global network of automatic telescopes can help supply these needs, which are global, spanning all countries. A new non-profit organization (GNAT, Inc.) has been formed to be the catalyst to develop and implement such a concept. We hope that many astronomers and organizations will become "members" of GNAT.

I. Introduction

There is an increasing amount of activity that could be classified as "global networking," including many of the reports at this meeting as well as at some of the earlier meetings devoted to automatic telescopes or photometry. People are already doing it. I won't go into any details here, except about the one I am most involved with, naturally. Let me just mention one or two (discussed in papers at this meeting) as examples: The Whole Earth Telescope program, the French-Morocco connection, the various supernovae search programs, several special variable star consortia (for a one shot deal or for longer activity), and the various APTS programs at Mt. Hopkins, including the link to Mt. Wilson. All of these (as well as just thinking about the issues) show that the topic is of increasing interest and viability. Let me again review the nature of the beast and describe a specific proposal to you (A Global Network of Astronomical Telescopes, GNAT, Inc.). I would like to challenge you to join GNAT.

II. Why a GNAT?

1. Research needs. All of us "have-nots" need more telescope time for our research. Besides the general there are special programs that would benefit greatly by a GNAT:

---

1. Operated by AURA, Inc. under cooperative agreement with the National Science Foundation.

long term coverage, monitoring, variable stars with periods near one day, variables needing continuous coverage, etc. There is no way for all the present and future astronomers to have enough observing time unless we find a way to increase significantly the number of telescopes available, and the only way to afford that is to do it with "small" telescopes. Small Science is still frontier science in astronomy.

2. Education needs. The world needs more and better science education, and astronomy can be a leader in helping to provide it, with real data available from and produced with a GNAT. GNAT can also be a source of many other materials, such as tutorials, data bases, and astronomical images.

3. These research and educational needs are global, spanning all countries and all peoples. Anyone, anywhere is a potential user.

4. It can be done at relatively low cost; new generation small telescopes are not astronomically expensive, and they provide a balanced complement to larger telescopes and space facilities. A GNAT can provide such facilities to many at a cost lower than everyone doing it themselves.

II. What and Where?

1. The first essential item is a computer "bulletin board system (BBS)," accessible to all. It will include a list of who is doing what with what, network contacts, many data bases, tutorials, digital images, and other valuable information. Structurally, it will be like many existing BBS's of all kinds, and like many of the resources available through Internet. It can and will be multi-lingual.

2. In addition, GNAT will have a complement of small, automatic telescopes at a number of world-wide sites. Other telescopes can and will be linked to the network, of course. These GNAT telescopes will be truly new-generation instruments, but they will be relatively low cost and will be essentially risk free to obtain and to put into operation. GNAT will not plan in the foreseeable future to offer real-time remote operation, but it will be able to produce high quality data for many people in a cost effective and timely manner.

3. A "Homebase" is required, to coordinate the operation, serve as a location for the BBS system, and as a resource of information to all in the network. There can and will be a number of regional or local nodes as well.

III. Who?

A new tax-exempt non-profit organization has just been incorporated to be the catalyst to develop and implement GNAT. "All of us" who can use and need information and observing time on small but powerful telescopes should be a part of GNAT. GNAT will be a network . We are all "have-nots." GNAT will not be a dictator, but a catalyst, complementary to all other programs and organizations. We all should become members of GNAT. Think of it as an "electronic scientific institution," with many un-paid, part-time members, world-wide. Currently, membership is free, though donations are most welcome. Just join up.

IV. How?

One needs funding, of course, and a Business Plan and funding proposals are being prepared by GNAT. Any organization or individual interested in supporting cost effective research and/or science education, especially on a global basis, are potential sources of such funding. Funding need not all come from one source or one country, of course. Your input and suggestions are most welcome.

V. When?

As soon as funding is found. There are essentially no problems with obtaining the hardware, software, or personnel necessary to develop and operate GNAT. It is should be truly a no-risk venture. A regular newsletter will begin shortly.

VI. A Few Other Aspects of Note:

1. GNAT should be a great help in insuring homogeneous and accurate data for all. All of us should benefit by uniform standards, systems, hardware, and software. Consider the gain to be made by the mass procurement of quality photometric filters.

2. The "80/20" law. GNAT will not normally be at the cutting edge of hardware or software technology development. It will have as a goal getting 80 percent or more of the performance of the frontier instruments at 20 percent or less of the cost.

3. The "1 +1 > 2" law. By working together, we can achieve much more than we can by doing it all in isolation, including achieving significantly lower over all costs and better quality facilities (such as matched filters and good detectors).

4. Balance: GNAT's small telescopes (quality ones) and facilities can and will complement large telescopes, space astronomy, radio astronomy (in fact, any multi-wavelength efforts), and university research everywhere. It is a viable extension of university research and education, for all.

5. A key item is "Value Per Cost," which should be maximized, in many ways, by a viable and effective GNAT. Astro-economics is an important and interesting subject. GNAT will be a full-time practitioner.

VII. GNAT Structure:

The GNAT Bylaws provide for a active Scientific Advisory Committee (SAC), meeting regularly mostly by Email, but occasionally in person. SAC will define the technical needs and advise how to meet them, as well as provide advice on all phases of the operation. SAC can and must be in close touch with active and potential users. The Bylaws provide also for "Members." I hope that all of you will want to become Members.

VIII. The Challenge:

There is no question in my mind of the need for a GNAT and of the fact that it can be done. However, there must be a clear cut demand shown for it. We must create that demand. Without it, there is little hope to get it funded in a timely manner. With it, I am convinced that we can get it funded. How to show the demand: Sign up now as a member. Write, let us know your thoughts about it and your need for it. Tell us how it seems attractive to you, why *you* need it.

IX. Summary:

GNAT is a viable and needed entity and concept. Join up!! Input advice at any time. Help us bring it to a reality, soon.

Note: This concept and proposal is not connected to NOAO or KPNO. It is strictly a creation of those individuals currently involved, a few of whom are staff astronomers at NOAO.

References:

Genet, R.M., and Hayes, D.S., Robotic Observatories, Autoscope Corp., Mesa AZ U.S.A., 1989.

Crawford. D.L., New Generation Small Telescopes, in New Generation Small Telescopes, Fairborn Press, Mesa AZ, U.S.A., pp.7-18, 1987.

Genet, R.M., Hayes, D.S., Geltmacher, H.E., and Crawford, D.L., Remote-Access Personal-Computer Astronomy, in Automatic Small Telescopes, pp. 1-10, 1989.

Crawford, D.L., Genet, R.M., and Hayes, D.S., A Global Network of Automatic Telescopes, in Automatic Small Telescopes, pp. 115-124, 1989.

Crawford, D.L., New Generation Small Telescopes: A Summary, in Automatic Small Telescopes, pp. 155-159, 1989.

Discussion

J. Baruch: *Already a couple of working groups have been started at Kilkenny and people are welcome to join. The objective of one of the groups - an interfacing and data structure working group is initially twofold: first, to agree where defined interfaces and datastructures would have more advantages than disadvantages, second, to pursue such agreement to show potential funding agencies that global agreement is possible. Please send your email address, or name and postal address, and offers to participate to John Baruch JEFB@UK.AC.BRADFORD Dept of Electrical Engineering, University of Bradford, UK.*

C. J. Butler: *At the Kilkenny meeting it was proposed that there should be a trial run which would enable observers to test their techniques and their ability to communicate. A forthcoming campaign in the MUSICOS series could fulfill this role and add to the astrophysical importance of the data.*

W. Tobin: *Does the MUSICOS programme have any connection with the UNESCO-sponsored World Astronomy Days being promoted as part of International Space Year?*

C. J. Butler: The MUSICOS programme will take place in early December, but I cannot tell you if this coincides with a World Astronomy Day.

A.R. Upgren: *I want to emphasize the extension of GNAT to existing telescopes for several reasons. As a sometime astrometrist, photometrist and spectroscopist I work on problems of stars that need observations in all of these fields. I see more contact needed between the positional and the photometric researchers. Furthermore membership in a worldwide network could buttress an observatory against the occasional attempt to close it down, and also act as a spur to the use of presently idle research-capable telescopes at developed sites and the securing of funds for this purpose. Telescopes in campuses and in cities have one advantage; they are accessible to large populations. Retention or resurrection of each of these greatly increases the educational element in GNAT's mix of goals.*

Crawford: Yes I think a viable GNAT can and should handle these needs.

J. B. Hearnshaw *The prime motivation for GNAT appears to be to increase the* **quantity** *of observational data. But does not progress in astrophysics at the present time depend far more on data* **quality** *and on its careful* **interpretation***?*

Crawford I think GNAT will offer both quality and quantity. We get it all. One will get excellent data whether we want it or not, because of better filters, standards, software, etc.

C.L. Sterken *I only would like to add that in variable star photometry, the only objective way to estimate the precision of your measurement is to compare results from multisite observations. (see Michel Breger's talk in this meeting). If GNAT would contribute to the quality aspect only it would already pay off. The increase in quantity of data is then a bonus.*

B. Anthony-Twarog *Can you share a larger list of the projected working groups? As a marginal 'have not' where one of the limiting quantities is time, many prospective contributors might be concerned about making a future, open-ended commitment of time and might* *like to carefully plan where they can best plan their contributed effort.*

Crawford Examples are, (there are others, and many subsets, of course):
1. Search for Extra Solar System Planets
2. Open Cluster Data Base
3. Variable Stars (many subsets)
4. Telescope
5. Filters
6. Systems
7. Communication
8. Software
9. BBS
10. Education
11. Summer Student Programmes
12. Post-Docs.
13. Optics
14. Sites
15. Meetings
16. Fund raising
17. Administration
18. Membership
19. Newsletter
20. Standard Stars
21. Databases and archiving
22. Fibres
23. Etc

R.M. Genet *The cost of a 2.5-metre system, as suggested earlier, is between US$2.5m and US$3.5m. This includes the telescope, several permanently mounted instruments, and the enclosure and its control.*

Crawford We must always be sure, in computing costs, that we compare 'apples and apples'. That is, include and exclude items and keep things comparable.

Progress with the Whole Earth Telescope

D. O'Donoghue[1] and J. Provencal[2]

[1] *Department of Astronomy, University of Cape Town, South Africa,* [2] *Department of Astronomy, University of Texas at Austin, USA*

Abstract

A summary of the results from seven global runs using the Whole Earth Telescope is presented, together with an evaluation of the scientific results obtained to date. Factors such as the distance of the target star from the equator, the nature and timescales of its intrinsic variability, etc. are shown to affect the value and quality of the results, as well as the traditional factors such as brightness and long-term coherent behaviour. Experience with the network shows that, taken as a whole, it enjoys far better weather than any one of its sites, and provides unprecedented 'uncluttered' resolution of time-series power spectra.

1. Introduction

The Whole Earth Telescope (WET) is a collaboration between astronomers from about a dozen nations who obtain, twice yearly on average, observing time on 1-2m class telescopes at about a dozen observatories in both hemispheres. The longitude distribution of these sites enables the observation of variable stars continuously, without any gaps in the data, for intervals of ~1-2 weeks. The scientific motivation for this capability is the elimination of the aliasing problem inherent in data collected at a single site. Aliasing confuses the interpretation of amplitude spectra of variable stars: instead of a single peak in the amplitude spectrum for each real frequency at which the star is varying, there is a retinue of peaks, as illustrated in Fig. 1 of Nather et al. (1990) (hereafter NWCHH). The amplitude spectra of multi-periodic pulsating stars often exhibit many overlapping 'aliasing patterns', and the disentanglement of the real signals from the aliases can be a hopeless task. The only way to overcome this difficulty is to obtain essentially continuous data and for this, a network of observing stations, well-spaced in longitude, is required. This paper describes progress with the WET; other descriptions appear in Nather (1989), NWCHH and Nather (1992).

2. Operating the Whole Earth Telescope

A world map showing WET sites as at 1989 March appears in Fig. 2 of NWCHH. More recent campaigns have added Kitt Peak, ESO, Poland, the Wise Observatory in Israel and Mt. John Observatory in New Zealand, although not all sites operate for every campaign. The minimum observing equipment required is a 2-channel high-speed photometer with offset guiding, and the ability to measure the guide star using the photomultiplier tube on the second channel. The preferred instrumental configuration is a 3-channel, portable photometer constructed by Prof. Ed Nather and collaborators at the University of Texas. The additional channel is used as a continuous monitor of the brightness of the sky background, measurement of which can only be done occasionally with a 2-channel instrument. Portability of the instrument, and its associated data acquisition laptop computer, is essential to allow observers to travel to distant sites without complicated instrumental handling procedures.

Each campaign concentrates on ~2 targets, each of which has an associated 'Principal Investigator' (PI), who has proposed the target for observation and who is expected to lead the analysis of the data. The PI's are present in the WET operations centre in Austin, Texas during the campaign. Observers are either attached to 'home observatories', such as McDonald, or travel to common user facilities, such as Kitt Peak or CTIO. The network is controlled by daily telephonic contact from the WET operations centre in Austin: each observer reports the weather conditions at the site and receives instructions about which target to observe. This 'real-time' operation is particularly needed at longitudes where there is more than one observatory: it ensures that the highest priority target is always observed (if possible), and that the most efficient use of telescope time is made. At the end of each observing session, the observer is expected to send the data just acquired back to Austin, either via email or telephone line using the Kermit data transfer protocol.

Table 1. Whole-Earth Telescope Observing Log

WET Run No.	Targets	Type	Date	Principal Investigator	Status
1	PG1346+082	Int. Bin.	1988 Mar	Winget/	In prep.
		White		Provencal	(Thesis)
	V803 Cen	Dwarfs		O'Donoghue	Published
2	V471 Tau	DA + K2V	1988 Nov	Clemens	Published
	G29-38	DAV		Winget	Published
3	PG1159-035	DOV	1989 Mar	Winget	Published
4	AM CVn	IBWD	1990 Mar	Solheim/	In prep.
				Provencal	(Thesis)
	G117-B15A	DAV		Kepler	Published
5	GD358	DBV	1990 May	Winget	Submitted
	GD165	DAV		Bergeron	In Prep.
	HD166473	Rap.Osc.Ap		Kurtz	Failure
6	GD154	DAV	1991 May	Vauclair	In analysis
	PG1707+427	DOV		Grauer	In analysis
7	1H0857-242	CV	1992 Mar	Buckley	In analysis
	PG1115+158	DBV		Barstow/	In analysis
				Clemens	
	G226-29	DAV		Kepler	In analysis
	WET-0856	Del.Sct.		Breger	In analysis
Forthcoming attraction:					
8	PG2131+066	DOV	1992 Sep	Kawaler	In anticip-
	G185-32	DAV		Moskalik	ation

The data are reduced as soon as they are received in Austin and a 'quick-look' analysis is performed, usually involving the calculation of Fourier Transforms. The PI's use the 'quick-look' analysis to evaluate the scientific return and debate the ongoing prioritization for each target. After the campaign is over, each PI receives a complete set of raw data to be rereduced and analyzed at a more leisurely pace. The PI is also expected to write the paper describing the scientific results obtained. All participants in the campaign are included as co-authors resulting in papers with 30 authors or more (e.g. Winget et al. 1991). Such a large list of co-authors is less than satisfactory from the point of view of many of the co-authors: authorities in their home institutions/countries evaluating their research productivity regard their contribution to such publications as far less significant than they were. From the point of view of the collaborators, the contribution of each member of the WET is vital for the success of the collaboration.

3. Results

3.1 WET History

Table 1 gives an observing record for the WET: most of the objects observed so far are white dwarf or pre-white dwarf pulsators (DAV, DBV, DOV). A few binaries showing oscillations have been observed and the Delta Scuti star observed in 1992 Feb was a serendipitous discovery: it was the 2nd channel/guide star chosen for 1H0857 and was not known to be variable prior to the campaign. However, it quickly revealed its pulsations soon after being observed for the first time. It is apparent that usually an interval of ~12-18 months elapses between the WET run and the production of the paper reporting the results.

3.2 Highlights

Undoubtedly the most spectacular success from the results obtained so far came from the observations of the pre-white dwarf PG1159-035 in 1990 March. These have been fully discussed in Winget et al. 1991 (hereafter W91), but it is worth emphasizing some features of the results which illustrate the power of WET to obtain new science. 264 hr of data were obtained during 12 nights giving over 90% coverage (Fig. 4 of NWCHH). This virtual elimination of aliasing from the power spectrum (Fig. 5 of NWCHH) enabled the number of reliably determined frequencies to increase from ~10, as found in previous studies, to more than 100 (see W91 and references therein). As pointed out by Nather (private communication), an improvement in measurement by a factor of 10 is not merely an improvement in precision, it enables a new 'kind' of science to be done. This potential was fully realized in the study of PG1159-035. Of the 125 frequencies obtained from the power spectrum, 121 were identified with $l=1$ or 2, non-radial g-modes. As described in W91, this identification could be verified using several independent checks, and enabled: (i) the mass to be determined to 3 significant digits; (ii) the rotation period of the star to be measured to similar accuracy; (iii) an upper limit to the magnetic field of 6 kG to be established; (iv) the pulsation and rotation axes to be determined to be collinear. Perhaps the most exciting seismological result from the study was the detection of a chemically-stratified envelope surrounding the stellar core, as expected from theory. This insight was gleaned from the slight departure of the frequencies in the power spectrum of PG1159-035 from the equal-period spacing expected from asymptotic theory applied to chemically- homogeneous envelopes. Clearly, impressive results can be expected from the analysis of WET data for other compact pulsators. Indeed, the analysis of the DBV star GD358 just being completed at the time of writing (September 1992) is bearing out this expectation (Winget et al., in preparation).

4. Factors affecting data quality

As expected for almost all astronomical targets, brightness is a major consideration in determining WET data quality. Fig. 7 of NWCHH provides a vivid illustration of the difference in signal-to-noise provided by a ~4-m compared to a ~1-m telescope for signals at the limit of detection. The inclusion of 4-m telescopes in the WET network is rare. Nevertheless, one of the less obvious advantages of WET is that it typically collects a very large number of photons on high priority targets resulting in much improved signal-to-noise ratios compared to data sets obtained by 1-2 m class telescopes at a single site.

Essential to the goal of ensuring no gaps in the data is a degree of overlap in coverage between sites at different longitudes. This overlap often occurs for large airmasses at both sites. Fig. 6 of NWCHH shows an example of the light curve of PG1159-035 observed by 3 sites simultaneously. The excellent agreement is apparent.

For the study of multi-periodic variable stars, the most important consideration for data quality is the elimination of all gaps in the data set. A number of factors have an important bearing on the viability of this goal, of which the completeness and redundancy of the longitude coverage is perhaps the most important. In the first WET run in 1988 March, there was no site in the network between Siding Spring, Australia and Sutherland, South Africa which resulted in significant 1 cycle per day aliases in the spectral window of even the highest priority target, PG1346+082. Even though the analysis of this star is still in progress, these aliases, combined with the complexity of its power spectrum (as apparent in Fig. 7 of Nather 1989) may defeat the attempt to decipher the frequency structure of this star.

As can be seen from the WET site map in NWCHH (Fig. 2), as updated above, there is significant redundancy for longitudes of Europe/Africa, and N. and S. America. But the longitudes from 60^oE to 120^oW include only 3 sites: Kavalur (India), Siding Spring (Australia) and Mauna Kea (Hawaii). This necessarily means that complete coverage is possible at any one time for only 1 target. Limited coverage of lower priority targets can, however, be useful: just three 2-hr light curves from Australia were sufficient to resolve the aliasing ambiguity in the 90-min photometric period of CP Pup during the first WET run (O'Donoghue et al. 1989). Nevertheless, such limited coverage is unlikely to produce 'ground-breaking' science of the kind demonstrated in the study of PG1159-035.

Another important consideration is the declination of the target. Obtaining complete coverage for objects far from the celestial equator is difficult because they are accessible only from sites in the corresponding hemisphere. This is especially true for southern objects because of the more restricted spread in longitude of southern hemisphere sites.

Finally, experience from the 7 WET runs to date has shown that the redundancy in longitude in the network has prevented bad weather from spoiling the coverage obtained for the highest priority target. WET weather is better than weather at any one site!

5. Resources required to run the WET

5.1 Manpower

There is a large administrative overhead in running the WET. Plans for each campaign have to be made well over a year before the campaign. Once the targets proposed by the PI's have been accepted, the PI's are expected to provide a 'generic' observing proposal to be supplied to each member of the WET collaboration who is going to apply for observing time. Members of the Texas group co-ordinate proposals for most of the common user observatories (e.g. La Palma, CTIO, Kitt Peak); other collaborators apply for time to their 'home' observatories. These applications have to be sent in before a very wide variety of deadlines. For example, observing time for the WET campaign of 1992 September had to be requested by 1991 Oct 15 for ESO and 1992 July 24 at Siding Spring, a difference of 9 months.

During the WET campaign, in addition to the PI's, the WET control centre in Austin is manned by at least 6 people (usually more) who work 8-hr shifts. These 'workers' phone each site daily to receive weather reports and assign targets, reduce the data that are sent back from each site, and perform the 'quick-look' analysis.

Each observing site is manned by at least one, and often two observers. Again, members of the Texas group usually travel to the common-user observatories accompanied by the 3-channel portable photometers mentioned above. The remaining observers are provided by members of the WET collaboration resident near the other observatories, using either 3-channel or 2-channel photometers obtained from the Texas group, or equivalent devices constructed at their home institutions. An important task at the end of each observing session is to communicate the data back to Austin for the 'quick-look' analysis. Not all sites enjoy this capability.

5.2 Money

The cost of running WET is very modest, both when considered against the scientific return, and when compared with other comparable networks (e.g. GONG). Perhaps the simplest way of reporting costs is to list the impact on the Texas NSF WET grant of each WET run. The cost of travel and maintenance for the two PI's to be present in Austin ($3500), 1 observer at Siding Spring ($3000), 2 observers at La Palma ($5500), 1 observer in Hawaii ($2300), 1 observer at CTIO ($2500) and 2 observers on two telescopes at McDonald ($2200) totals $19000. This accounting does not include the costs to collaborators at their 'home' observatories: no estimates of these costs are available but they are likely to be significantly less than the above. Finally, it the goal of WET to equip each site with identical instrumentation. The cost of a 3-channel, portable photometer from Texas is $20000. Funding is being sought to provide such an instrument to those currently lacking it.

6. Conclusion

The WET is a working global network of telescopes of modest aperture which operates twice per year on average. From the 7 campaigns that have been conducted, 5 papers in refereed journals have been published along with some progress reports in conference proceedings. Analyses of other data sets are at an advanced stage and should soon be in preprint form. Although administratively onerous and manpower intensive, the WET network requires very modest financing. The study of PG1159-035 has demonstrated that WET is capable of increasing the number of frequencies measurable in the power spectra of compact pulsators from ~10 to more than 100. With the exception of the Sun, this improvement in high-resolution power spectroscopy of variable stars is unprecedented. The resulting asteroseismological capability has opened up new horizons in the exploration of the interiors of pulsating stars.

Acknowledgements

We gratefully acknowledge the contributions of all members of the WET collaboration, especially Prof. R.E. Nather who provided useful ideas and information for this paper.

References:

Nather, R.E., 1989, in Int. Astr. Un. Coll. 114: White Dwarfs, ed. G. Wegner, Springer Verlag, Berlin, Heidelberg and New York, p109.

Nather, R.E., 1992, in Proceedings of the 8th Europoean Workshop on White Dwarfs, ed. M. Barstow, Kluwer, Dordrecht, in press.

Nather, R.E., Winget, D.E., Clemens, J.C., Hansen, C.J., Hine, B.P., 1990, Astrophys. J., 361, 309 (NWCHH).

O'Donoghue, D., Warner, B., Wargau, W., Grauer, A.D., 1989, Mon. Not. R. astr. Soc., 240, 41.

Winget, D.E., Nather, R.E., Clemens, J.C., Provencal, J., Kleinman, S.J., Bradley, P.A., Wood, M.A., Claver, C.F., Frueh, M.L., Grauer, A.D., Hine, B.P., Hansen, C.J., Fontaine, G., Achilleos, N., Wickramasinghe, D.T., Marar, T.M.K., Seetha, S., Ashoka, B.N., O'Donoghue, D., Warner, B., Kurtz, D.W., Buckley, D.A., Brickhill, J., Vauclair, G., Dolez, N., Chevreton, M., Barstow, M.A., Solheim, J.E., Kanaan, A., Kepler, S.O., Henry G.W., Kawaler, S.D., 1991, Astrophys. J., 378, 326 (W91).

Discussion

C.L. Sterken: *You analyse the data, and publish the scientific results of your analysis, but what do you do with the data? Do you publish these data* in extenso *or do you submit to the data archives?*

O'Donoghue: The data enter the public domain after 1-2 years. Manpower shortage prevents the data from being sent to an archive like Strasbourg.

J. Baruch: *The World Telescope is a trivial result of the GNAT programme. Many continuous observing programmes will be viable. GNAT will also be suitable for observing the active galactic nuclei and searching for chaotic behaviour.*

O'Donoghue: When do you expect to get data?

M. Breger: *The problem is that some observers need publications in which they are first authors. In the Delta Scuti Network the problem was addressed by observers publishing their own data in their national publications. An example can be found in China. The main paper with the astrophysical analyses is still published in the major journal with everybody as co-author.*

R.M. Genet: *A long list of co-authors that are observers can be a problem. In the main, robotic telescopes do not require robotic co-authorship. Presumably this would change if robots took over the world.*

W. Tobin: *Does WET have enough redundancy? Doesn't having a second priority target complicate administration by requiring global real-time decisions as to which star to observe, and also degrade the window function you finally obtain?*

O'Donoghue: No a compromise on achieving maximum coverage on the highest priority target is made. The first WET campaigns have gaps in longitude coverage.

D.L. Crawford: *Those interested in these issues should also be aware of the excellent work done by the Global Oscillation Network Group (GONG) for solar research and its potential relative to networked astronomy.*

Globalizing Observations: Prospects and Practicalities

E. Budding

Carter Observatory, Wellington, New Zealand

Abstract

To promote fuller awareness we should aim to globalize observations. In a more immediate sense, this addresses the planetary distribution of observing facilities and their access. What components in such facilities occur in 'global networks'? What factors optimize the growth and scientific viability of such a network? What kinds of progress can be expected?

1. Introduction

Astronomers sometimes consider, or encounter, general strategies for improving observational data-acquisition, given built-in limitations of location, diurnal cycle, weather, human capabilities and associated logistics, as well as other factors of the present environment. Prominent among these are the rapidly growing electronics and telecommunications 'revolution', and greater production of manufactured optical and mechanical components of high quality at lower real cost for standard items. The present paper outlines some thinking points — mainly in the area of networking observational (photometric) facilities, though other aspects arise. A fuller version will appear elsewhere (Budding, 1992).

Many related issues were discussed at the Strasbourg Conference on *Coordination of Observational Projects* (eds. C. Jaschek and C. Sterken, 1988), as well as *Multiwavelength Astrophysics* (ed. F. Córdova, 1988). Also bearing on this article are a group of papers appearing in *Automatic Small Telescopes* (eds. Hayes and Genet, 1988), and redeveloped at the JCM on automated telescopes at the IAU General Assembly of 1991 (ed. S. Adelman).

2. Global networks and questions they raise

Numerous global networks, of different types and degrees of complexity exist, eg. the world post-mailing system, meteorological services or the organization of public lending libraries. The recent growth of *computer networks* is ushering in a highly relevant new era of effective communicability (Benn, 1990). Remote logging into a computer or database becomes a standard operation, and from that remote control is foreseen, particularly with automatic (photometric) telescopes (APTs). These represent one of the simplest examples, with telescope and photometer controlled by a computer, which can be remotely accessed, perhaps by a modem and telephone line. But how are things organized?

Demand, supply and efficiency

We should consider first the demands promoting observational networks. Undoubtedly, more data is a primary demand. APTs demonstrate the potential for higher levels of data — not just more, or 'ungapped' data-sets, but data of a fundamentally higher quality (Young *et al.*, 1991). Another key reason, mentioned by Sterken (1988), is basic attestability of observations. If two (or more) facilities independently record the same events then a vital element of proof is added to the result. Of course, this depends on appropriate quality control of separate facilities purporting to observe the same, or similar, objects. Crawford *et al.* (1988) have also commented on educative and developmental functions of robotic observatories.

We can model 'quality' scientific output as a function of the aperture size of used telescopes. Let us write

$$q(r) = f n_{ag}(r) r \sqrt{t(r)}, \tag{1}$$

where q is the rate of scientific output measured by published (observational) information; n_{ag} is the number of astronomers who get observing time; r is telescope aperture size (proportional to S/N, and therefore data quality); t is the average allocation time for each 'getting astronomer'; and f is an adjustment factor, used to estimate aspects of quality science which may be hard to specify exactly. Since $t \propto 1/n_{ag}$ and $n_{ag} \propto n_t$, the number of telescopes of aperture size r, we have

$$q(r) \propto f r \sqrt{n_t(r)}. \tag{2}$$

Now $n_t(r)$ can be set $\propto r^{-m}$ (Figure 1, cf. also Krisciunas, 1988, Fig. 8.6). We can also relate m with m' of an aperture-cost law $C \propto r^{m'}$ (Abt, 1980). This leaves f, which, to account for 'prestige' effects, we set $\propto r^p$. We then find

$$q(r) \propto r^{1+p-m/2}. \tag{3}$$

A value of $m \approx 2.1$ was obtained from data on 186 optical telescopes of aperture >0.5m listed by Kuiper and Middlehurst (1960). A slightly lesser value (1.9) holds for 282 telescopes detailed in the *Astronomical Almanac*, 1981-4. Abt's (1980) values for m' were about 2.4 initially, and 2.1 in the longer term. The general result implies near unitary demand elasticity m'/m, for larger (professional) facilities. Elasticity in supply, however, associated with technological progress and durability, is likely to considerably boost the numbers of smaller facilities.

The value of p is inevitably a matter of speculation, but it clearly has to outweigh the negative effects of both availability $(m/2)$ and cost (m') to justify the proposition that best *value* observational astronomy comes from telescopes of largest size. The value productivities (v) of large telescopes were already questioned on this basis by Abt (1980), without respect for the networking possibility.

If ν networked telescopes of aperture r' can be regarded as equivalent to one of size r where $r/r' = \sqrt{\nu}$, we can compare quality science rates at r and r' as:

$$\frac{q_\nu(r')}{q(r)} = \nu^{m/4} e^{-a(\nu-1)}. \tag{4}$$

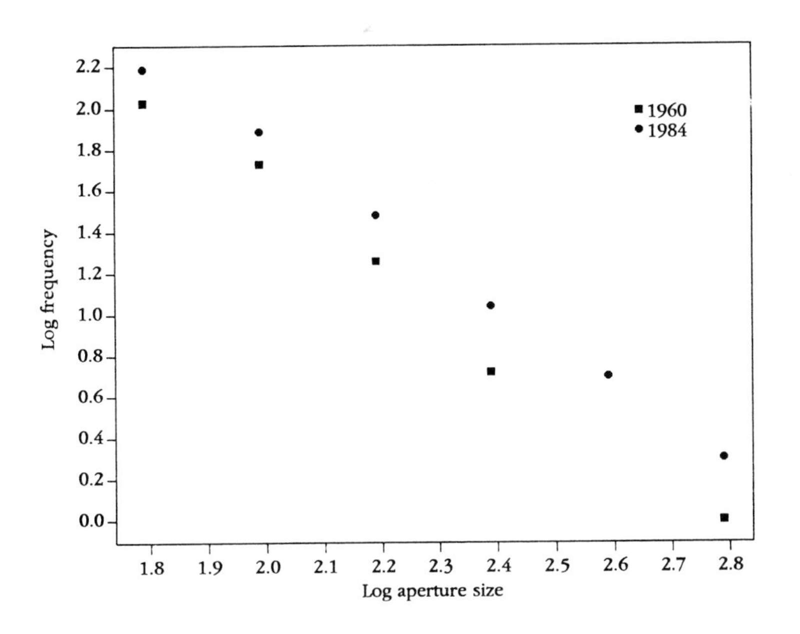

Figure 1 Larger size optical telescope log(frequency) versus log(aperture).

The 'prestige' of ν networked telescopes should be equivalent to that of one of the same effective area, however, the exponential term reflects the communicational difficulty of co-ordinating many smaller telescopes, which exponentiates with increasing ν. On this model we find an optimal group size $(m/4a)$ for effective networking. The cost advantage in networking ν telescopes at a given effective size scales like $\nu^{m'/2-1}$.

Since we are borrowing terms from economists, we may also refer to 'diminishing returns' effects. Two come to mind... the increasing demand for radio frequencies to serve telecommunication needs actually encroaches upon the prevention of radio-pollution, a cause well known to radio-astronomers. Another potential problem is information over-supply — the flooding of computer memory areas with unsolicited data, or the software explosion (Lawden, 1992).

System specification

Planning for a computer network involves a spelling out of key factors such as node interrogation and response rates. One point is the large difference in bandwidth and message time combination for human and machine processors. Modern areal observation involves data collection rates of $\sim$ 10Mbd, though intelligent reduction implies bringing this to a final rate comparable to that of typical human awareness — $\sim$ 10 bd, or perhaps the information rate in scientific papers — $\sim 10^{-3}$ bd (per astronomer). In this way, transmission speeds required for various operations can be estimated. A thousand active users could justify mean network transmission speeds of $\sim$ 10 Kbd, which allows sufficient time for front-end processing.

Real time or batch mode operation

Many astronomers know of remote operation through observing by satellite. The IUE retains real time investigator presence during observations, though this is not strictly necessary (cf. Wamsteker, 1988). Batch-mode working suggests sometimes frustrating 'bureaucratic' delays for active users. Faster and more plentiful processing power has tended to push general usage towards on-line timesharing. An implication of this for observations is many small 'intelligent' telescopes, rather than one centralized super-device. One can envisage batch-mode requests to an international observing network agency (perhaps like the proposed 'GNAT' — Crawford, 1992), for new data on an object — in a similar way to seeking out existent, but obscure, data via inter-library loans. The relevant observational task should not be a complex challenge. A sufficiently detailed request is put in, and, some time later, new data comes down an e-mail connection.

Batch-mode operation rules out the sometimes vital element of on-line feedback and correction, as well as lowering the chance of serendipitous discovery. Observation-oriented research astronomers may not favour it, but for routine-type data-acquisition it seems appropriate.

Human issues

A number of other issues, generally connected with the way humans organize themselves, are raised by this subject. These seem inherently complex: some are referred to in Budding (1992).

3. Three levels of global network for astronomy

In time, various categories of electronic linkage to distributed observational facilties may well develop — for present purposes we consider three possible levels of global network, designated simply as high, medium and low.

High

Under this heading comes dedicated equipment involving 'centre-of-excellence' type institutions. There is an implication of highly competitive usage on well specified proposals. A powerful network centre directs proceedings. Project costs can be estimated to be $\sim\$10^6$ (cf. also Wisniewski, 1988). The 'Whole Earth Telescope' (WET), at least in its fully planned form (Nather *et al.*, 1990), may be an example.

WET is a going concern: it has already produced impressive long continuous data runs. A comparable facility, still at the planning stage, is that of a lunar APT network (Zeilik, 1991). This facility is intended as an add-on to the science centred around a large lunar base. Unlike the WET, it is not controlled in real-time by a Principal Investigator. Rather, it has intelligent controllers which schedule and process received observational requests in batch-mode.

MEDIUM

Most progress in networked observations hitherto has been made in medium-level, internationally co-ordinated, generally *ad hoc* observation campaigns with very specific targets. Many of these were reviewed in the Strasbourg Conference of 1988 (*op. cit*). A number of organizations, often centering around certain active professionals, have formed themselves at this level to help propagate campaigns on particular stars, usually with the aim of combining data from multi-site multi-wavelength facilities. One example was the 1990 campaign on AB Dor, which gave rise to thought-provoking correlations between the radio, optical and X-ray 'light curves' (Budding, *et al.*, 1992).

In medium level projects existing equipment is used, so costs are more related to the movements of observers and data-processing. Individual campaigns might involve gross outlays of $\sim\$10^4$, but an organization setting itself up to deal with such projects could arrange for several in a year.

LOW

A potentially significant development of recent times is the the 'PCO', a low-cost, PC-controlled backyard observatory (Hudson *et al.*, 1992). Local groups of individuals with such facilities can network via public bulletin boards. Genet (1992) has envisaged a community of participating (largely amateur) astronomical observers with: a) a large number of users, b) small telescopes, c) an educational function, d) a coordinated approach to observing which can have applications to certain research goals (cf. also Shibata, 1991).

To fix ideas, a sum of say $\$10^3$ buys a telescope of around 20cm aperture. Such a telescope receives photons over a broad-band filter's range from a 5th mag star at around 1MHz, and in terms of photon noise, would be adequate for millimagnitude surveillance of bright (naked-eye) stars, or providing more continuous records on peculiarities of well known but enigmatic objects, eg. η Car.

4. Inferences and final remarks

We can use the formulae of section 2 to make a few general comparisons across the project levels of section 3. Let us suppose that the WET group size is near optimal, so that a in (4) $\sim$ 0.05. We similarly adopt, for exploratory purposes, m=2, $m' =$ 2.5, p = 2. The results are indicated in the table,

On these figures it would appear that quality science is still preferentially generated by high level projects, though medium level work gives comparable value productivity. Low-level work seems unlikely to match the output of these groups, unless it is much easier to organize very large groups than we surmise. On the other hand, the cost advantage of low-level installations will considerably enhance their value productivity if they can form compact active networks.

Projecting here how further developments might be fostered, I offer some suggestions about possible functions of a 'GNAT'-type organization as follows: (a) assist

Table 1: Quality science rate and value productivity for different network groups.

	H	M	L(a)	L(b)
$\bar{r}$	1	0.5	0.2	0.2
ν	10	40	250	30
q	10^{-3}	4.5×10^{-4}	3.1×10^{-8}	7.6×10^{-5}
v	1	0.63	6.9×10^{-5}	1.4

with communications of high-level programmes; (b) encourage observatory directors to allocate facilities, particularly with regard to batch-mode 'services'; (c) propagate information on remotely controlled observing; (d) help develop smaller, underused or remote facilities; (e) encourage and educate observing amateurs and students; (f) record existence of low-level contributors, receive and communicate their data; (g) process requests for certain types of data from astronomers.

References:

Abt, H. A., 1980, *Publ. Astron. Soc. Pacific*, bf 92, 249.

Benn, C. R., 1990, *Electronic Mail, Guide 1*, Royal Greenwhich Obs. publication.

Budding, E., 1992, (preprint — available on request.)

Budding, E., Slee, O. B. and Stewart, R. T., 1992, in *Proc. IAU Symp., 151*, (preprint)

Crawford, D. L., Genet, R. M., Hayes, D. S., 1988, in *Automated Small Telescopes*, (eds. D. S. Hayes and R. M. Genet), Fairborn Pub., (hereafter referred to as *AST*), p7.

Crawford, D. L., 1992, Private communication.

Genet, R. M., Hayes, D. S., Geltmacher, H. E., Crawford, D. L., 1988, *AST*, p17.

Hudson, R., Hudson, G., Budding, E., 1992, (elsewhere in this volume).

Krisciunas, K. 1988, *Astronomical Centers of the World*, Cambridge Univ., Press.

Kuiper, G. P., and Middlehurst, B. M., 1960, *Telescopes*, Univ. Chicago Press.

Lawden, M., 1992, *Starlink Bull.*, No. 9, 1.

Nather, R. E., Winget, D. E., Clemens, J. C., Hansen, C. J., Hine, B. P., 1990, *Astrophys. J.*, **361**, 309.

Shibata, K. 1991, *Proc. 5th Meeting Astron. Educ. Soc. for Teaching and Popularization of Astronomy*, (Japan), p2.

Sterken, C., 1988, in *Coordination of Observational Projects in Astronomy*, (eds. C. Jaschek and C. Sterken), Cambridge Univ. Press, (hereafter *COP*), p3.

Wamsteker, W., 1988, *COP*, p37.

Wisniewski, W., 1988, *COP*, p245.

Young, A. T. *et al.*, (7 authors) 1991, Preprint – available from Fairborn Pub.

Zeilik, M., 1991, *Proc. 3rd New Zealand Conference on Photoelectric Photometry*, (eds. E. Budding and J. Richard) Fairborn Pub., p73.

Discussion

W.Z. Wisniewski: *The network should be financially independent. Even with the best will, directors of observatories cannot easily allocate telescope time unless participants get their own money.*

Budding: I believe that the proposed global network (GNAT-type) organisation provides some financial incentive to directors to help achieve the required function.

C.L. Sterken: *I have a worry about the PCO concept. It will not only increase the diversity of instruments applied, there is also the introduction of an extra parameter, which is the necessary motivation to push the measurements to the level of precision the networks needs. I have no experience with PCO, but I had some experience in multisite campaigns involving personal observations, which in fact are progenitors of PCO. I found it extremely difficult to persuade such participants to push their performances to the millimagnitude level of precision.*

Transformations and Modern Technology

Jaap Tinbergen

Sterrewacht Leiden, Kapteyn Sterrenwacht Roden, Netherlands

Abstract

Transformations are a central issue in making a global network do more than simple monitoring of low-amplitude variability. I explore the approach of observing at a much narrower *instrumental* bandwidth than is required for the *scientific* problem; such an approach would have the following advantages:

- transformations can be handled in standard fashion at the instrumental level; at the scientific level, they can be avoided entirely,
- users have almost complete freedom in specifying the shape of the scientific passbands, hence comparison of observational data with stellar atmosphere models can be maximally effective,
- standard star observations can be used repeatedly, for programmes running concurrently in *different* scientific photometric systems,
- an observer can use existing standard-star observations to create his own specially-tailored photometric system *from scratch,*
- all-sky homogeneity of the instrumental system can be tested against space photometry such as that provided by HIPPARCOS; this will benefit other scientific systems synthesized from the same instrumental system.

Key components in the hardware will be array detectors with low readout noise and a calibration lamp system designed specifically for this application. A data base provides the link between the observations (in the *instrumental* system) and the results (with *scientific* passbands defined by the end user).

Introduction

Young (1992) is the latest author to show how tricky transformations can be, even between photometric systems *of roughly the same bandpass.* This is a matter we shall have to face in depth when operating global photometric networks. Experience shows that in the course of years passbands always change a little. Equally, *nominally* identical photometric systems at different sites always differ from each other in at least minor ways. Extensive standard star observations are the only insurance against having to conclude after a number of seasons that the system changed while we were not looking and that, *unfortunately,* we are not quite sure how it changed. If, with considerable effort, we construct a global photometric-telescope network, should we not spend some thought and engineering effort in deciding whether modern technology can help us avoid such pitfalls in future? In this paper I examine what I believe to be key issues in one approach, that of using array detectors to observe at very much narrower bandwidth than the scientific problem really requires, which will allow us to adjust the scientific passbands as we think fit *after* we have completed our observations. This is in fact spectrophotometry, integrating or convolving to whatever resolution is optimum for the scientific problem in hand; spectroscopists are extending their craft in this direction, but they tend not to emphasize the 'photo' part as much as we should like (e.g. they 'rebin' when convenient, which amounts to a transformation from one photometric system to another).

Observing at narrower bandwidths is not going to solve all our problems. In particular, we should not expect to do better than the best 'monolithic-filter' photometry in absolute flux calibration. However, there would seem to be excellent prospects of limiting transformations to the standard, *instrumental,* stage of the proceedings (e.g. correcting for equipment drift or for known Doppler shifts). There are bound to be other examples of what can be improved and what cannot. The all-important question is: how much more science can we do by making use of what new technology offers? The answer will depend on how well we can control the transformations from instantaneous to standard *instrumental* system.

I shall assume that funds and a modest engineering laboratory will be available to a global network consortium (cf. VLBI); to me, there would be no point at all in organising expensive networks of 'cheap and nasty' instruments.

System outline

In this global-network discussion, I consider a photometer operating on a single star (probably sky simultaneously), but making use of modern technology (see my review of New Techniques):

- Fast telescope of 1-metre class, run by computer programme. Observing schedule determined by 'committee', but up to 50% of the observing time reserved for standards and calibration lamp exposures, to ensure good quality on behalf of all users.
- Spectrophotometer using a nominally 10-arcsec input diaphragm, a multi-passband filter for passband definition, prism spectrometer for physical separation, an array detector to observe N channels simultaneously (N of order 1500), the light of each wavelength channel being spread over M detector pixels (M of order 100; possibly achieved by input scrambler).
- Calibration input simulating the telescope beam, but coming from a local lamp laboratory via a fiber link. The instruments in this lamp laboratory will be crucial to the quality of the spectrophotometry.

— Archival data base containing, in the instrumental system:

- raw data of programme stars, standards, calibrations, auxiliary quantities,
- data reduced for known *and understood* instrumental effects, and transformed to an above-atmosphere standard instrumental system.

— Distant scientific user of the data base, feeding back to the 'committee' above.

Transformations

At the scientific level, there will be no transformations, as passband characteristics will be determined by suitably weighted combinations of the instrumental-system data and can be kept constant. However, the problems noted by Young (1992) will return to plague us at the instrumental level; as Young shows, they do not disappear as we narrow the bands. But at least they can now be tackled by a specialist in a standard and well-documented way, *for the benefit of all users* (cf synthesis radio-telescopes); this is one potential gain of great significance.

If we have (as we must) a fully-sampled spectrum, we possess all the information needed to correct that same spectrum for out-of-band and high-order-moment contributions. To use the information, we need *relative* flux calibration of some source or other at all wavelengths within the response of the system. As contributions arise from increasingly distant wavelengths, calibration becomes more uncertain, so the design of the photometer must include sufficiently high blocking for far-out wavelengths. Closer to the central wavelength of a channel, however, the calibration is more certain. This will make things easier compared to doing the transformations in the much wider scientific passbands, in much the same way as the transformation to outside-atmosphere is more reliable for narrower bands. This is another potential gain, of even greater significance (or so I maintain).

Finally, at the instrumental level, extinction treatment will benefit enormously from simultaneous measurements in narrow bands.

Limiting magnitude

Objections commonly levelled against observing at much narrower bandwidths than required are A) that it will take for ever and B) that limiting magnitude will suffer badly. These are assumptions, which must be proved and may be disproved.

If we reduce bandwidth by a factor n, it will take n times as long to collect the photons, there will be n bands to observe and photon-limited observations would indeed take very long with a single detector. By observing all bands simultaneously with an array detector, we remove one factor of n. Many observations are not photon-limited at all, so that the problem is much less severe than it might seem; it should be assessed in detail and set against the advantages of the present approach, such as multiple use of standard star observations.

Deterioration of limiting magnitude also appears to be an unconscious fear rather than solid fact. For accurate point-source spectrophotometry, limiting magnitude *(given sufficient photons)* is determined by the signal from darkest sky through a focal-plane diaphragm of at least 10 arcsec, or about 17th magnitude; fainter objects rapidly get swamped by photon noise in the sky signal. Let us estimate what combination of telescope aperture, integration time and spectral bandwidth will yield the sufficiency of photons we have just assumed. For a 1-metre telescope, 17th magnitude corresponds to about 1 incident photon per Angstrom-second. If we are to record such a signal without degrading it appreciably by detector readout noise, we should collect of order 100 *incident* photons per detector pixel (a state-of-the-art CCD has readout noise of about 3 electrons and quantum efficiency of about 50%). Assuming that the light in one passband is spread over some 100 pixels for photometric stability, we should integrate for 10000 Angstrom-seconds. Taking 1000 seconds (cosmic-ray upper limit) for limiting magnitude, we see that minimum bandwidth (FWHM) is of order 10 Angstrom.

This is a very interesting result. Most scientific passbands are of order 200 Angstrom or wider. Since a wavelength range about twice the FWHM contributes to a band and fully sampling the spectrum means spacing the elementary passbands at intervals of half the FWHM, of order 80 instrumental passbands will contribute to a single scientific passband. This is excellent for freedom in shaping the scientific passbands, as in suppressing a conspicuous but irrelevant spectral feature (digital equivalent of a 'notch' filter) or in narrow-band photometry such as H-beta.

Integrating for 1000 seconds at limiting magnitude may sound excessive. But how often does one observe limiting magnitude? Certainly only around New Moon, and certainly NOT for standards and lamp exposures. I do not see this as a serious problem; the technique of 'charge binning' on the CCD chip could probably eliminate the problem altogether while preserving the photometric stability (with a 30-arcsec diaphragm or 1-electron readout noise, the problem also ceases to exist).

Calibrations

A large fraction of the engineering effort may have to go into devising suitable calibration systems for our new-fangled photometer. We need two basic kinds of routine calibrations for stability of the system:

— wavelength stability: are the instrumental passbands where we think they are?

— 'gain' (flux sensitivity) of the individual instrumental channels relative to each other (and, less urgently, the time variations).

A suitable selection of standard stars may have to be the final reference for *gain*;

between observations of stars on the basic standards network, lamps will have to provide interpolation. It is worth emphasizing that the lamp signal need not stay constant, *as long as it changes only in a known way (monitoring at the photometer input by filtered photodiodes?);* there is a need for some creative engineering.

The approach to routine *wavelength* calibration which I should like to see thought out in detail would use a solid Fabry-Perot etalon as calibration component. As in the case of the birefringent filter, only one or two parameters determine its properties; the effect of change in those parameters will be visible in many instrumental channels and the pattern of change will to some degree be specific for each parameter. In such an overdetermined system one can expect substantial gain in correcting for wavelength drifts. The approach is similar to the use of pointing models to improve large telescopes and similar gains may be expected here from creative use of computers. Such a calibration situation could comprise:

— lamp spectrum: 2 parameters (blue/yellow, yellow/red)

— birefringent filter: 1 parameter (phase shift by the basic crystal slice)

— solid F-P: 2 parameters (finesse, spacing)

Our hypothetical system would thus have 5 parameters and regular observations in hundreds of channels, so would be extremely over-determined. In the light of experience we would probably wish to add a few parameters to represent slow changes in CCD spectral sensitivity or flexure of the instrument. The key to good results is a sensible form for the parametrization of the system.

The aim of calibration is to relate the actual instantaneous instrument to a virtual instrument with constant parameters; ideally, the global network will have just one (virtual) reference instrument to which all actual instruments relate. For global network application, portability of the calibration system would be a great advantage. It seems possible to design the lamp and its monitoring diodes, and the F-P, in cabin-luggage format. The other components mostly belong to the photometer and should be left in place. A portable standard CCD system does not seem beyond the realm of possibilities and would allow separate calibration of the spectrophotometer optics and the detector at each site. 'Portable' calibration, if successful, would reduce the network to a single instrument *without making use of standard star observations;* these would then be available for independent consistency checks.

To determine now and again the exact passband shapes and blocking factors for all the channels, a high-purity monochromator will have to be part of the site installation. We *measure* simultaneously the relative response of all channels to one wavelength at a time, later we *use* the response of one channel at a time to all the wavelengths. Since we must use some flux-calibrated source to transform the measured stellar signal at one wavelength into a stray contribution at a channel nominally recording some quite different wavelength, the stray contributions from more distant channels are increasingly uncertain and should therefore be blocked sufficiently well for this increased uncertainty not to matter.

It is essential that the lamp beam looks exactly like a beam from the telescope. Spectroscopists know this and designs for suitable optical systems are available.

Operations

As I see it, a scientific user of the system would lobby for his observing programme to be first accepted and then scheduled optimally. When the semi-automatic system has obtained the data, the user will inspect the (instrumental-system) data corrected (i.e. 'transformed') for extinction and for equipment changes detected by the calibrations. He will then gather the corrected standard star observations of 2 or 3 seasons and set up his private photometric system *ab initio* by defining the passbands and determining (above-atmosphere) standard-star differences in magnitude and colours, then looking for closure errors and other systematic effects. Having decided what further (partly ad-hoc) corrections to apply, the user will obtain highly accurate results in his private photometric system.

Of course, there will be quicker routes to less-than-optimal results. The distinction will be in how we use the archival data and we do not throw away all possibilities of 'the best' results whenever we decide to optimise for something else rather than accuracy. If only we can agree on what the passbands of the standard systems 'really' are, widespread use of private photometric systems for specialised purposes will not stop us using the same data again for standard systems. We can eat our cake and still have it; paraphrasing one of the other speakers: while peasants guzzle coarse bread, Marie Antoinette can nibble at her Gâteau des Rois, *both confections created from the same trough of dough.* And that would be a Revolution.

Conclusions

I have suggested that, for point-source spectrophotometry, it may be possible to avoid the problem of 'transformations' at the scientific level, and to control it more effectively at the instrumental level, by implementing an old concept. The essence of the technique is *a posteriori* passband control, which seems practicable with the newest generation of CCDs. Many engineering details will need to be solved before one can confidently plan a global-network instrument on this basis, but if we don't set our sights on progress why are we here at all? VLBI provides a valid precedent of organising a single but multi-site facility for a variety of users.

References:

Young, A.T. 1992, Astron. Astrophys. 257, 366-388

Discussion

R. Florentin Nielsen: *In your calculations of feasibility you use a read-out-noise of 3 electrons. This is not really the state-of-the art today. CCDs with multiple non-destructive read-outs can have a read-out-noise of < 1 electron (F. Fanesick).*

Tinbergen So much the better, if detailed engineering shows this to be practicable, it will reduce the 1000s maximum integration time to 100s, short enough to monitor extinction on a nearby star between 100s partial exposures. The data-handling problem will not improve, however.

E. Budding: *I just wanted to make a comment about your expression "cheap and nasty", which may reflect a certain heretical tendency when set against the leading orthodoxy of market-related policy making. Perhaps we should also be thinking about cheap and attractive(?).*

Tinbergen: I was referring to a simple classical photometer, which is cheap but generally produces nasty transformations. I am not against simplicity when it helps.

Archiving of photometric data

B. Hauck

Institut d'astronomie de l'Université de Lausanne, Switzerland

Abstract

Lausanne University's Institute of Astronomy has been collecting stellar photometric data since 1970. Our collection contains data for some 170,000 stars in 78 photometric systems. This database has been used in various applications: comparison of systems, Hipparcos Input Catalogue, studies of star samples, etc. With CCD detectors we are facing a new challenge, since a great deal of information is obtained, mainly for star clusters. But unfortunately, a policy for archiving these data has not yet been established. This is an urgent problem to solve, all the more so as in many cases only a part of the data is published. A solution to this problem is discussed here.

1. Introduction

Since the first work of Stebbins at the beginning of the century, photoelectric photometry has faced many changes, the latest being the use of CCD detectors. Technical improvement has not only made it possible to obtain more precise data and to observe fainter stars, but it is also responsible for a large growth of data. Growth of photometric data is also due to the increase of photometric telescopes, particularly in the southern hemisphere. In a study on the growth of data in stellar astronomy, Jaschek (1978) has shown that photometric data increase exponentially. Confirmation of this is to be seen in Table 1.

Table 1. Growth of photometric data

Strömgren system:		Geneva system:	
1965	$1.2\ 10^3$	1964	$3.4\ 10^2$
1973	$7.5\ 10^3$	1966	$6.8\ 10^2$
1975	$9.3\ 10^3$	1971	$1.4\ 10^3$
1977	$1.4\ 10^3$	1976	$4.6\ 10^3$
1978	$1.9\ 10^4$	1978	$1.3\ 10^4$
1985	$4.0\ 10^4$	1980	$1.5\ 10^4$
1990	$4.5\ 10^4$	1988	$2.9\ 10^4$

Before the 1970s, photometric data were dispersed in various publications and no systematic efforts were made to archive them. This situation encouraged us to collaborate with the Strasbourg Data Center and to compile all stellar photoelectric photometric data, with the exception of light curves for variable stars (they are now collected by Dr M. Breger at Vienna Observatory). Our database now contains data for about 166,000 stars. With CCD detectors we are facing a new challenge, since a great deal of information is obtained but unfortunately no policy for archiving these data is currently established.

2. The general catalogue of photometric data

For nearly twenty years we have been compiling all the photoelectric photometric data for stars. At the end of 1989 our database contained nearly 410,000 measurements for 166,000 stars. No fewer than 78 photometric systems were found! This database is the result of the merging of the various catalogues we have established (e.g. Hauck and Mermilliod, 1990, for the Strömgren system, Mermilliod & Mermilliod 1992 for the UBV). Our catalogues are formed in two parts: the first contains the original data and the second the weighted mean values. Both parts are available from the CDS, while only the second is published in some cases. In order to establish the second part, it is necessary to make a critical evaluation of the data. To do that, we follow the method described by Nicolet and Hauck (1977). Each list of new data is compared to a reference list. On the basis of this comparison we can define a weight for the new publication and eventually a systematic correction to the data of the new list.

Our photometric database runs on a Vax machine. Its master index is our General Catalogue of Photometric Data, which can be obtained from the CDS. The contents of this catalogue is for each star an identification number (Mermilliod 1978), equatorial coordinates (if available), V magnitude and photometric keys. Each photometric system has its own key. For small samples it is possible to send us by e-mail (mermia@obs.unige.ch) a table with a list of identification numbers according to Mermilliod. Thus with such a catalogue we can know in which system a star is measured. We can also obtain various samples, for example all stars having UBV, Geneva and Vilnius data.

3. CCD Data

The future of photometry lies undoubtedly in the use of CCD detectors. Observations with a CCD camera are currently made with the Strömgren, Geneva, UBV, Washington and Cousins systems. With the development of this new kind of detector we are facing an important problem: what to archive and where? Huenemoerder (1992) and Mermilliod (1992) have already briefly discussed this matter and have made some interesting suggestions. Huenemoerder has quoted that

> ... "photographic plates can be stored in a vault for posterity. The primary archive is not then a question of medium but of organization. With CCDs or other electronic data, both the storage media and organization of data are open issues".

Huenemoerder is right. However, even in the case of photographic plates it is difficult to know where the information is archived. An attempt was made some years ago by the working group on astronomical data of IAU Commission 5 (Hauck 1982a,b) to centralize the information about plate vaults. Unfortunately, it was not possible to give a good follow-up to this initiative. An attempt is now being made for stellar spectra by the members of a working group of IAU Commission 29. This working group is chaired by Dr E. Griffin. Concerning the photometric data, we can assume that at least nearly all photoelectric data for stars are published somewhere and that our database reflects the situation reasonably well! But with CCDs we are facing a very disturbing problem. In many papers none of the data, or only some, are published. Data are used for plotting the diagrams which are printed, but this excludes access to the primary data! We can easily understand the reason for this situation, but we should be concerned about the archiving of these data and their accessibility, not only now but also in the future, even in a century from now! But what should we do? It seems that a good policy would be the following:

1. After reduction data have to be archived in an astronomical data centre. A description of the reduction method must be a part of the archive.

2. Each data centre should establish a list of fields archived, with an indication of the central position, types of filters, ... An updated version of the list should be published regularly, or at least available by e-mail. The list is communicated to the other data centres.

3. Before accepting for publication a paper based on new CCD data, the editors of the journal have to check, or at least to receive the assurance that point 1 is fulfilled.

Now we have to persuade the ADC to accept points 1 and 2 and the editors to accept point 3! An interesting attempt has been made by Acta Astronomica, who do not publish voluminous tables of CCD data, but make the file available by FTP through a computer. This seems to be a very attractive solution, but journals must have a financial equilibrium. Thus I doubt whether this solution could be generalised and that publishing companies will spend time and money on archiving astronomical data. An alternative solution would be to increase the subscription price! But we can forget that immediately as librarians already have enough problems today raising finance.

4. Conclusions

The situation is fairly satisfactory for photoelectric data and the astronomical community will surely appreciate the new catalogue of UBV data prepared by J.-C. Mermilliod and M. Mermilliod and published by Springer-Verlag. Our database is deposited in various locations and is easily accessible. Our main problem is that we cannot assume that all the published data are included in the database. That is the reason we appreciate it when authors send us their data directly in computer-readable form. However the position is rather bad concerning the CCD data and I fear that we are losing primary data. It is becoming urgent to find a solution not only for photometric data obtained with a CCD camera, but for all CCD data. I.A.U. Commission 5 (Documentation and Astronomical Data) will be discussing this matter and I would appreciate receiving suggestions.

References:

Hauck B. 1982a, in *Automated Data Retrieval in Astronomy*, C. Jaschek & W. Heintz (eds.), D. Reidel, Dordrecht, p. 217.

Hauck B. 1982b, in *Automated Data Retrieval in Astronomy*, C. Jaschek & W. Heintz (eds.), D. Reidel, Dordrecht, p. 227.

Hauck B., Mermilliod M., 1990, AAS, **86**, 107.

Hauck B., Nitschelm C., Mermilliod M., Mermilliod J.-C. 1990, A&AS **85**, 989.

Huenemoerder D. 1992, in *Highlights of Astronomy*, Vol. 9, J. Bergeron (ed.), in press.

Jaschek C. 1978, Bull. Inf. CDS **15**, 212.

Mermilliod J.-C. 1978, Bull. Inf. CDS **14**, 32.

Mermilliod J.-C. 1992, in *Highlights of Astronomy*, Vol. 9, J. Bergeron (ed.), in press.

Mermilliod J.-C., Mermilliod M. 1992, Springer-Verlag, in press.

Nicolet B., Hauck B. 1977, in *Compilation, Critical Evaluation and Distribution of Stellar Data*, C. Jaschek & G.A. Wilkins(eds.) D.Reidel, Dordrecht, p.121.

Discussion

A. T. Young: *The problem of archiving CCD data is very similar to the problem NASA has had in archiving old spacecraft data. The descriptions of the processing used to produce the archive have often turned out to be inadequate. I suggest that the actual computer programme used be stored as part of the archive.*

Hauck: I think you are right and you made a good suggestion. However the problem with CCD data is more complicated because the data are in many locations.

J. Baruch: *The Robotic Telescope we are building at Bradford is designed for profile fitting faint object CCD photometry. It will extract photometric data from faint objects in crowded fields and generate target acquisition assurance from pattern recognition algorithms. It will also produce indices of image quality. The original image will not generally be stored.*

S.B. Howell: *I would just like to make a point in agreement with the above. A CCD light curve time-series can generate 200-400, or more, Mbytes/night. I forsee that many of us, as well as many of the APT CCD telescopes, will perform real time data reduction and not keep the image themselves.*

Hauck: Light curves are perhaps a special case and, if you keep them, (not the images) and archive them in a data centre they are saved!

D.L. Crawford: *Another example is the Search for Extra Solar System Planets, which will generate an immense amount of CCD data, most of which it is not possible to archive.*

We should always offer all the encouragement we can to those who are archiving, such as the groups at Lausanne and Strasbourg.

W. Tobin: *If one wishes to archive all CCD data acquired, it will be necessary for unreduced data to be stored too, because the fact is (1) not all potentially useful frames are reduced, and (2) when a frame is reduced, not all objects in it may be reduced. At Mt John we have no plans to archive our CCD data until the hardware problem of a suitable archival medium has been solved. To add to Dave Crawford's comment, a friend who works for an insurance company tells me their company archivists spend their time deciding what to throw away from the archive, not what to save!*

Hauck: I agree with you it could be a huge problem, but I think we should archive at least all magnitudes, colours (and coordinates) we have obtained. In many cases we have only plots. The storage of all frames is now a problem but I think it will be resolved and we have to be ready!

Russ Genet shares his birthday cake with Phil Hill and Gene Milone

Session 6

Photometry with CCDs

Photometry with CCDs

Alistair R. Walker

Cerro Tololo Inter-American Observatory, Casilla 603, La Serena, Chile

Abstract

Characteristics of telescopes, photometers, CCDs and data acquisition systems relevant to achieving high quality photometry are discussed.

1. Introduction

In the recent past, the area of the typical CCD available for astronomical use has increased by an order of magnitude. Second generation CCD controllers, essential to efficiently handle large CCDs with multiple readouts, can also provide relatively seamless integration with reduction software and telescope control environments. Computer power has increased greatly, and comprehensive data reduction packages have been written. CCD photometers have become more complex and are often used with auxiliary optics or combined with spectrographs. Active optics and low order adaptive optics, together with attention to telescope collimation and heat management, have been demonstrated to dramatically improve image quality. All these factors have a direct effect on both quality and quantity of CCD photometry.

2. Telescopes

When CCDs first started to be used a decade ago, frequently on relatively moderate sized telescopes, the inadequacies of the guiding, tracking and pointing provided often became painfully obvious. Somewhat more recently it has been generally realized that at good observing sites telescopes and their surroundings frequently limit the image quality by thermal mismatch of telescope and dome with the ambient air, or by the optical quality of the telescope mirrors, or by limitations in mirror support systems. The ESO NTT has demonstrated the dramatic gains possible given the latest advances in telescope and enclosure designs, while the CFHT is an example of how a more traditional telescope and enclosure design can be improved. Since much CCD photometry, independent of telescope size, is done in crowded fields, good image quality is a critical parameter. As an example of a program to improve an already existing telescope, at the CTIO 4m the thermal environment is being improved by moving out of the building all functions not directly concerned with operation of the telescope, by moving the control room to the ground floor, by cooling the oil used in the RA drive bearing, by controlling the temperature of the primary mirror and cell and by drastically venting the dome. At the same time the f/8 secondary mount is to be modified to allow rapid collimation, the secondary itself is to be refigured, and active control of the primary support structure is to be implemented. In addition, a Shack-Hartmann image analyser is to be permanently installed at a side port of the RC focus to allow regular monitoring of the image quality.

Telescopes designed for wide field photography are not always optimal for use with CCDs since they usually require a field flattener in close contact with the plate. For small CCDs the field flattener can generally be left out with negligible effect on the image quality, however CCDs are now approaching the size of small photographic plates, so without the flattener images can be expected to vary as a function of field position. For Cassegrain telescopes coma at field corners can intrude, as an example the CTIO 0.9m f/13.5 telescope provides a 13 arc min square field with a Tektronix 2048 CCD, but even in moderate (1.4 arc sec) seeing the fwhm are 10 percent larger at the corners of the field. Fitting psf functions to stars on such frames requires software which can build a psf which varies with position on the CCD (eg DAOPHOT) and many psf stars. An alternative is to design a suitable corrector, for a 10 cm field at f/10 or slower these can be rather simple, and with sufficient back-focus to allow a normal (plane) window for the CCD dewar. In dedicated dewars the field flattener can serve dual purpose as the dewar window; this is common practice for fast spectrograph cameras.

Difficulties with psf fitting can also arise if the CCD is not flat. TI 800^2 CCDs, which are a floating membrane of silicon supported only at the edges, are notorious for this problem and many are so warped that they cannot be used in fast beams at all. Hopefully, the Reticon 1200J is the last of this genre, but even supposedly flat CCDs are not (eg Tek 2048), and CCDs are not neccessarily aligned in their packages. For prime focus photometers it is a good idea to provide a facility for aligning the CCD with respect to the optic axis, preferably by being able to rotate and tilt the dewar slightly.

In principle, baffling (Young 1967) for the oft-used Cassegrain telescopes should be optimized for the size of CCD. This is often not possible due to the necessity of using an off-axis guider, but in any case the baffles should be well designed with multiple traps in the "chimney" (N. Caldwell, private communication). A related problem is preventing reflections off CCDs and their surroundings returning to the CCD via reflections of filters etc. Dewar windows, filters and any auxiliary optics should be AR coated. It is a good idea to mask the filters with square apertures of a size just large enough not to cause vignetting.

3. Improving the Resolution

We have already mentioned passive and active optical means of improving resolution. Some improvements can be made a posteri (eg maximum entropy sharpening) which are beyond the scope of this review and have in any case received much recent attention in the context of HST images. In most cases it seems that resolution is improved at the expense of photometric integrity (eg in the cases where the resolution improvement is proportional to S/N).

Adaptive Optics methods attempt real time correction or partial correction of wavefront errors. Instruments such as HRCAM, DISCO and MARTINI improve resolution by a combination of removal of star image wander due to seeing inhomogeneities, wind buffeting and telescope drive errors, by closing a fast shutter to exclude the worst seeing, and by using individually corrected sub apertures. Reductions of fwhm by a factor two have been reported (Racine & McClure 1989, Tanvir et al. 1991). Isoplanicity is maintained over a field of a few arc minutes square only (McClure et al. 1991), and even within the corrected field the psf will change with distance from the reference star. As emphasised by Devaney (1992), a reduction code allowing the fitting of variable psfs will be required. Gaussians are a poorer and Lorentzians a better fit than they are to unsharpened images. The above-mentioned instruments are all quite complex. It is possible in principle (T. Ingerson, private communication) to build a very simple instrument on the basis of moving the CCD itself, using one quadrant of a quad-amplifier CCD as the guider. Although such an instrument would not allow the easy use of sub-apertures, the other advantages mentioned above remain. It may be preferable to use a CCD which does not use bond wires in order to avoid mechanical stresses and microphonics.

4. Instruments

CCD photometers, particularly those used at Cassegrain foci, tend in general to be rather simple instruments, often containing no more than one or two filter wheels, an electronic shutter, and an illumination source for providing a preflash exposure. The latter device has become more use to verify correct operation of the CCD since most recent CCDs do not require preflashing. Guiding facilities normally are provided via a guide box to which the photometer is attached.

Photometers used in faster beams, such as at the prime focus of a 4m telescope, tend to be more complex instruments, at least in part due to the restrictive 8-10 cm backfocus behind the final corrector element which decrees that the guider be integrated with the photometer. Correctors in use are often those left over from the days of photography, but newer designs incorporating atmospheric dispersion compensation are feasible. Two such correctors are being built for the CTIO 4m and the La Palma 4.2m telescopes; these will provide 0.25 arcsec images over almost a degree diameter field from 3000-10000Å and correct for dispersion up to 60 deg. from the zenith, and thus will provide high quality imaging even for very large mosaics. Maintaining a sensible pixel scale is the motivation for installing CCDs at the prime foci of large telescopes. The conflicting requirements of obtaining a reasonable sized field and at the same time sufficiently sampling star images in the best seeing have been relaxed with the availability of larger CCDs. Some reduction programs (eg DOPHOT) work optimally with fwhm $\sim$ 2 pixels, however better sampling is required for work in crowded fields. Focal reducers (eg Aldering & Bothun 1991) can alter the pixel scale and are particularly useful where well sampled stellar images are not required. Similar motivation has led to the popularity of CCDs on Schmidt telescopes.

We mention only briefly integrated instruments, those that can do both spectroscopy and photometry. Given that these instruments are of necessity optimized for spectroscopy there are some trade-offs for photometry. UV throughput is normally poor (although new glasses and coatings have improved matters), field size is limited preventing use of the largest CCDs, while it is difficult to prevent low level ghost images. The major advantage is the ease of changing from spectroscopy to photometry, and since these instruments have a parallel beam many modes of operation are possible. Although EMMI (d'Odorico 1990) is a recent example of such an instrument which has few of the mentioned disadvantages, its complexity and high cost suggests that for pure imaging applications the simple instruments are generally to be preferred.

5. CCDs

Within the last 2-3 years, after several years of relative paucity, a number of excellent devices have become available. It has become de rigueur to use a CCD with 1024^2 pixels for photometry while 2048^2 CCDs are desirable for many programs. At the time of writing, the most popular CCDs for photometry are the Tektronix 512^2, 1024^2, 2048^2 family of back-illuminated CCDs. Also common are the Thomson 1024^2, EEV (eg 1242x1152) and the Loral (ex Ford Aerospace) 512^2, 1024^2, 2048^2, all of which are front-illuminated, and are often laser-dye or phosphor coated to improve blue and UV QE. Even larger devices (4096^2 in both 7.5 micron and 15 micron pixels sizes) have been fabricated by Loral, while Thomson, EEV and Reticon all have R&D programs for thinning large CCDs. Thinning of foundry (usually Loral) CCDs has also been accomplished by SAIC and M. Lesser (University of Arizona). In addition, Kodak produces small-pixel CCDs up to 2048^2, while older devices (TI 800^2, RCA, EEV) still see much use. Long term commitment to producing scientific grade CCDs by at least some of these companies seems assured. Janesick & Elliott (1991) have provided a fascinating in-depth look at the history of CCD evolution including recent developments. Table 1 summarizes the most popular mega-pixel CCDs.

These latest devices are not only larger than their predecessors but are improved in several important respects. The number of amplifiers provided is often two or four per CCD, offering the possibility of reducing readout time for the mega-pixel devices to that of the smaller CCDs. Unfortunately, if you want extra working amplifiers, there is often a price premium. Cosmetics for the top grade devices can be perfect or nearly so, charge transfer efficiency has improved with the incorporation of sculptured channels ("minichannels") which expose small charge packets to less silicon, read noise is now just a few electrons, while full-well per given pixel area has increased. Quantum efficiencies have generally improved with the perfection of thinned, anti-reflection coated CCDs chiefly by Tektronix (although RCA successfully pioneered this technology more than a decade ago). We shall now consider some of these properties in more detail.

Process improvements have led to a general increase in yield which has resulted in the virtual elimination of hot columns and fully blocked columns from grade one devices. Even image area charge traps, which if not too numerous are more of a nuisance to be avoided rather than a catastrophe, are now not common. Achieving

Table 1: Some Megapixel CCDs useful for Photometry

Manufacturer	Type	Size	Pixel size	Amps	Comments
Tektronix	thinned	1024^2	24	4	AR coated, implant
Tektronix	thinned	2048^2	24	4	AR coated, implant
Thomson	thick	1024^2	19	4	
Loral	thick	2048^2	15	2,4	foundry
Loral	thick	4096^2	15	4	foundry
Reticon	thinned	1024^2	13.5	4	1992?
Reticon	thinned	2048^2	13.5	4	1992?
EEV	thinned	1242x1152	22.5	2	
EEV	thick	2186x1152	22.5	2	1992?

Table 2: Examples of UBVRI filter transformations with various CCDs

Filter	Ingredients	Band	Tek 1024 + Harris	Tek 1024 + Tek	TI + Harris	Thomson + Harris
U	UG2/1, 5mm 25% $CuSO_4$	: V	-0.03	0.02	-0.04	-0.01
B (Harris)	BG12/1, BG39/2, GG385/1	: B-V	1.21	1.05	1.09	1.08
V (Harris)	GG495/2, BG39/2	: U-B			0.98	0.95
R (Harris)	OG570/2, KG3/2	: V-R	0.98	0.98	0.95	0.96
I	Interference	: V-I	0.97	1.01	0.95	1.00
B (Tek)	BG1/2, BG39/2,GG385/1[1]	:-------------------------------- ----------------				
V (Tek)	GG495/2, BG38/2	:Eqns. of the form V = v + coeff. x (B-V)				
R (Tek)[2]	OG570/3, KG3/2	: and B-V = coeff. x (b-v), etc.				

Notes: [1] GG375/1 would improve the nominal match but there is a danger that unwelco effects would occur due to the confluence of the Balmer lines. This filter has 20% m throughput than B (Harris) and 30% more than B (Bessell).

[2] Bessell (1990) suggests OG570/2, KG3/3 but this filter has only 78% of the transmiss of the two R filters here.

uniformity of response on scales ranging from pixel-to-pixel to 100's of pixels is very important. It is obviously much easier to flat-field a CCD that is intrinsically flat to start with than one which has enormous variations on every scale (such as the TI 800^2 CCDs). Although front illuminated CCDs have traditionally a flat response, thinning technology has advanced so that large back illuminated CCDs are fully competitive in this respect. Tektronix 1024 CCDs have typically 1 percent pixel-pixel variations and 2 percent variations at lower spatial frequencies. The number of low pixels (QE less than 90 percent of their neighbors) is under 0.01 percent. Ten years ago, the first generation RCA CCDs could approach these figures, and it is to be hoped that further improvements can be made.

Full well capacities are now typically more than 250000 e^- for 15 micron pixels and more than 500000 e^- for 24 micron pixels, less by a factor ~ 2 for MPP (multi-pinned phase) CCDs. The latter have dark generation several orders of magnitude less than conventional devices at room temperature (typically they have the same dark rate as a non-MPP CCD operated 30C cooler) but in order to reduce dark to just a few e^-/hour they have to be cooled to much the same temperature (typically -100C). For most ground-based astronomical applications the non-MPP versions, with larger full-well, are preferable. Noise floors can be as low as 2-3 e^- rms, thus the full dynamic range is typically 100-400 Ke^-. To digitize this range completely 17-19 bits of resolution are needed. We will return to this point below when we discuss CCD controllers. Most CCDs appear to be linear to at least 0.1-0.5 percent from very low signals to at least 80 percent of the full well capacity. The low noise figure in itself is less important for broad-band photometry, particularly on large telescopes, than it is for narrow band imaging where sky background signal is often small.

High quantum efficiency is vital to the user. Some examples are shown in Figure 1. As will be discussed elsewhere at this conference, there are particular difficulties in reproducing UBVRI passbands with CCDs. This is for two reasons; CCDs have an extended red response even compared to Ga:As photocathodes, and the blue-UV response often falls very steeply. The latter makes U band photometry difficult or impossible, the former imposes stringent red leak requirements on all filters (see Stetson 1988). The B passband is very difficult to reproduce using available glasses without sacrificing throughput, and the I band is best reproduced with an interference filter. It is also possible to imitate the B, V and R bandpasses with interference filters (eg the "Mould sets") but caution must be taken with the red leak specification, which as Stetson (1988) has pointed out, even a very small red leak integrated over several thousand Å of bandpass can cause severe problems. Different types of CCDs have very different response shapes in the blue and UV and filters which work well for one type may not work well for another. Some examples are given in Table 2. UV flooding (eg Oke et al. 1988, Janesick & Elliot 1992) still appears to produce the best UV response in thinned CCDs on which a flash oxide has been grown, but experiments (particularly by Reticon) with flashgates and biased flashgates are encouraging. The use of boron implants (eg Tektronix and EEV) has the advantage of providing a stable UV QE. Advances in anti-reflection coatings on thinned CCDs have resulted

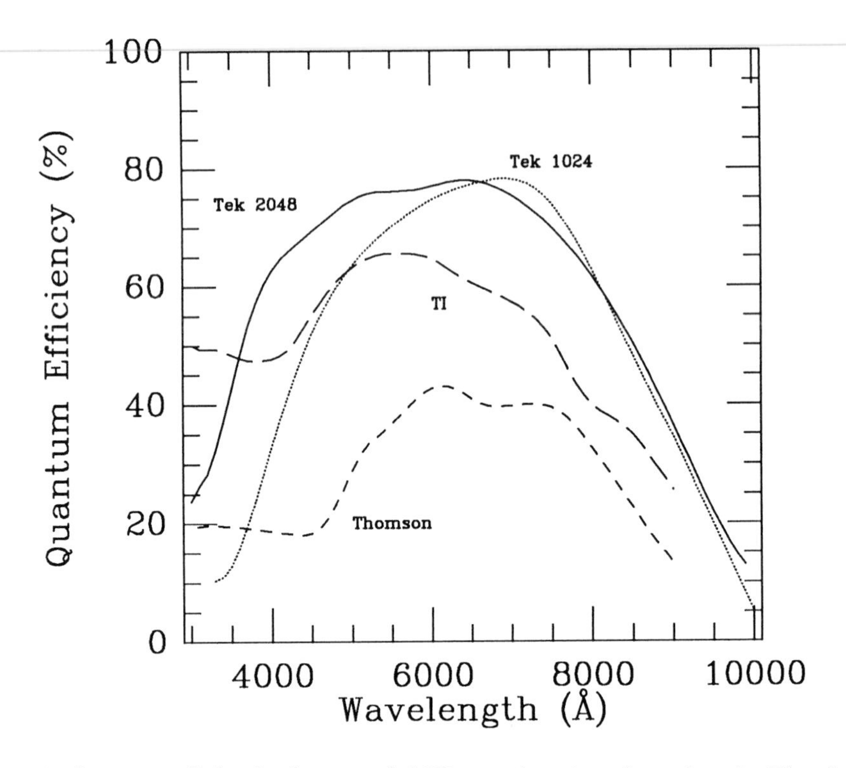

Figure 1: Quantum efficiencies for several CCDs as a function of wavelength. That for the Thompson CCD is typical for front-illuminated CCDs that have been coated to improve the response below 4500Å. The TI curve is typical for a thinned, UV flooded CCD, while an AR coating is responisble for the high peak QE of the thinned Tektronix CCDs.

in excellent QE's peaking near 80 percent, and even better can be expected in the future since M. Lesser (JPL CCD newsletter no. 3) has coated thinned Loral CCDs with Hafnium oxide and by tuning the coating thickness can obtain over 90 percent QE over a wide wavelength range including the UV.

6. CCD Controllers and the Data-taking environment

Second generation CCD controllers are beginning to appear, replacing the controllers used for the past decade. There are many motivations, including improving reliability by using modern techniques and components, improving performance, providing facilities for more efficient optimization, lowering power consumption, integrating cleanly with instrumentation and telescope control systems, etc. Controllers can be given enough computing power to allow such features as a real-time display capable of scaling and displaying the image as it is read out, with zero overhead. Pre-processing can be incorporated so that a single controller architecture can be used for both optical and IR arrays. User interfaces can run in modern environments (such as X-Windows) to better help the observer obtain and reduce his or her data. Three aspects will be considered in more detail.

Firstly, it is now feasible to read out CCDs with multi-amplifiers in parallel, and to build mosaics of CCDs. This directly decreases the fraction of time that data is not being taken. Several recent designs use architectures based on Inmos transputers which are designed for parallel processing applications and thus are particularly suitable. Enough computing power can be incorporated in the controller so that the time needed for unscrambling the data from multi-amplifier CCDs can be just a small fraction of the readout time. There is then little excuse for skimping on (short exposure) standard star observations, particularly since CCD-sized fields are now available containing standards with a wide range in color (Landolt 1992). Secondly, the utility of the data from the CCD depends critically on the faithfulness with which the analog signal from the CCD is amplified and converted into digital form. There are now much improved analog-to-digital (ADC) converters available, driven by the audio market. Further improvements are likely in the whole area of analog and digital signal processing, driven by medical imaging requirements. The latest generation of ADCs use less power, are faster, have no missing codes and offer impressive linearity. Very importantly, they are self-calibrating so that this level of performance can be maintained indefinitely. They can also be very small. Examples are the Crystal CS5101, Datel ADS-930 (both 16-bit) and the Analogic ADC5030 (18 bits). Even with the 16 bit units and gain set at several electrons/adu, these ADCs, particularly the Crystal unit, have such high accuracy that it is feasible to work with the least significant bit set equal to half the readout noise without ADC errors and digitization noise dominating. This is of most important in situations where there is little background.

Finally, if efforts are made to optimize the data flow right through into the data reduction environment then it is practical to process data while observing. Often it is best to have two observers (one taking the data, the other reducing it – networking and high bandwidth satellite links can mean that one or both observers can be remote from the telescope). Although it is not viable to completely reduce all types of data this quickly, it is certainly possible to carefully evaluate data, and to reduce standard star observations in this way. With readily available computing power, it is possible to read a large quad-amplifier CCD, unscramble the data and remove the instrumental signature in under 30 seconds. Representative figures for a Thomson 1024^2 CCD and the CTIO ARCON controller are: readout 11 seconds; reformat 8.5 seconds; trim, de-bias and flat field 15 seconds (reformatting and subsequent processing using a Sparcstation 2 with a 12 ms SCSI disk – this is 1991 technology; faster processors and disks are now available for much the same price). The raw data can be inspected as readout occurs on a display with 1152x900 pixels. Features such as rapid field preview can be built into the hardware and the software.

7. Calibration and Test Procedures

Test procedures are those which evaluate whether the instrument is functioning correctly, whereas calibration procedures include removal of the instrumental signature from the data followed (for photometry) by determination of response sensitivity usually from observations of standard stars. Massey & Jacoby (1991) show how to recognize "good data".

More stringent tests to verify operation with "second generation" controllers are desirable given that the data path from the detector to the data reduction computer is generally more complex than before. Probably the best way to accomplish this is by incorporating an Artificial Data Generator in the controller itself, at the start of the digital part of the signal chain. The read-out sequence for this data is exactly the same as for the CCD itself, and at the end of the readout the data can be compared with that expected. Telemetry of clock and bias voltages can be checked against a table of expected values. ADCs can be self-calibrating. Beyond these system checks, correct operation of the shutter (and the measurement of the usual offset from the integral number of seconds selected for the exposure time), plus various CCD parameters such as read out noise, gain calibration, linearity, etc. can be checked with a suitable series of exposures.

Another valuable technique is to plot the histogram of counts on a frame. The most suitable frame is one containing a ramp of counts ranging from the zero level up to the maximum allowed by the ADC (or full well, whichever is the lowest). A short exposure of a globular cluster is a good approximation. The histogram, after trimming of the CCD edges and any major defects, should show a smooth distribution with no "bumps" or missing codes.

Removal of the instrumental signature is some variant of subtracting a zero-level (bias) frame and then divison by a flat field frame, with removal of defects by interpolation, or simple flagging, as necessary. Dark frames are generally not needed except for very narrow band work, since most CCDs at temperatures of -100C have a dark rate of only a few e^-/hour. In practice several frames of each type are combined together, in order to improve the Poisson statistics and to allow removal of cosmic ray events. The generation of a truly "flat" flat field is something to be aimed for, but unfortunately never achieved. Djorgovski (1984) and Djorgovski & Dickinson (1988) have discussed flat fielding procedures in some detail, and only a few points will be made here.

Dome flats (exposures of an illuminated white spot inside the dome) are very popular since they do not occupy any observing time. Therefore many can be accumulated and statistical counting errors can be very small. Thus pixel to pixel QE variations can be accurately removed. Lower spatial frequencies are less well removed and there are several unfortunate effects which can conspire to make dome flats work poorly. The following are some guidelines: 1) Have a well baffled telescope, else illumination onto the CCD can be very different from the night sky. Sharp edges (dust on filters, window) may not divide out correctly, due presumably to illumination differences; part of this may be due to specular reflections off the white spot which should be made from suitable materials (see Jacoby et al. 1991). 2) You should use lamps with as a high a color temperature as possible, quartz halogen lamps with color temperatures in excess of 3000K are suitable (3300K is the upper limit, and these lamps have typically only 100h lifetime). Since this color temperature is still some 3000K too low, the lamps should be used together with a color balance filter of value -200 mred for BVRI dome flats (not U). Schott BG34 in 2mm or 3mm thickness is suitable. The Melles Griot filters are not suitable as they have high transmission in the far red, as shown in Table 3. Metal halide lamps have color temperatures of 6000K and may be an alternative; their spectra consist of many densely packed lines (rare earth elements) but they do not seem to have strong lines. 3) Your filters must not have red leaks (see Stetson 1988).

Table 3: Color Balance Filters

Band	Effective Wavelength (Ångstroms)	Transmission BG34/3	MellesGriot -180mred
U	3659	0.10	?
B	4382	0.60	0.62
V	5448	0.18	0.20
R	6407	0.05	0.07
I	7982	0.035	0.15?
	9000	0.045	high
	10000	0.07	high

Twilight sky flats can be used alone, or smoothed and used as illumination corrections to rectify the incorrect lower spatial frequencies of dome flats. Twilight and dawn are fleetingly short when one is waiting for a 2048^2 CCD to read out, and the color match is not exact; but if one is using these frames as illumination corrections only then the CCD can be suitably binned to speed read out. Other types of flat fields impact on observing time. Multiple exposures of "star-less" fields can be used as illumination corrections, after removal of the few remaining stars, or, as a variant, some or all of the object frames for a night (or the whole run) can be combined. The latter will not work well if all the frames contain images of globular clusters (see Kuhn et al. 1991 for an opposing view) but can be very successful if many relatively star-free frames can be used. Tyson (1988) has used this technique to reduce his systematic flat field errors from 20 to 0.03 percent. Drift scanning techniques (continuously reading out the CCD while moving it in synchronization) makes flat fielding a 1-D problem but sky brightness, transparency and seeing are then not constant over the whole frame. Careful reduction techniques can minimize these variations, and the method is most valuable when surveys of a large area of sky are required. A variant on this method is short-scanning, in which the CCD is read out only a few (20-100) rows during the exposure. The CCD is moved in synchronization via a precision stage (eg KPNO) or else the TV guider (and thus the telescope) moves on a precision stage (eg CTIO). As pointed out by Djorgovski & Dickinson (1988) this method should have wide application, but has seen little use.

References:

Aldering, G.S., and Bothun, G.D., 1991, PASP, 103, 1296

Bessell, M.S., 1990, PASP, 102, 1181

Devaney, M.N., 1992, A&A, 257, 835

Djorgovski, S., 1984, in: *Proceedings of the Workshop on Improvements to Photometry*, eds. W. Borucki & A. Young, p152

Djorgovski, S., and Dickinson, M., 1988, Highlights of Astronomy, 8, 45

D'Odorico, S., 1990, ESO Messenger, 61, 51

Jacoby, G.H., Schoening, W., and de Veny, J., 1991, NOAO Newsletter, 28, 25

Janesick, J., and Elliot, T., 1991, in: *Astronomical CCD Observing and Reduction Techniques (A.S.P. Conferences Series*, ed. S. Howell, in press

Kuhn, J.R., Lin, H., and Loranz, D., 1991, PASP, 103, 1097

Landolt, A.U., 1992, AJ, in press

Massey, P., and Jacoby, G.H., 1991, in: *Astronomical CCD Observing and Reduction Techniques (A.S.P. Conference Series*, ed. S. Howell, in press

McClure, R.D., et al., 1991, PASP, 103, 570

Oke, J.B., Harris, F.H., Oke, D.C., and Wang D., 1988, PASP, 100, 116

Racine, R., and McClure, R.D., 1989, PASP, 101, 731

Stetson, P.B., 1988, Highlights in Astronomy, 8, 635

Tanvir, N.R., et al., 1991, MNRAS, 253, 21P

Tyson, J.A., 1988, AJ, 96, 1

Young, A.T., 1967, Appl. Optics, 6, 1063

Discussion

M. Zeilik: *What is done at Cerro Tololo to archive CCD frames, especially the large format ones?*

Walker: At present, nothing.

A.T. Young: *Be careful with twilight flats. The twilight spectrum is heavily mutilated by ozone absorption, so it contains almost no UV and has a big dip around 6000 Å due to the Chappius bands.*

R.P. Edwin: *Would you please comment of the spectral variation of flat field calibrations. Pixel to pixel variations are maybe 1% but what is the spectral dependence, typically, in this variation?*

Walker: Photons with shorter wavelengths get converted into electrons nearer the surface of the CCD than do photons with longer wavelenths. U-band photons get converted very near the surface and the resulting U-band quantum efficiency depends critically on surface conditions. On the other hand, I-band flat fields depend little on surface properties. Thus there is little correlation between (say) U-band and I-band flat fields, even though both may have similar pixel-pixel and lower spatial frequency errors. Indeed, thickness variations in thinned CCDs may cause an *anti-correlation* between short and long wavelength flat fields.

D.H.P. Jones: *On La Palma we have been experimenting with running our CCDs 40 ^{o} hotter than normal in order to extend their red response. In this way we have made successful observations of the HeI 10830 A line. Also, on the 1-meter telescope on La Palma, we have a drift-scanning arrangement where the CCD is driven by a lead screw. Every time the lead screw advances the appropriate amount one row of the CCD is read out. This is used to extend the size of the CCD.*

Walker: Referring to your first point, some CCDs now are of MPP design and have a lower dark rate than non-MPP CCDs. These would seem to be particularly suitable for being operated at elevated temperatures.

I. van Breda: *What are the limitations on the type of chip to which you can apply UV flooding?*

Walker: The CCDs must be thinned and have a 'steam oxide' layer grown.

T.J. Kreidl: *I think it is impressive to see the new developments in hardware at large institutes. Many smaller establishments are forced to buy commerical systems which often have older technology. One hopes many of the new developments will eventually trickle down to smaller observatories, and it's important that experiences gleaned from observatories with major development programs are made widely known within the astronomical community.*

Walker: Instrumentation at National Observatories tends to be very well documented, and there is no reason why detailed information cannot be supplied to anyone who requests it.

R. L. Hawkins: *Where is your colour correction filter for dome flats located? Also, how do you then deal with dust spots and irregularities in the colour correction filter which affect your flats?*

Walker: In answer to your first question: in the optical path above the other filters. With regard to the second point: we try to minimize those as much as possible, but I agree that it is a problem.

Further Progress in CCD Photometry

Peter B. Stetson[1]

[1]*Dominion Astrophysical Observatory, Canada*

Abstract

I comment on some of the steps required to convert raw instrumental magnitudes, derived by profile-fitting or synthetic-aperture photometry from CCD images, to final calibrated photometry on a standard system. The status of the DAO program to obtain homogeneous *BV* photometry for star clusters and nearby galaxies will also be discussed.

1. Introduction

Photoelectric photometry is easy. If you want to measure some stars photoelectrically, take your list of coordinates and some finding charts to the dome. Point the telescope, identify the program star, center it in the little hole, and write the star's name down on the strip of paper that comes out of the printer. Integrate for a while through each filter, then move the star out of the hole and integrate on the sky. Reobserve that star a few more times if you want, or move on to the next star in your program. If you are fast you can observe a hundred stars or more in a night. Later, divide the star and sky counts by the integration times, subtract the skies from the star observations, and −2.5 times the logarithm of what's left is your instrumental magnitude; the difference between magnitudes in different filters is an instrumental color. You can do it with a piece of paper and a slide rule, although a pocket calculator is even better. Compare your instrumental magnitudes and colors for some stars observed at several airmasses and derive an extinction correction. Then take your extinction-corrected instrumental magnitudes and colors for some standard stars, compare these to the published magnitudes and colors, and derive transformation equations. Apply these to your program objects. Publish the results.

If you want to measure a lot of stars in a cluster you can do it photographically, which is even easier. At the telescope take several plates of your field. Once your plates are developed and dry, make a large-scale print of the best one. On this, next to the image of each star write a little number. Then take your plates into a darkened room, and place them one by one on the stage of an iris photometer. Consulting your finding chart, steer the machine to each star, center it up, turn a knob to make a needle go to zero, and write down the star's ID number and the number from the readout on the front of the machine onto a piece of paper. Later, plot up machine index *vs* published photometric magnitude for a sequence of stars that have been observed photoelectrically, sketch or compute a smooth curve through those points, and run the machine indices through this curve to provide magnitudes on the standard system. Publish the results.

Nowadays, we have the charge-coupled device (CCD). This remarkable chip combines the advantages of the photomultiplier and the photographic plate: it is linear and repeatable, so a photometric system can be accurately transferred from standard stars in one part of the sky to program stars elsewhere; it is an area detector, so many stars can be recorded with each integration. Observing with a CCD is not much more difficult than observing with a photomultiplier or plates: there may be a few more calibration observations to be made around sunset and sunrise, but on the other hand you don't get eyestrain trying to decide whether some faint puff of light is centered in the aperture, and you *certainly* don't have to slop around in chemicals next morning.

No, the price you pay for enjoying the benefits of the CCD is not in the observing, it's in the analysing. Not only is the analysis more complex with CCDs than with photomultipliers and plates, there are also immensely larger volumes of data. No longer are we talking about one measured datum per star per plate for a few hundred stars in a cluster, or a half dozen data per star for a hundred or so stars observed in a night with a photomultiplier. These days it's possible to obtain a hundred CCD frames in a night, each frame potentially containing of order 10^4 stars and of order 10^6 individual data-numbers. The process of converting these immense mounds of raw data into a publication comes in three basic stages, which may be called "preprocessing", "processing", and "postprocessing."

Preprocessing includes bias-subtraction and flat-fielding, and does not decrease the volume of data from an observing run; what it does do is to ensure that the data are on a consistent photometric zero-point and scale. Dozens of manuals and guidebooks and papers (including several at this meeting) have been written on how to preprocess your images. Perform these steps properly, and you now have images that contain accurate, linear representations of the brightness of objects in your program fields.

Processing — converting the rectified CCD intensity arrays into lists of positions and magnitude indices for the stars contained therein — is another matter. If your CCD frames contain few stars, it is often possible to simulate photomultiplier photometry by simply defining some geometric region as an "aperture": sum up the photons detected within an "aperture" centered on a star, sum up the photons detected within a star-free region of the same area elsewhere in the frame, and call the difference between these two numbers the signal from the star alone. Readout noise in the CCD and the flux contributed by the diffuse light of the nighttime sky make this method very imprecise for faint stars, and for crowded fields, as in star clusters, it may not be possible to define an aperture which is large enough to contain all of one star's flux while entirely excluding the flux of neighbors. For dealing with faint stars against noisy backgrounds and stars in crowded regions, model-profile fitting offers the best possible recovery of photometric information for stellar objects. A half-dozen or so profile-fitting packages written by astronomers originally for their own use have become fairly widely distributed in the field. Since the problem is complex, the various programs are quite individualistic, and adopt an interesting variety of approaches to each of the hurdles that must be overcome (see, *e.g.*, Stetson 1990a,b, 1991a, 1992a for previous discussions and references). The bottom line, however, is that each package reduces the two-dimensional data array representing an input CCD frame to an output list of presumed *bona fide* detections, their positions, and their apparent brightnesses on an instrumental system. At this point, my story today begins.

2. The many and various steps of postprocessing

To begin, let us assume each astronomical image obtained during our CCD run has been reduced to a list of stars: their positions within the coordinate frame of that image and their apparent instrumental magnitudes relative to some zero-point for that image. How do we get from here to the part about publishing the results? I illustrate the procedures required by describing a typical post-processing reduction path, using software with which I am personally familiar.

a. Aperture corrections

First, we note that there may not be a consistent photometric zero-point from frame to frame. While we expect that the CCD is inherently a consistent, stable, linear detector of photons, our measuring process may have mislaid some of that consistency. In profile-fitting photometry, the instrumental magnitude for a star comes from the height of a model point-spread function ("PSF") for that frame scaled to the intensity values recorded inside the star's image. Since the instrumental magnitudes for each frame are defined relative to a *different* PSF, it is difficult to be certain at the sub-percent level that the zero-points for all frames are identical. One solution to this problem has been discussed at length elsewhere (Howell 1989; Stetson 1990a,b), so I will summarize it here in a few words. Synthetic-aperture photometry is performed for several selected, bright, isolated stars in each CCD frame; if necessary, the profile-fitting results are used to subtract unwanted stars from the frame beforehand, to prevent their contaminating the aperture photometry. Either the selected stars are measured through an aperture large enough to contain a constant fraction of their total flux, independent of seeing and guiding differences from frame to frame, or they are measured through a series of apertures of increasing size, and these results are extrapolated on a frame-by-frame basis to an aperture large enough to contain a consistent fraction of their flux. We then rely on the stability of the CCD to provide a consistent relationship between the total number of photons detected from a star, and its true brightness as perceived through the atmosphere, telescope, and a filter, just as one does with a mechanical aperture and a photomultiplier. The integrated total instrumental magnitudes for several stars in each frame are stored in a so-called "total magnitude" file.

b. Cross-identification

Next, since we will eventually want to compare repeat measurements of a given star, whether to beat down noise, to measure variability, or to determine colors from observations through different filters, we must cross-identify re-observations of the same star in different frames. The telescope may have been shifted between successive exposures, and we may want to combine data from different cameras or different telescopes, so we can not expect a given star to reappear at the same position in every image. While a human being armed with a finding chart *could* display each CCD frame and note the position of every star of interest, this could take forever. We want a computer program which, when presented with several lists of positions and magnitudes, can intercompare the lists, recognize prominent asterisms, cross-identify the corresponding stars, and estimate the geometric transformation equations that interrelate the coordinate systems of the various frames. The program must be able to cope with arbitrary translations, rotations, scale changes, and flips, and it must be effective when the overlap between the two lists is incomplete, as when some stars in one frame lie outside the area or beyond the magnitude limit of another.

See Murtagh 1992 for an extensive catalog of automatic pattern-matching algorithms which have been implemented, but let me describe mine briefly here. "DAOMATCH" attempts to recognize triangles of stars, exploiting the fact that while their positions, orientations, and sizes may change, triangles won't change *shape* as they are translated, rotated, and scaled. It considers brightest stars first, since these are the most consistently detectable objects in the field. For each target field, the astronomer selects one frame to serve as the "master" frame — normally this will be the deepest one, or among a set of partially overlapping images it may be the one closest to the center of the pattern. The program reads the star list from the master frame and sorts the stars by apparent instrumental magnitude; it then does the same for the next frame of that field. Taking the brightest three stars from each list, it encodes the shape of each triangle in two numbers: the ratio of the length of the second-longest side to the longest side, b/a, and of the shortest side to the longest side, c/a. These ratios are clearly independent of translations, rotations, and scale changes, and each triangle defines a unique point in two-dimensional $(b/a, c/a)$-space. If the two points in $(b/a, c/a)$-space coincide within a certain tolerance, the three stars may be provisionally cross-identified: the star opposite the longest side in frame 1 corresponds to the star opposite the longest side in frame 2, and so on. Transformations are estimated from the three presumed cross-identifications and are presented to the user for approval. If the scale and rotation factors look about right (*e.g.*, scale ≈ 1, rotation ≈ 0 for frames taken with the same equipment), the user may accept these transformations and move on to the next frame. If, however, the first two triangles do not have the same shape, or if the user wants transformations based on more than three provisional cross-identifications, the program will take the fourth brightest star from each list and compute the three new triangles formed by this star with each pair of the previous three. If any of these have consistent shapes, new cross-identifications are made and the implied transformations are offered to the user. If these are still unacceptable, the fifth brightest star in each frame and six new triangles are taken under consideration. When the user is satisfied that the cross-identifications are correct and the geometric transformations are adequate, the coefficients of the transformations are written to a disk file (a "match file"). The user then enters the name of the star list for the next image of that field, and the process is repeated. In my experience, this routine works reliably and without human help provided the degree of overlap between the input lists is greater than about 25%: if at least one-fourth of the stars which appear in list 1 also appear in list 2, and *vice versa*, the correct transformation constants will be found in a matter of seconds.

The number of possible triangles can be huge ($O(n^3)$), so it is not possible in general to cross-identify *every* star in the field in this way. Therefore another program has been written ("DAOMASTER"), which accepts the approximate transformation equations provided by DAOMATCH, reads in all the star lists for that field, and cross-matches *all* stars by spatial proximity: if after transformation to the coordinate system of the master frame, a star lies within a specified distance of a star in the master list, it is provisionally identified with that star; if it lies near *no* star in the master list it is added to the list as a possible new detection. As subsequent lists are considered, their stars may be identified with ones in the original master list or with stars added from other frames, or may themselves be added to the master list if no such identification is possible. After all possible cross-identifications have been made, the augmented master list is cleaned of "insignificant" detections, on the basis of user-supplied acceptance criteria. Then the current set of acceptable cross-identifications is used to refine the geometric transformation constants, and all available positional determinations for each star are averaged to provide a New, Improved! master list. The process is repeated with an increasingly stringent tolerance for positional agreement: at the beginning a rather lax agreement criterion is used because some allowance must be made for positional mis-match caused by errors in the provisional transformations provided by DAOMATCH; as the transformations are refined, a more strict criterion is used to minimize the number of false cross-identifications. Several passes of increasing strictness can be performed on several dozen star lists for a given field in a few minutes.

Once the user is satisfied that the geometric transformations are accurate and the master list is as complete as possible, DAOMASTER offers several types of output, such as: a copy of the complete master list with average positions and magnitudes on the system of the master frame (a "master file"); a file containing the new, more accurate transformation equations (a new "match file"); and a simple transfer table of cross-identifications (a "transfer file"): star 1 in input list 1 is the same as star 37 in list 2, it doesn't appear in list 3, it is star 196 in list 4, ... The output files can serve a wide variety of purposes. For instance, the master file contains an index of the frame-to-frame scatter in a star's magnitudes, which may be used to identify variables if all frames were taken through the same filter. The new match file also contains an estimate of the magnitude zero-point differences among the frames; taken together, the geometric transformations and photometric offsets allow the user to combine partially overlapping CCD frames in a single montaged image covering a larger area of sky (see, *e.g.*, Stetson & Harris 1988, Figs. 1–6.) The transfer file will be used later in the final photometric reductions.

c. Real-world identifications

At this point each star is still only a few entries in some lists. Although we can now identify each star of interest among all frames of a given field, we have yet to associate particular entries in the master list with real-world names, such as the stars' identifications in published standard-star lists or in previous photometric investigations of the cluster. This may be done with a routine called FETCH, which was once a FORTRAN program interacting with the DAO's VICOM image display; it has since been reincarnated as an IRAF script. The user displays the master frame or a montaged image of the target field. Then, having placed the cursor on a star, the user enters the star's real-world name; the routine then searches the master file for the star lying closest to the cursor's position. A "fetch file" is created which contains the stars' real-world names and the corresponding entry in the master list — in particular, the stars' (x, y)-coordinates in the system of the field's master frame. This user-interactive process needs to be done only once per field.

d. Preparing for the photometric transformations

We now have almost everything we need to derive the photometric transformations between our instrumental system and the standard system: we can identify a given star's real-world name with a a particular (x, y) position in the master system for its field; either through the geometric transformations derived in §b above or using the transfer file from §c we can then find a named star in *any* frame of that field. Now we must provide a file containing the ancillary data for each exposure: the filter, the time of mid-exposure, the airmass, and the exposure time. This is not difficult. Now we are ready to collect the observed data for named stars from a night's worth of observing into a single file.

It is easiest for me to describe the process by following the series of steps taken by my computer program, "COLLECT." The user begins by specifying the name of the file with the ancillary data (filter, airmass, exposure time, etc.) for all the CCD frames of the night. Then, frame by frame, the user enters the name of the file with the large-aperture photometry for selected stars in that frame, which was obtained as described in §a above. It reads in both these data and the profile-fitting magnitudes for *all* stars the same frame from the output of the profile-fitting routine. It identifies the stars with aperture photometry among the profile-fitting results by positional coincidence, and uses the comparison of these data to determine the additive correction required to place the profile-fitting magnitudes onto the system of the aperture magnitudes; this allows a consistent comparison with data from other exposures. Next the user enters the name of the fetch and match files for the field. Named star by named star, the program takes the star's master-frame position from the fetch file, uses the geometric transformations to determine the star's position in *this* frame, and searches for the star among the input lists of aperture and profile-fitting photometry. If the named star has both aperture and profile-fitting results, the program considers the standard error of each: the large-aperture magnitude may be uncertain because of read noise and sky noise in the large aperture, or because it has been extrapolated from a much smaller aperture by means of a poorly defined growth curve; the profile-fitting magnitude may be uncertain because the correction from the profile-fitting zero-point to the large-aperture zero-point may be badly determined for that frame. Whichever of the two total-magnitude estimates is less uncertain is written to the output file, which I call an "observation file," along with the star's real-world name, and the filter, time, airmass, and exposure time of the observation. If no concentric-aperture photometry is available for that particular star, then it is the profile-fitting magnitude corrected to the system of the large-aperture photometry that is written out. The process of building up the observation file proceeds as fast as the user can type a file name followed by several carriage returns; a night's data for stars with real-world names can be assembled in around a half hour.

e. Deriving the transformations

I have described my transformation procedure elsewhere (*e.g.*, Stetson & Harris 1988, Stetson 1992b), so here I will provide only a brief summary. The observed magnitudes on the (large-aperture) instrumental system are fitted by robust least-squares to transformation equations consisting of terms involving the standard-system magnitudes and colors of the standard stars, and the airmasses and times of observation of the frames. For instance, for BV photometry the equations may be of the form

$$\begin{aligned} v_{ij} &= V_j + F_1(V_j, (B-V)_j, X_i, T_i) \\ b_{ij} &= B_j + F_2(V_j, (B-V)_j, X_i, T_i) \end{aligned}$$

for a measurement of standard star j in frame i, where maybe

$$F_1 = A_0 + A_1 \cdot (B-V) + A_2 \cdot (B-V)^2 + A_3 \cdot X + A_4 \cdot (B-V) \cdot X + A_5 \cdot T \cdot X + \ldots$$

I want to stress that the *observed, instrumental magnitudes* are fitted by functions of the *standard photometric indices* and the time and airmass of the observation — no attempt is made to derive or employ instrumental colors. Not only is this proper least squares, since the errors of the fit are dominated by uncertainty in the observed instrumental magnitudes, but it is also more convenient: unlike the case with photoelectric photometers, where it is easy to cycle quickly through the filters (*e.g.*, *VVBBBBVV*), so each observation can be made symmetric about its midpoint and both filters therefore have the same effective airmass, it is difficult to define consistent instrumental colors with a CCD. The airmasses of the B and V exposures of a given frame pair will necessarily be somewhat different, and the difference between the two will change depending upon the declination, hour angle, and duration of each exposure. The *standard-system* magnitude and colors are constant attributes of a (non-variable) star, and the transformation equations can be consistently expressed in terms of them.

The equations above are suitable for photometric nights, where the throughput of the system depends only upon the color of the star and the airmass of the observation (with perhaps an allowance for a secular time-variation of the atmospheric transparency — the $X \cdot T$ terms). On cirrusy nights, the transparency may fluctuate unpredictably from frame to frame. We nevertheless expect that the effective exposure for different stars *in the same frame* will be the same, and we may define analogous transformation equations: *e.g.*,

$$F_1' = A_{0,i} + A_1 \cdot (B - V) + A_2 \cdot (B - V)^2 + A_4 \cdot (B - V) \cdot X + \ldots$$

Here, $A_{0,i}$ is not a constant applying to all frames for the night, but is rather an instantaneous photometric zero-point which applies to the i-th frame alone. The other constants A_k have precisely the same meaning as in the clear-sky reductions. Provided at least some of the frames taken during the night each contain several photometric standards with a range of colors, the color-transformation terms can be computed; once they have been established, any frame which contains even one photometric standard can have its individual zero-point determined.

My program to perform these solutions ("CCDSTD") requires a file containing a standard-star library, where each star is identified by its real-world name accompanied by its published photometry on the standard system, a short "transformation file" expressing the form of the desired transformation equations in a simple code, and the observation file created in §d above. It then computes the least-squares transformation equations, whether in the clear-sky or cloudy-sky mode. The encoded transformation equations and the derived values of the transformation coefficients are written out to a "calibration file."

f. Applying the calibrations

Once the photometric calibrations for a night, or an observing run, or a series of runs have been derived, they can be applied to program stars in an inverse sense, to infer their values of V and B–V (in the current example) from their observed values of v and b. This is done by Newtonian approximation:

$$\begin{aligned} V_j &= v_{ij} - F_1^{(i)}(V_j, (B-V)_j, X_i, T_i) \\ B_j &= b_{ij} - F_2^{(i)}(V_j, (B-V)_j, X_i, T_i) \end{aligned}$$

This is another reason why the transformation equations are expressed in terms of *standard-system* magnitudes and colors: although we do not yet know these values for our program stars, we can utilize *all* the v and b observations we ever made of the star — from different frames, different nights, different runs, different telescopes, whatever — to determine them. Each observation has a corresponding set of calibration constants A so, having assembled all the observations, the program merely postulates a crude guess at the standard-system values of V and B–V for star j, and computes new, least-squares estimates of V and B from the observed residuals $\langle v - F_1 \rangle$ and $\langle b - F_2 \rangle$. The new values are fed back into the right sides of the equations and the process is iterated to convergence, which happens very quickly. By the end of the process, each observation has been calibrated with the *best possible* estimate of the star's true magnitude and colors. In contrast, had we tried to define the transformation equations in terms of the *instrumental* colors (*e.g.*, $V = v + f_1(v, b-v, X, T)$) then one night's observation of the star would have been calibrated using that night's value of b–v (which might be ill-determined), another night's observation would be calibrated using a different, comparably poor, value of b–v, and data from a night on which the star was accidentally observed in only one filter would have to be discarded. Not so with the present scheme.

There are two situations to which this method may be applied: (1) to stars with real-world names which appear in the observation file created in §d above — these could be standard stars whose magnitudes and colors you want to redetermine from your own observations, or stars with names of your own devising without prior photometry; or (2) stars in some cluster or other field, most of which have no names yet, but which exist as a set of cross-references obtained by DAOMASTER (§b). In essence, these situations may be thought of as photomultiplier-like and photographic-like reduction modes. The approaches taken in the two cases differ.

In the first situation the user simply runs a program named "CCDAVE" and types in the names of all the observation files to be included in the analysis. The program reads in the observed data and the corresponding transformation constants from the appropriate calibration files. It then assembles all the available data for each star — identified by real-world name — and performs the inverse transformation by Newtonian approximation as described above. The program produces a file with the final, averaged photometry and its standard errors, as well as an index of the goodness-of-agreement which can be used to help spot variable stars.

The second situation is slightly more complex. A program named "FINAL" accepts the transfer file generated by DAOMASTER, which contains the names of the profile-fitting photometry files for all the CCD frames of the field, and the transfer table of cross-references for all observations of stars in the final master list. It then accepts the final averaged photometry for named stars produced by CCDAVE above. Next the user specifies the name of the fetch file for the field; from this the program learns the real-world names of local standards in the field, and recovers their standard-system magnitudes and colors from the averaged-photometry file. It also uses these stars' positions as recorded in the fetch file plus the cross-references provided by the transfer file to recover each star's instrumental, profile-fitting magnitude from each CCD frame. Finally, it reads in the photometric transformation constants for each frame from the calibration file for the night on which the frame was obtained. Now we are ready to fly.

First, FINAL redetermines the photometric zero-point of each frame from the local named stars with standard-system photometry generated from CCDAVE above: for each frame small correction to A_0 is estimated from

$$\hat{A}_0 = \langle v - V - F_1 \rangle$$

The reason for this step is precision. There are any number of reasons why any given

CCD frame may depart systematically from the optimum photometric calibration for its night by a few hundredths or a few thousandths of a magnitude: uncertainty in the growth-curve correction, a contrail drifting past the telescope during the exposure, a touch of mist on the telescope mirror, ... Such frame-to-frame jitter in the photometric zero-point can introduce extra scatter in a cluster's color-magnitude diagram, since not all stars will necessarily be measured in all frames. By recorrecting all frames to a single consistent zero-point defined by a specific sequence of well-measured local standards, one does not place the average cluster photometry on the standard system with any greater accuracy, but one does produce sharper sequences in the color-magnitude diagram.

The rest is easy. The set of cross-references contained in the transfer table allows FINAL to assemble all observations of each star in the master list for the program field, and the photometric calibrations — complete with the revised zero-points — are used to correct these observations to the standard system by Newtonian approximation, as before.

Now you may publish the results.

3. Current status of the DAO cluster-photometry program

Since 1983, a number of us at the DAO have been collecting observations in the *BV* photometric system for open clusters, globular clusters, and nearby galaxies. Contributing observers include Roger Bell, Mike Bolte, Bill Harris, Jim Hesser, Bob McClure, Linda Stryker, Nick Suntzeff, Don VandenBerg, and me. Many of these observations have been reduced — mostly by means rather more crude than described above — and published. But in bits of time scattered between other projects, I have been trying to rereduce these data using the latest software and consistent reduction techniques. As of this moment, early August 1992, data from 16 observing runs have been reduced. These comprise a total of 52 nights (33 photometric, 19 partly cloudy) on five telescopes (CTIO 4m and 0.9m, KPNO 4m and 0.9m, CFHT 3.6m), from which 2,556 useable CCD frames of astronomical objects have resulted.

The central pillar of our standard system is the work of Landolt (1973, 1992), but additional standards have been taken from a number of sources: Graham 1982 (E regions); W. E. Harris, unpublished (M11 cluster); Christian *et al.* 1985 and L. E. Davis, unpublished (M92 and NGC 7006); Heasley and Christian 1986 (M92); Baade and Swope 1961 and Stetson 1979 (Draco); Hawarden 1970 and Anthony-Twarog *et al.* 1979 (Mel 66); Stetson 1981 (NGC 1851); Tifft 1963 (field near NGC 121); and Graham 1981 (field near NGC 300) bring the total number of "external" standards which we have used to date to 255, most of them observed by us many times. Through the intermediary of our own observations we have derived zero-point corrections to place the results of these latter studies on the system of Landolt's work — in the mean, at least — after the fashion of Stetson and Harris 1988 and Stetson 1991b.

From our own observations we are also able to define a number of new secondary "standards", which serve two important purposes. First, since they have been reobserved a number of times on different nights and with different telescopes, adding well-observed program stars to the standard list strengthens the internal homogeneity of our various reduced data sets. Second, the secondary standards in program fields will be the "local standards" used to refine each field's photometric zero-points in the second mode of applying the transformations as described above. In addition to the 255 primary standards, we have so far been able to define 2,325 secondary standards in a total of 95 CCD-sized fields scattered over the sky. The secondary standards all satisfy the criteria $N(V) \geq 4$, $N(B) \geq 4$, $\sigma(V) \leq 0.02$ mag, $\sigma(B-V) \leq 0.03$ mag, and $\chi < 2$, where χ represents the ratio of the observed, external, frame-to-frame scatter in the results for the star to the scatter expected from the photon statistics, read noise, quality of the profile fits, and quality of the nightly transformation solutions; a large χ value could indicate intrinsic variability or crowding/seeing problems, either of which would disqualify a star as a useful standard.

The work continues, and may even see publication some day.

References:

Anthony-Twarog, B. J., Twarog, B. A. & McClure, R.D. 1979, ApJ, 233, 188

Baade, W. & Swope, H. 1961, AJ, 66, 300

Christian, C. A., Adams, M., Barnes, J. V., Butcher, H., Hayes, D. S., Mould, J. R., and Siegel, M. 1985, PASP, 97, 363

Graham, J. A. 1981, PASP, 93, 29

Graham, J. A. 1982, PASP, 94, 244

Hawarden, T. G. 1970, MNRAS, 174, 471

Heasley, J. N. & Christian, C. A. 1986, ApJ, 307, 738

Howell, S. B. 1989, PASP, 101, 616

Landolt, A. U. 1973, AJ, 78, 959

Landolt, A. U. 1992, AJ, 104, 340

Murtagh, F. 1992, PASP, 104, 301

Stetson, P. B. 1979, AJ, 84, 1149

Stetson, P. B. 1981, AJ, 86, 687

Stetson, P. B. 1990a, in *CCDs in Astronomy II: New Methods and Applications of CCD Technology*, eds. A. G. D. Philip, D. S. Hayes & S. J. Adelman, L. Davis Press, Schenectady, p. 71

Stetson, P. B. 1990b, 102, 932

Stetson, P. B. 1991a, in *3rd ESO/ST-ECF Data Analysis Workshop*, eds. P. J. Grosbøl & R. H. Warmels, European Southern Observatory, Garching, p. 187

Stetson, P. B. 1991b, AJ, 102, 589

Stetson, P. B. 1992a, in *Astronomical Data Analysis Software and Systems I*, eds. D. M. Worrall, C. Biemesderfer & J. Barnes, Astr. Soc. of Pacific Conf. Series, 25, p. 297

Stetson, P. B. 1992b, J. Roy. Astron. Soc. Can., 86, 71

Stetson, P. B. & Harris, W. E. 1988, AJ, 96, 909

Tifft, W. G. 1963, MNRAS, 125, 199

Discussion

C. Morossi: *How do you take into account shutter problems in your data reduction procedure?*

Stetson: See the *Highlights in Astronomy* article referred to by Alistair Walker. The problem can be largely circumvented by avoiding the shortest exposure times – less than, say, 10 seconds.

M.J. Stift: *In profile-fitting one has to be aware that the shapes of the profiles can depend on colour when using broadband photometry. Already, at airmass 1.2, asymmetric profiles (positive or negative skewness) are possible. Taking the maximum intensity of a profile can lead to systematic errors in magnitude.*

Stetson: I believe that that would be a multiply differential effect. To first order, the profile fits conserve profile volume, so for isolated stars it is a second-order effect, and for faint stars inside the profile of brighter stars it would be a first-order problem. The model profile is derived from several stars in each field, and in some sense it is appropriate to a star of average colour; I think only a star of extreme colour could be seriously affected. Finally, the problem would affect standards just as it does program stars, so it would come out in the colour terms of the calibration.

R.M. Genet: *How much human time is involved in the reduction? Can it be totally automated?*

Stetson: I spend of the order of 10-15 minutes of my time per CCD frame. It is totally automated now, if you don't want to do intermediate visual checks.

A. Walker: *Are programs such as DAOGRON etc. available for the general user in much the same way as DAOPHOT is?*

Stetson: I've never refused a copy to anyone who asked me nicely.

T. von Hippel: *Do you have plans to add astrometry to your package?*

Stetson: People at the U.S. Naval Observatory are extracting astrometric information from DAOPHOT reductions. The process of correcting for scale, rotation, and other transformation terms requires another post-processing package analogous to the one I've described for photometry.

A.J. Penny: *Do the programmes take into account the change in atmosphere extinction due to the change in the zenith distance across a frame?*

Stetson: No. For small fields this is negligible if observations are made at reasonable air masses.

S.B. Howell: *I'd just like to point out that timing errors in the shutter are a big problem when using brighter standards such as the original Landot stars. Getting many sets of* *fainter standards is important.*

A. T. Young: There must be angular effects in CCDs, which should show up in defocussed images. The response depends on angle as well as position. A pixel in an out-of-focus image receives light from only a small part of the pupil. But in flat fields, it sees the whole pupil. These effects, at least, should vary according to the Fresnel reflection coefficients in back-illuminated CCDs, and should be larger due to shadowing by electrodes in front-side illumination.

Stetson: I remain to be convinced that these effects would cause photons to be systematically lost or gained at the 0.5% level in slightly defocussed images.

Problems of CCD flat fielding

William Tobin

Mount John University Observatory & Department of Physics and Astronomy, University of Canterbury, Private Bag 4800, Christchurch, New Zealand

Abstract

CCD flat fielding using an illuminated dome screen is discussed. Rings of reduced response caused by dust specks on the cryostat entrance window indicate the required uniformity of screen illumination but may not divide satisfactorily because of telescope flexure. Illumination colour is important for shorter-wavelength, broad-band filters. Flat fields can be distorted significantly by light scattered off telescope baffles. Special precautions are needed for uv flat fielding. The Mt John TH 7882 CDA chip is slightly sensitive to polarisation. Dome flat-fielding accuracy of 0.3% would seem achievable.

1. Introduction

The system sensitivity is nonuniform across an astronomical CCD image because of factors such as interpixel sensivity variations intrinsic to the chip, shadows cast by dust specks on the cryostat window and other optical surfaces, filter inhomogeneities, streaky or bubbly optical cements, and telescope vignetting. A flat-field (sensitivity-map) image attempts to represent the combined effect of these variations. Before photometry can be extracted from CCD science images they must be corrected to uniform sensitivity through division by the appropriate flat-field image. (The process of flat fielding cannot of course correct for any intrapixel sensitivity variation.)

Correct flat fielding is crucial for the observer who seeks precise and accurate photometry. This paper discusses aspects of flat-fielding attempts made using the Photometrics Ltd. cryogenic CCD system which has been in use at the Mt John University Observatory (MJUO), Lake Tekapo, since late 1989 (e.g. Tobin, 1991). Because MJUO is a private facility, it has been possible to conduct more tests than is practical for a visiting observer at an international facility. While I have yet to achieve the ~0.3% flat-fielding accuracy desirable for my scientific programme (photometry of eclipsing binaries in the Magellanic Clouds), I hope my experience so far will provide useful guidance to others, especially those at smaller observatories. The reader should also study the articles by Djorkovski (1984), Djorkovski & Dickinson (1989) and Stetson (1989), and Chapter 5 of the book by Buil (1991).

To obtain a flat-field map it is only necessary to image a uniform, extended source. The twilight sky may be a suitable source for the small fields of view of many CCDs, but twilight flats are difficult to obtain because the sky must be unambiguously clear

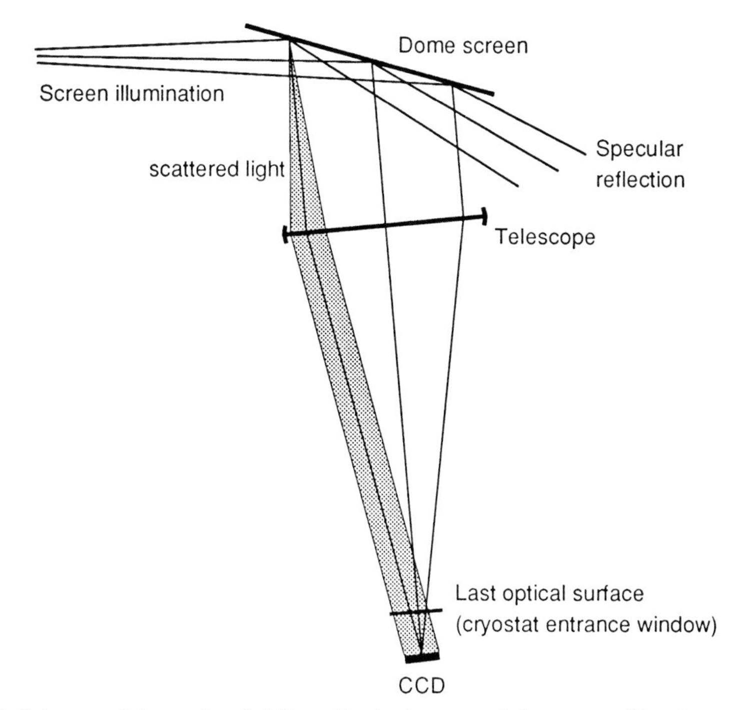

Figure 1 Schema of dome flat fielding. Each element of the screen illuminates the whole CCD, and each point of the CCD is illuminated by all elements of the screen.

and the signal strengths are appropriate for only a few minutes. Another, easier option is to use an illuminated screen within the dome: with the telescope focused at infinity, the claim is that the screen is so far out of focus that the telescope focal-plane illumination is nevertheless uniform.

At MJUO resident technician-observers do much of the variable-star observing. They cannot routinely be asked to extend their working night by the 1–2 hours needed to obtain sky flats (nor can clear twilight be assumed in New Zealand). On a nightly basis dome flats must be used at MJUO, so this paper concentrates on the question of dome flatting.

The tests reported here have been made with either the 0.61-m or the 1-m cassegrain telescopes at MJUO. The CCD is an overcoated 384×576-pixel Thomson TH 7882 CDA chip. On both telescopes its field of view is ~3.8×5.8 arcmin.

2. Dome flat fielding.

Can dome flatting be satisfactory? I think so, if properly arranged. Figure 1 sketches an arrangement. So long as one avoids angles at which the dome screen scattering function is changing rapidly (i.e. close to specular reflection) each elemental area of

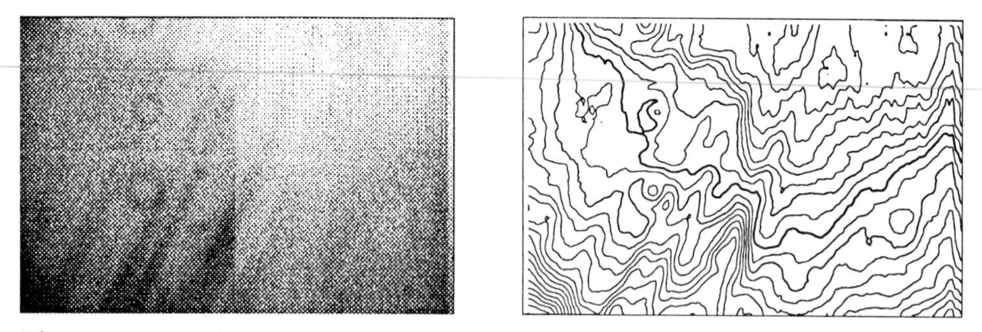

Figure 2 A MJUO V flat field represented *left* as a positive intensity-coded image and *right* in smoothed form as a contour map in which the heavy line represents the mean and the contour spacing is 0.5%. The r.m.s. variation of the flat field is 3.1% with a range of ~12%. In U, B and I the r.m.s. variations are respectively 3.6%, 3.5% and 2.5%.

the dome screen provides uniform illumination within the few arcminute spread of the ray directions which the telescope focuses to differing points on the CCD. The chip is thus uniformly illuminated (and its sensitivity variations correctly mapped) whatever the illumination pattern of the dome screen.

The uniformity of the screen illumination is however important for mapping the throughput variations caused by optical elements in front of the detector. The most obvious of these variations are rings of reduced response resulting from the occultation of the hollow core of the telescope beam by dust specks on the cryostat entrance window. Three examples are clearly seen in the sample MJUO flat field shown in Figure 2, which also shows sensitivity variations on all spatial scales. If the telescope beam is not uniformly filled because the screen is unevenly illuminated, the shadow ring is modulated and differs in intensity from that caused by a star. At MJUO these dust shadow rings depress the level of the flat field by ~3%, and this furnishes an indication of how uniform the scattered light from the dome screen must be. For 0.3% flat fielding the scattered light across the board must be uniform to ~0.3%/3% or ~10%. This can probably be checked with a light meter.

The relative insensitivity of MJUO dome flats to the details of the screen illumination is illustrated in Figure 3 which presents a ratio image in which a dome flat obtained with only *half* the screen illuminated has been divided by a dome flat obtained with a fully-illuminated screen. This severe modulation of the illumination generally only affects the flat field at the ~4% level. These variations principally derive from the glue that holds together the individual glasses in the filter some 50 mm in front of the CCD chip. The dust rings, being caused closer to the chip, are more severely affected and show more obviously the effect of the half-illuminated beam.

So long as the telescope is properly focused at infinity MJUO dome flats seem reproducible at the ~1% level irrespective of telescope orientation and small differences of screen illumination (e.g. illuminating lamp to east *vs.* lamp to west). However,

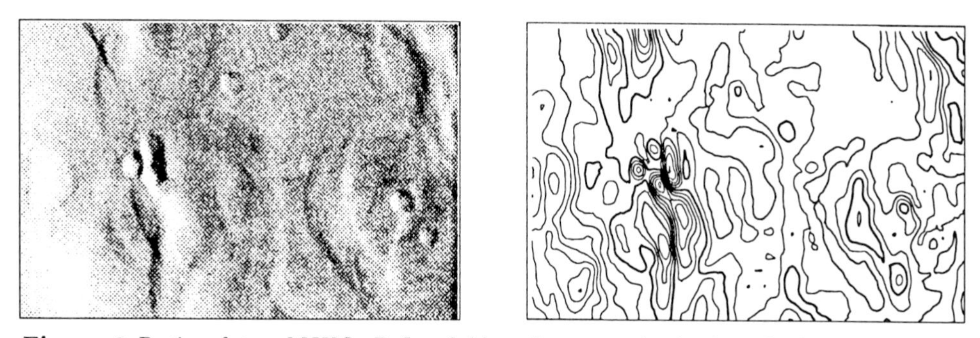

Figure 3 Ratio of two MJUO *B* flat fields. One was obtained with the dome screen illuminated in the normal way, the other with only half the screen illuminated. The ratio image is presented in the same two ways as in Figure 2.

because of flexure, the position of the dust rings on the chip varies with telescope orientation. For the best photometry dust rings should not impinge upon target stars.

Sky flats in *B*, *V* and *I* obtained from the dusk sky were compared with dome flats. The high spatial frequency variations agree very well, but the (sky flat/dome flat) ratios show large-scale systematic effects in all three bandpasses at the 1.0-1.5% level (Figure 4 gives an example). The filter bandpasses are weighted differently by the red illumination of the dome flats and the blue illumination of the sky flats. To investigate whether this could be the cause of the (sky/dome) differences, dome flats were obtained with the regular *UBVI* filters, and then with cut-on or cut-off filters which bisected the *UBVI* bandpasses placed in front of the screen-illuminating lamp. Linear combinations of the full and bisected *UBVI* responses were made which more-or-less matched the responses to sky illumination, and corresponding synthetic (sky flat/dome flat) ratios were calculated. These show systematic differences of ~1.5% in *U*, ~1% in *B*, ~0.5% in *V* and ~0.1% in *I*. Figure 4 presents the results for *B*. The predicted (sky/dome) variation is sufficient to explain the observed (sky/dome) variation in *B*, but not in *I* (with *V* an intermediate case). Conversely, colour effects are likely to be important in the shorter-wavelength broad-band filters at the ~0.5% level in *V*, increasing to ~1.5% in *U*. Appropriate colour-balance filters are desirable to match the dome screen illumination to the colour of the stars being observed for short-wavelength, broad bands.

With dome and sky flats agreeing at the ~1% level one might think that MJUO flat-field images are flat at this level. To check this, observations were made of two E-region standards which are ~1 arcmin apart on the sky. The magnitude difference between the stars varied in a reproducible way by ~4% depending on their location on the chip. Despite their similarity, the dome and sky flats are warped.

I believe the cause of this systematic error is scattered light. The telescope baffles prevent sky light from reaching the CCD chip directly, but as Figure 5 shows, considerable light is scattered off the inside of the primary mirror chimney baffle. The

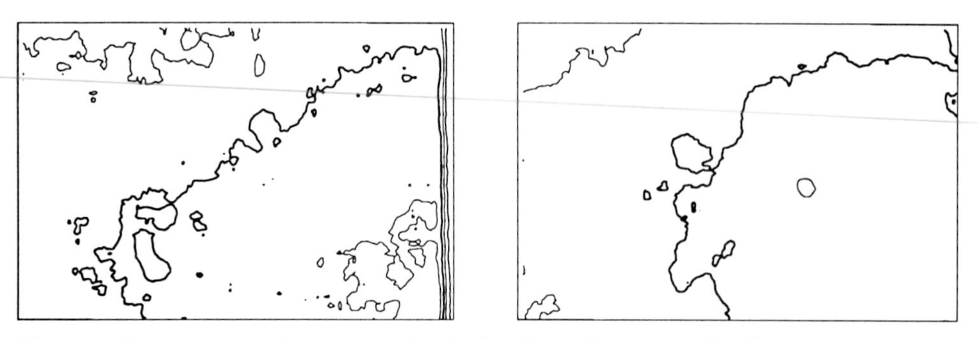

Figure 4 Contour plots of smoothed B (sky flat/dome flat) ratios (0.5% contour intervals). *Left* observed ratio, *right* model ratio synthesized from dome flats obtained with full and bisected bandpasses. The plots are very similar.

dome and sky flats agree because both derive from an extended source (the sky or the rectangular dome screen and reflective pale green dome interior), but both are wrong. Support for this explanation comes from dome flats obtained after the dome screen had been painted down with black paint to the size of the telescope beam. These flats are curved by several percent compared to earlier dome flats.

Annuli, which are very effective at curtailing scattered light, will now be installed in the baffles. Non-flatness of dome and sky flats can probably better be tested using not a pair of stars but observations of an extended open cluster taken at various different chip positionings. Pairwise magnitude differences can then be fitted to a 2D polynomial to check for and to model any remaining curvature of the flat field images, as suggested by Djorgovski (1984).

With the understanding of flat-fielding now obtained, I believe that the prospects are good for achieving the desired 0.3% flat-fielding accuracy.

3. Screens and lamps for dome flat fielding

Until recently slide projectors were used to illuminate rectangular screens of white-plasticized wood composite. In future I wish to extend observations to Strömgren u, and for this the spectral reflectance of the plasticization has been checked since many modern white materials contain 'optical brighteners'. These fluorescent dyes absorb from 300-400 nm and re-emit at longer wavelengths. The reflectance of the MJUO board drops by ~50% from 420 nm to 380 nm, which may cause colour-balance problems not so much for uv filters as for B and Strömgren v. Teflon sheet would seem a possibly more suitable screen material.

A uv-rich lamp is required to illuminate the screen. Mercury lamps show strong emission lines. Deuterium lamps are spectrally smooth and very blue from 300–400 nm, but are intrinsically faint. Xenon lamps have good and smooth uv output but necessitate precautions against explosion, ozone—and sunburn. Because of its safety, simplicity, low light spill, filter slots, easy availability and low cost a 1200 W theatrical

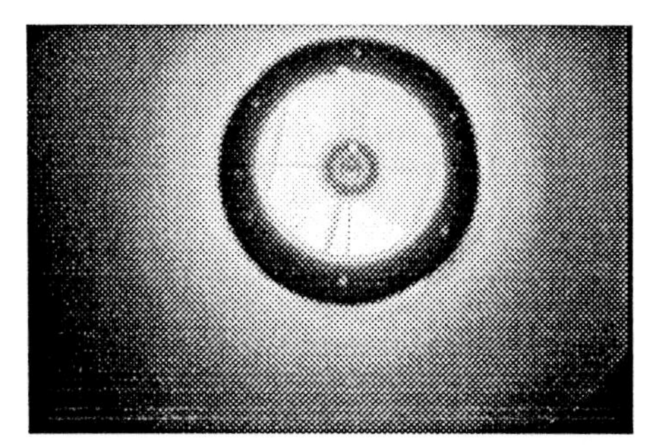

Figure 5 Scattered light in the chimney baffle of the MJUO 0.61-m Boller & Chivens cassegrain reflector. This image was acquired by placing a 50-mm camera lens in front of the CCD. The scattered light in the image is 11% of the light in the beam.

spotlight with mains-operated quartz-halide bulb has now been purchased for MJUO dome flat illumination. Its glass fresnel lens has been removed and a tube collimator installed to restrict its beam to the dome screen (rings within this tube dramatically tightened the beam). Using this spotlight, a dome flat with 0.5% statistical noise can be obtained in U in 40 s, and, it is estimated, in u in 4 min. For colour balancing or attenuation theatrical filters are readily available and can stand proximity to a 1200 W bulb. Further, filter sample books now include spectrophotometric tracings (though only to ~700 nm). The 1200 W spotlight is far too bright for red and ir flat fielding, so a 5 W lamp is under construction for this.

4. Polarization sensitivity

An attempt has been made to check the commonly-made assertion that CCDs are insensitive to polarization. Polaroid sheet was placed in a 1-m diameter rotatable holder located between the telescope and the dome screen. Dome flats obtained at different angles of polarization were compared for U, B and V. (Polaroid sheet is ineffective at I.) For a 90° rotation of the plane of polarization there is evidence of pixel-to-pixel sensitivity variations of at least 0.7%, but less than 3%.

Acknowledgments:

I thank Stephen Duncan, Murray Forbes and John Pritchard for assistance with various aspects of this work.

References:

Buil, C., 1991, *CCD Astronomy*, Willman-Bell, Inc., Richmond.

Djorgovski, S., 1984, in: *Proceedings of the Workshop on Improvements to Photometry*, eds. W.J. Borucki & A. Young, NASA CP-2350.

Djorgovski, S., Dickinson, M., 1989, Highlights of Astronomy **8**, 645.

Stetson, P.B., 1989, Highlights of Astronomy **8**, 635.

Tobin, W., 1991, Proc. Astron. Soc. Australia **9**, 164.

Discussion

T.L. Kreidl: *I would contend that sky flats are particularly important when weather conditions are variable. In certain bandpasses, the night sky lines can be quite variable and hence, affect the flat fielding significantly.*

Tobin: This applies for sky-limited images. It will be less of a problem if you are observing brighter stars.

A.T. Young: *The gradient pictures you showed, in comparing flats taken with different pupil illuminations, illustrate those angular effects I mentioned after Peter Stetson's talk. The stray light is in part due to scattering from dust on optical surface.*

W. Tobin: Quite so.

R. Florentin-Nielsen: *What kind of data acquisition system do you use, i.e where do the data go from the CCD controller.*

A. Walker: The data go into a SUN computer. This is presently a VME-bus SUN, but we are changing to use S-bus SUN's.

R. Florentin-Nielsen: *Did you try to let daylight illuminate your dome board for your dome flats.*

Tobin: No, but I may, once I've reduced the scattered light.

J. Tinbergen: *The last three talks bear out what I said (hoping it was true): Imagers are receiving sufficient attention. However, even better results can probably be achieved by experimentation, such as described by Tobin, but with an optical engineer looking over your shoulder. In particular, improvements might be expected from a pupil image within the photometer, by putting filters and a shutter in (cf. Walker, review) and a properly designed calibration light system to supplement the sky flat field information. It is worth looking at the TAURUS II optical system (La Palma Observers' Guide or AAO equivalent).*

Time-Series Photometry: CCDs vs. PMTs

T.J. Kreidl

Lowell Observatory, Flagstaff, Arizona 86001 U.S.A.

Abstract

With the ability to obtain simultaneous photometry of many objects, CCD time-series photometry is a potentially powerful method for obtaining data, even under non-photometric conditions. In particular, the ability to utilize one or more comparison stars on the same frame without the need to move the telescope to a different field makes for a higher duty cycle than conventional photoelectric photometry. In addition, the ability to determine the local sky in a variety of ways plus the ability to use more complex analysis techniques such as profile fitting and curves of growth permits a variety of analysis options. Some of the advantages of utilizing CCDs and the techniques used in time-series photometry of compact objects are discussed. With the flexibility of modern CCD control systems, possibilities for real-time or near real-time data analysis using readily available computer technology are stressed. Brief discussions of periodicity analysis considerations and other aspects of the data acquisition are presented.

1. Introduction and history

Time-series photometry is basically an extension of classical differential photometry, except that with computer-controlled data acquisition systems and telescope control, it has become possible to work considerably more efficiently. The evolution from a two- or multi-star photometer to a CCD as a multi-object photometer was a natural consequence. It did, however, require that CCDs and their control systems reach a certain maturity in terms of achievable precision as well as adequate field coverage, not to mention reasonable readout times and adequate data storage space.

Some of the first such experiments were carried out by Howell and Jacoby (1986). An on-line system was described by Stover and Allen (1987). As soon as computers and disk storage became reasonably inexpensive, the use of CCDs for time-series photometry exploded. Once CCD systems came down in price and became more commercially available, their use on on small telescopes became possible even for observatories with modest resources. Before discussing some of the characteristics of CCDs, a brief review of photoelectric photometry (PP) is appropriate.

2. Photoelectric photometry

For PP of a single star, "bright" sources are dominated by scintillation noise and by sky measurement errors due to the need to use large apertures. Changes in seeing, periodic drive errors, or poor repetition of centering can have drastic consequences for small apertures. Precision is *not* limited by shot and scintillation noise. In general, the inability to see (and hence center on) very faint objects is itself a problem, and in addition, faint stars that can contaminate the aperture are often invisible or nearly so.

For differential photometry (DP), better results than for PP can be obtained in some cases. In particular, for studies of variability on time scales roughly greater than 20 minutes, DP is preferable. Problems can result from drifts in transparency on time scales similar to the sampling rate, and from poor centering or repeatability of aperture placement on the sky. Also, if large differences between star brightnesses exist, the poorer photon statistics of the fainter object degrade the end product. Finally, even with modern systems, there is an inherent loss of duty cycle switching between variable, comparison star(s) and sky measurements.

3. CCDs and some of their characteristics

CCDs aren't the panacea for photometrists, either, but do have some characteristics that make them very nice tools for time-series photometry. For CCDs, scintillation noise usually dominates for shorter exposures, but since one often wants to study faint objects, getting high total counts is more of a concern; exposure times tend to be minutes in length. The higher quantum efficiency of CCDs plus their excellent linearity make them almost ideal detectors. Since they can act as a multi-star photometer and obtain many objects on the same detector, the transparency is essentially identical for all the objects in the field (which is typically around a few arcminutes in diameter) and thus the observations are independent of transparency variations caused by imperfect observing conditions (even fairly thick clouds). For frequencies below about 1 mHz, a CCD is generally better than a PMT for this reason.

So, why, one may ask, do observers even bother using PP any more if the CCD has so many advantages? Let us turn now to some of the problems, since the real world is anything but ideal. For one, CCDs suffer from flat fielding errors; this may be one of the biggest error sources, since the characterization of a CCD is a time-consuming task and its characteristics may not all remain stable over long periods of time. In addition, we are taught that we should take flats every night as there can be instrumental effects that differ night-to-night. Given the fact that each pixel has an associated error with it and the two-dimensional flux distribution not only moves around, but in addition, is not evenly distributed over the same pixels for every exposure, the problem becomes more obvious. Imperfect guiding and changes in seeing cause such an effect. There are also inherent hardware error sources in CCDs, such as charge transfer efficiency (due to the CCD itself, or to improper electronic settings that affect the readout process), cosmetic defects, hot pixels, cosmic ray events, and hysteresis (in some older CCDs). A/D converters can also be sources of problems: some suffer from linearity problems at extreme levels, have "sticky" bits, and can change characteristics with age. Not only is the CCD chip itself a source of errors, but so are its electronics settings and how the observer uses the system. For example, if the gain exceeds the readout noise, one undersamples the read noise and introduces additional errors; see e.g. Massey and Jacoby (1992) for a discussion of this and other related problems.

Still, it is possible to achieve rather good results, and above all, for certain types of projects, precision can be achieved that cannot be attained from ground-based telescopes with PP. Fig. 1 shows the theoretical S/N that can be achieved for a CCD vs. a PMT (assuming unrealistically that the PMT has the same quantum efficiency as the CCD and that no overhead exists for either system). Such an algorithm is easy to construct and valuable for estimating what to expect before observing. The expected signal-to-noise (S/N) ratio for PMT and CCD data can be readily predicted, and is discussed for PP in Henden and Kaitchuck (1990), and for CCDs in Howell (1992) and in Kjeldsen and Frandsen (1992).

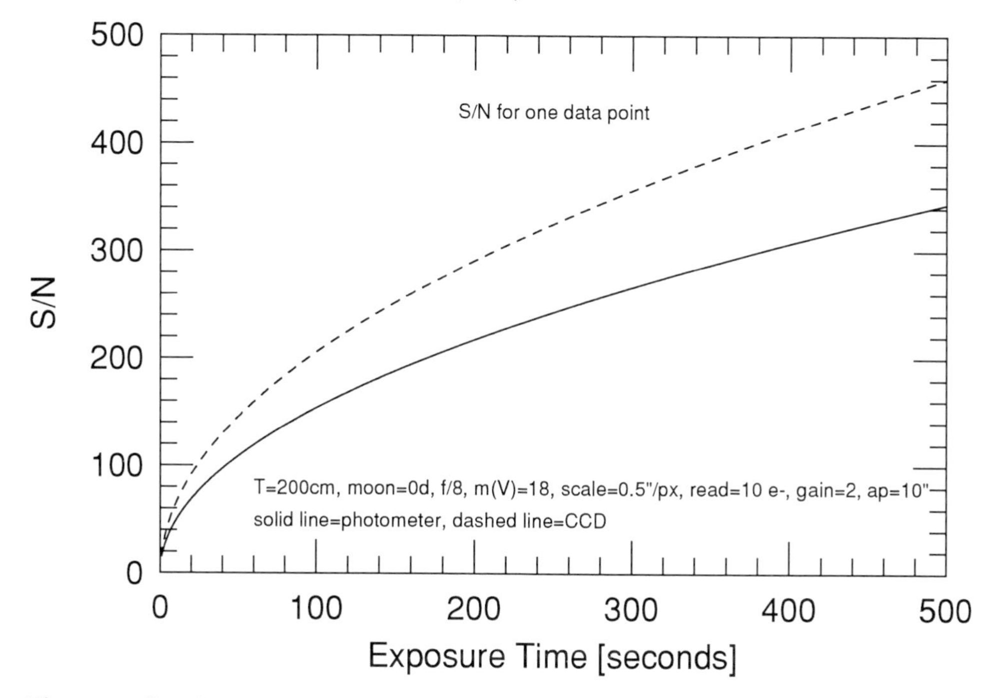

Figure 1 Signal-to-noise ratio compared for a single measurement with a CCD and a PMT. Note the unrealistic assumption that the QE is the same for both detectors. It is also assumed that a sky measurement of equal length was obtained independently with the PMT system. Here, the CCD sky determination used 10× the number of pixels as used to integrate the brightness of the star.

Clearly, the CCD beats the PMT for all but the very brightest objects; the use of sufficiently many CCD pixels for sky determination is important. Fig. 2 illustrates a typical data set obtained over nearly six hours; these particular data were reduced via IRAF's apphot. Fig. 3 shows the results of a Fourier analysis of data taken with different CCD chips of constant stars of quite different magnitudes. Using 300-second integrations, we have even managed to get reasonable light curves of stars as faint as 20 mag in the red with a 1.8-m telescope (Howell et al. 1990).

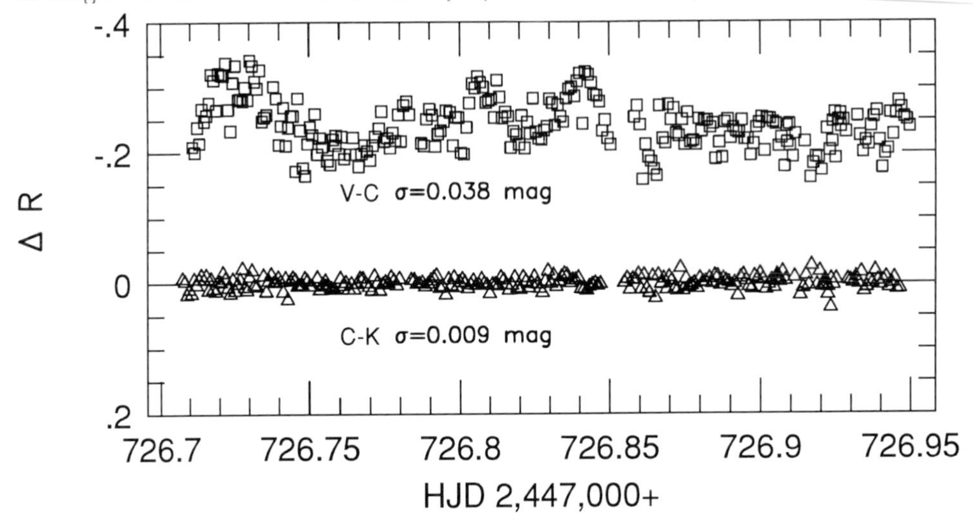

Figure 2 Differential CCD magnitudes of the cataclysmic variable V404 Cyg (V) and two comparison stars (C and K). Note the degree of activity of V-C and the constancy of C-K.

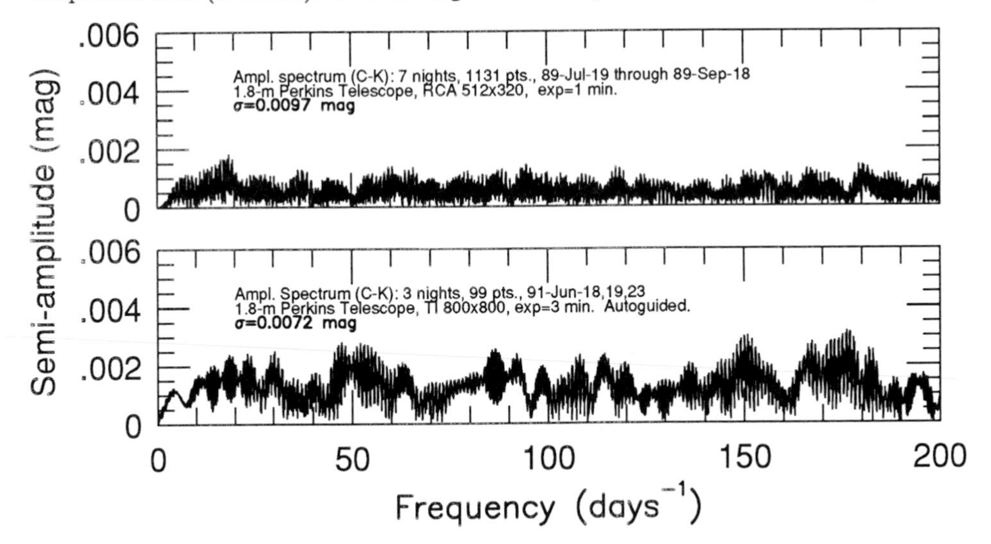

Figure 3 Amplitude spectra of two different, combined time series of comparison stars. C and K in the lower panel are about 3 mag fainter than C and K in the upper panel. The noise level in the amplitude spectrum depends not only on the σ of the data, but the number and spacing of the data and the period of time over which they were obtained.

4. Considerations for data acquisition and reduction

Understanding the characteristics of the CCD itself is important both for data taking and reduction. The gain, range of linearity, readout noise, uniformity of response from pixel-to-pixel, stability, bias pattern, dark current buildup, and image scale (pixel oversampling rate) need to be considered.

A number of steps can be undertaken to improve upon simple aperture photometry. The use of comparison stars of similar brightness and color as the variable star is important; a quick look using a filter of different bandpass can be used to pick good comparison star candidates. The integration times or the filter width can be optimized (see Kjeldsen and Frandsen 1992). One must be careful not to undersample the expected period of variability. For a sinusoidal variation, the amplitude "suffers" as a function of $sinc(\nu t)$ where ν is the frequency of the period and t the data spacing (cf. Martinez 1989). For 1% precision in one period, 14 points/cycle are needed, and at the Nyquist frequency (2 points/cycle), the amplitude is sampled only at the .637 level. The most important point I'd like to make is that the algorithm used for extracting magnitudes is critical. Simple aperture photometry generally will *not* do as good of a job as a more sophisticated algorithm, such as DAOPHOT (Stetson 1987), ROMAPHOT (Buonanno et al. 1983), DoPHOT (Mateo and Schechter 1989), MOMF (Kjeldsen and Frandsen 1992), or aperture correction methods (e.g., DaCosta et al. 1982; Howell 1989). It can be argued that, for non-crowded fields, this can be "overkill" to some extent, but there are potentially large differences in precision at stake. Clearly, for uncrowded or semi-crowded fields, the astronomical community needs some easy-to-use routines that improve on the simplistic approach of aperture photometry. MOMF seems to be a step in the right direction. It is equally as important to make sure users understand the algorithms utilized and don't just use programs as "black boxes." There are clearly many such routines in existence in the astronomical community, but few have been published.

In this regard, it should be noted that some reduction software packages assume pure Gaussian noise for the background, which for very low sky levels is not always the case (cf. Newberry 1992). Note also that a common estimate of the mode as $3\times$median $- 2\times$mean is not always close to being correct.

Mention should be made of the ensemble photometry method (Gilliland and Brown 1992). This offers the most precise method available for searching for periodicities with amplitudes of a few tens of micromagnitudes in long time-series data, provided that sufficient numbers of high S/N stars are present in the CCD frame.

This brings us to the topic of analysis. Periodicity analysis is a major subject in itself. Planning of the observations (S/N, sample rate, gaps in the data, etc.) will affect the analysis one way or the other, and hence should be done carefully. The use of Fourier analysis, PDM or other periodicity routines should be followed up with an investigation of the window function, the significance of detection (see, e.g. Scargle 1982; Horne and Baliunas 1986; Stellingwerf 1978) as well as tests for variability, such as Howell et al. (1988).

5. Future CCD systems

The future holds much promise for CCD time-series photometry. We will see less pixel-to-pixel sensitivity variation (perhaps eventually obviating the need to flat-field). Some chips are already constant to the 1-to-2% level (see the paper by A. Walker in this volume). Many CCD characteristics, such as quantum efficiency and linearity, are already nearly at theoretical levels and many newer detectors have substantial full well capacities and low readout noise, hence the need for better A/D conversion (20-24 bits and greater stability). Faster readout rates while maintaining good CTE would also be beneficial for time-series work. The ability to do a partial frame readout, or make use of multiple simultaneous readouts or the frame transfer mode available on some chips, can also save substantial time. Having powerful computers to do some analysis at the telescope can be very valuable; decisions can be made on whether or not to change observing strategies, keep taking data on an interesting object, etc. Even if a more precise analysis is performed later, some of us have found quick feedback to be extremely useful.

Finally, one ought to consider other means to get the information desired. For example, some stars can best be studied via RV variations where the characteristics of the CCD come in with a different weight than when used as a "pure" photometer.

I would like to thank S. B. Howell, R. M. Wagner, C. A. Gullixson, J. A. Holtzman, G. W. Lockwood, and countless other colleagues for many stimulating discussions.

References:

Buonanno, R. *et al.*, 1983, Astron. Astrophys. **243**, 160.
DaCosta, G.S., Ortolani, S., Mould, J., 1982, Ap. J. **257**, 633.
Gilliland, R.L., Brown, T.M., 1992, Pub. Astron. Soc. Pacific **104**, 582.
Henden, A.H., Kaitchuck, R.H, 1990, *Astronomical Photometry*, William-Bell, Inc., Richmond, Virginia.
Horne, J.H., Baliunas, S.L., 1986, Ap. J. **302**, 757.
Howell, S.B., 1989, Pub. Astron. Soc. Pacific **101**, 616.
Howell, S.B., 1992, in: *A.S.P. Conf. Series*, Vol. 23, p. 240.
Howell, S.B., Jacoby, G.H., 1986, Pub. Astron. Soc. Pacific **98**, 802.
Howell, S.B., Mitchell, K.J., Warnock III, A., 1988, Astron. J. **95**, 247.
Howell, S.B., Szkody, P., Kreidl, T.J., Mason, K.O., Puchnarewicz, E.M., 1990, Pub. Astron. Soc. Pacific **102**, 758.
Kjeldsen, H., Frandsen, S., 1992, Pub. Astron. Soc. Pacific **104**, 413.
Massey, P., Jacoby, G.H, 1992, in: *A.S.P. Conf. Series*, Vol. 23, p. 240.
Mateo, M., Schechter, P.L., 1989, in: *Proc. First ESO/STECF Data Analysis Workshop*, ESO, Garching, p. 69.
Martinez, P., 1989, Mon. Not. R. Astron. Soc. **248**, 439.
Newberry, M.V., 1992, A.S.P. Conf. Series, Vol. 25, p. 307.
Scargle, J.D., 1982, Ap. J **263**, 835.
Stellingwerf, R.F., 1978, Ap. J. **224**, 953.
Stetson, P.B., 1987, Pub. Astron. Soc. Pacific **99**, 191.
Stover, R.J., Allen, S.L., 1987, Pub. Astron. Soc. Pacific **99**, 877.

Discussion

D. O'Donoghue: *Did I understand you to say that the S/N ratio is always better for CCD's rather than photomultipliers, even for bright stars.*

Kreidl: For very bright stars, photon statistics dominate the error, and precision should be essentially the same. The important point is that for time series, CCDs eliminate transparency variations, so, in particular for variability at frequencies below one or two millihertz, one can often do better with a CCD. This assumes that other sources of error can be adequately minimized, of course (which can be difficult).

For faint stars, the determination of the sky background increasingly becomes extremely important, so the CCD will always have the potential to do better than a photoelectric photometer.

CCD Time-Series Photometry of Faint Astronomical Sources

Steve B. Howell

Planetary Science Institute, Tucson, Arizona 85719 U.S.A.

Abstract

Using CCDs to obtain time-series light curve information is an increasing area of interest in astronomy. For brighter, high signal-to-noise sources, the data collection and reduction procedures are very robust and easy to use. However, for fainter, low signal-to-noise objects we must resort to new methods. These include the use of optimum data extraction techniques, a fuller understanding of the CCD itself, and a more complete error model. This paper will provide a brief introduction to CCD time-series photometry and then explore the above new methods in relation to real observational situations.

1. Introduction

The recent astronomical literature contains many papers in which CCD time-series observations have been used. Howell and Jacoby (1986) were the first to provide a recipe for this type of work and they alerted the user to some of the parameters to consider when using CCDs as time-series photometers.

The basic idea of time-series photometry is to collect a temporally contiguous set of CCD frames of a particular source or sources, and to do this with as short a time resolution as possible. Typical values for current time resolutions are 5-30 seconds up to 5-10 minutes for sources of 14th to 22nd magnitude when using 1 to 2.5-m telescopes. The set of CCD frames are then reduced in the standard way (e.g., Gilliland 1992), and is likely to include bias subtraction, flat fielding, etc.

The reduced frames are now ready for photometric reductions to be performed. Any number of 2-D aperture photometry methods can be applied at this point. A few of the most popular are discussed in Adams (1980), Stetson (1987), and Howell (1989). Errors can then be assigned to each datum and the photometric results corrected for extinction and color terms, i.e., standard photometric reduction procedures. Finally, light curves can be produced (either absolute or differential) and the analysis of these data for variability or periodicity performed. Howell (1992) gives a detailed review of time-series CCD photometry.

2. Bright vs. Faint Sources

The signal-to-noise (S/N) for a point source imaged on a CCD is given by (Howell 1992),

$$\frac{S}{N} = \frac{N_*}{\sqrt{N_* + n_{pix}\left(1 + \frac{n_{pix}}{n_{BG}}\right)\left(N_S + N_D + N_R^2 + G^2\sigma_f^2\right)}} \tag{1}$$

We can define a bright source as one in which the following condition is true,

$$\text{if } N_* \gg n_{pix}\left(1 + \frac{n_{pix}}{n_{BG}}\right)\left(N_S + N_D + N_R^2 + G^2\sigma_f^2\right) \tag{2}$$

and a faint source is one for which

$$N_* \ll n_{pix}\left(1 + \frac{n_{pix}}{n_{BG}}\right)\left(N_S + N_D + N_R^2 + G^2\sigma_f^2\right) \tag{3}$$

These two definitions are guides for us in our usage of the terms bright (high S/N source) or faint (low S/N source). They are however, relative terms which depend on the magnitude of the source, the telescope used, the CCD characteristics, the sky brightness, etc. The user must decide for herself which objects are bright and which are faint. Figure 1 shows example models (see Merline and Howell 1992) of a high S/N source and a low S/N source. Figure 2 shows the range of S/N compared with the ratio of N_* to p where $p = n_{pix}\left(1 + \frac{n_{pix}}{n_{BG}}\right)\left(N_S + N_D + N_R^2 + G^2\sigma_f^2\right)$ for a 1-m telescope. The other specifics about the telescope, sky, and CCD are listed in the figure. The models in Figure 1 used these same listed parameters. Note that in Figure 2, near 15th or 16th magnitude, there is a slope change in the curve which may be useful in deciding between bright and faint sources in this particular plot.

3. Optimum Data Extraction and Error Contributions

If one carefully examines equation (1) given above for the S/N of a point source, it is apparent that for a given telescope-CCD combination, the only parameter directly under the user's control is the value n_{pix}. Thus, if the user can decrease the relative contribution of n_{pix}, the S/N of a given observation will be increased. Howell (1989) discusses this at length and finds that while standard 2-D aperture extraction uses an extraction radius of ~3 FWHM, an optimum radius for data extraction occurs at near 0.5 FWHM. This radius is optimum in that it provides the largest S/N measurement for a point source, although caution must be exercised to assure the source is not undersampled and the extraction software handles partial pixels in a flux-conserving, correct manner. Increases to the S/N of 20-50% are realized by use of these optimum apertures. DaCosta (1982) and Stetson (1990) also discuss small aperture data extraction.

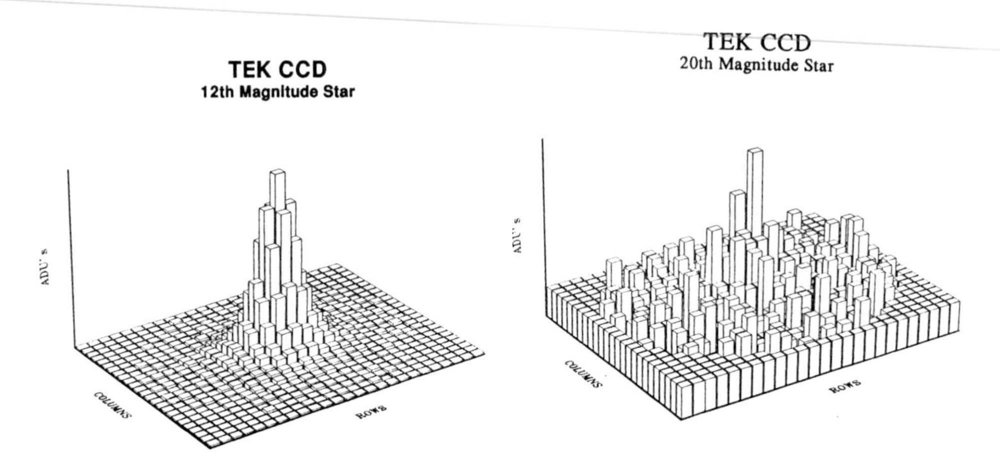

Figure 1: Models using a TEK CCD and a 1-m telescope of a bright (high S/N) and a faint (low S/N) point source. The techniques described in the paper are aimed at obtaining good photometric information on time-series CCD observations of these fainter type sources.

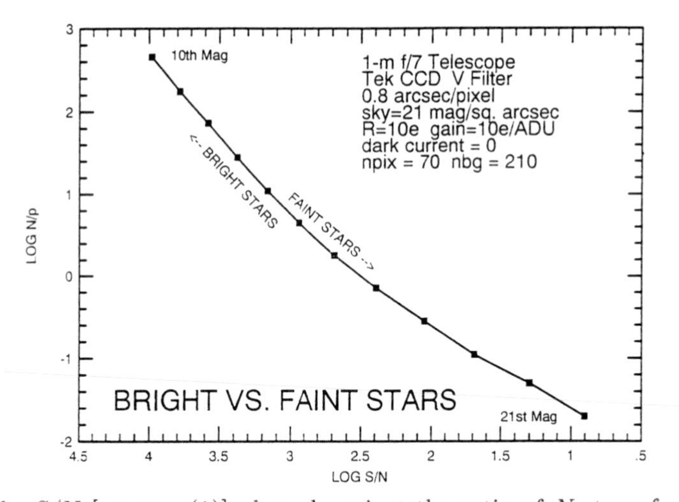

Figure 2: The S/N [see eq. (1)] plotted against the ratio of N_* to p for a range of point source brightness. The telescope/instrument characteristics are given on the figure.

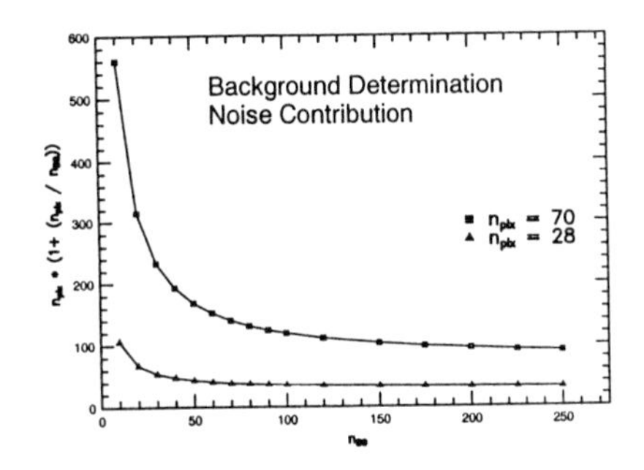

Figure 3: The relative error contribution of the number of background pixels used. Note how the optimum case requires fewer pixels and always has a smaller error contribution.

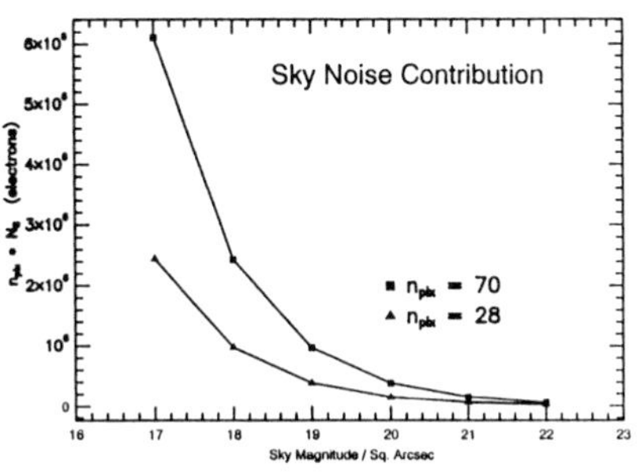

Figure 4: The relative error contribution of the sky noise. For less than ideal dark sites on nights when the moon is present, the optimum aperture case provides far less of an error contribution.

The three error terms, which we collectively called p in the last section, will now be examined separately. In each case, we compare the error term contribution of standard 2-D aperture extraction to that obtained by the use of optimum 2-D aperture extraction. All the plots in this section are based on the CCD and telescope characteristics listed on Figure 2.

First is the determination of the background. This is usually done by extracting some number of pixels which appear free of any faint stars, galaxies etc. and away from any obvious CCD imperfections. Figure 3 shows the relative contribution of the term $\left(1 + \frac{n_{pix}}{n_{BG}}\right)$ over a range in the number of background pixels used. In general, using $n_{BG} \geq 3n_{pix}$ provides essentially the best possible background determination. In the optimum extraction case, the use of 60% less background pixels yields, in all cases, a smaller noise term.

We next examine the sky noise contribution to the error budget. Figure 4 shows this noise term over a range of sky brightness. Again, we see that the optimum extraction case provides a significantly lower error contribution, particularly for brighter sky backgrounds.

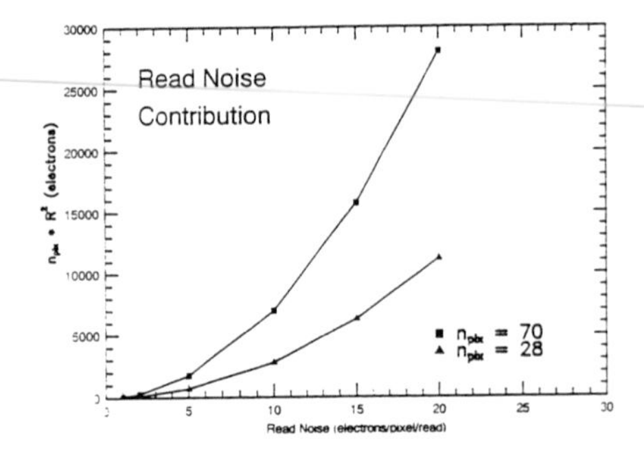

Figure 5: The relative error contribution of the CCD read noise. Most modern CCDs have very low read noise, thus this term becomes less important.

Finally, the read noise contribution is shown in Figure 5. This term is simply a scaled factor based on the read noise of the particular CCD used. For the newer, low read noise CCDs, the difference between standard and optimum data extraction is small for this term.

4. Magnitude Error and Observational Example

Using equation (1) given above for the S/N of a point source, the expected variance in flux for a source can be written as (see Howell et al. 1988)

$$\sigma_F^2 = \frac{N_* + p}{t^2} \left(e^-/\text{sec}\right)^2 \tag{4}$$

and the error, in magnitudes, for a given observation will be

$$\sigma_{mag}^2 = (2.5 \log e)^2 \frac{\sigma_f^2}{f^2} \tag{5}$$

where $f = N_*/t$. Therefore,

$$\sigma_{mag} = \frac{1.0857\sqrt{N_* + p}}{N_*} \tag{6}$$

As an example of the use of optimum data extraction and the use of the magnitude error equation above, Figure 6 shows an example CCD time-series light curve of a V=19.7 variable star. The data were obtained with a 1.8-m telescope and a TI CCD. Each point represents a measurement from a single CCD frame of integration time 297.6 μFortnights (i.e., 360 seconds). The top curve was extracted using a standard aperture of radius 2.4 FWHM while the bottom curve used an optimum aperture of 0.64 FWHM. The bottom curve was shifted downward by about 1 magnitude for clarity. The average 1 σ error for the standard extracted curve is 0.087 mag while the optimally extracted data have an average 1 σ error of 0.055 mag.

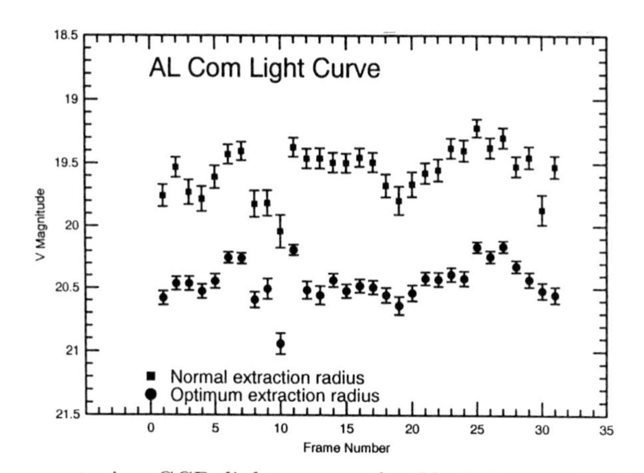

Figure 6: A representative CCD light curve of a V=19.7 magnitude variable star. See the text for details.

References:

Adams, M., Christian, C., Mould, J., Stryker, I., and Tody, D. 1980, *Stellar Magnitudes from Digital Pictures* (Tucson, AZ: Kitt Peak National Observatory).

DaCosta, G.S., Ortolani, S., and Mould, J. 1982, *Ap.J.*, **257**, 633.

Gilliland, R.L. 1992, *Astronomical CCD Observing and Reduction Techniques*, ASP Conference Series, Volume 23, 68.

Howell, S.B., and Jacoby, G. 1986, *PASP*, **98**, 802.

Howell, S.B. 1992, *Astronomical CCD Observing and Reduction Techniques*, ASP Conference Series, Volume 23, 105.

Howell, S.B., Mitchell, K.J., and Warnock III, A. 1988, *A.J.*, **95**, 247.

Howell, S.B. 1989, *PASP*, **101**, 616.

Merline, W.H., and Howell, S.B. 1992, in *Astronomical Data Analysis Software and Systems I*, ASP Conference Series, Volume 25, 316.

Stetson, P.B. 1987, *PASP*, **99**, 191.

Stetson, P.B. 1990, *PASP*, **102**, 932.

Discussion

E.F. Milone: *Variable star observers cannot always be choosey about the air mass of their observation. Do you worry about effects of image refraction at larger air masses? Also, do you select the optimum aperture at the time of your observation?*

Howell: I have not seen any effects from air mass for my differential photometry using objects on the same CCD frame. One can pick an optimum aperture for each object on a frame by frame basis.

D. O'Donoghue: *What is your limiting magnitude on a 1-m telescope with a few minute exposure?*

Howell: In a one minute exposure at a good S/N your limit will be something like 17th magnitude. If you are willing to increase your integration time a bit, you'll be able to get to 20th magnitude or so.

E. O'Mongain: *How do you center your apertures?*

Howell: For standard 2-D aperture data extraction, I use simple X-Y centroiding. For the optimum aperture extraction, I use more sophisticated schemes being very careful about under-sampling.

W.H.S. Monck: *You may be aware that Stephen Dixcon and I made the first electrical measurements of starlight in 1892, in Dublin, using George Minchin selenium photovoltaic cell. It is certainly impressive to see what can now be obtained with CCD photometry. Therefore, could you give me an example of just how sensitive a CCD is compared with my selenium cell?*

Howell: Well, to observe the same star which you did on your telescope, one would need to attenuate the incoming flux by first passing the light through roughly a pint of Guinness!

Extended Strömgren Photometry with CCD's

B.J. Anthony-Twarog, B.A. Twarog

University of Kansas, Lawrence, Kansas, U.S.A.

Abstract

CCD's have made possible the extension of intermediate-band photometric systems, including the Strömgren *uvby* system, to larger, fainter and cooler stars, with successful applications in old disk and globular clusters. Some of the applications in globular clusters demonstrate the ability to remove foreground stars from photometric diagrams and have enabled a re-evaluation of the evolutionary correction needed for distance modulus determination for metal-poor stars. We have developed an additional index based on measurement of the Ca II H and K lines which retains sensitivity to metallicity changes for extremely metal-poor stars. Finally, we are testing the utility of CCD Hβ photometry in the unusual and old open cluster Melotte 66.

1. Introduction

Despite its enormous success, the Strömgren photometric system has suffered from a few limitations. The spectral range for which the system was originally developed and calibrated did not extend to cooler stars and therefore excluded much of the galaxy's history. Both theoretical models (Gustaffson & Bell 1979) and empirical applications (Eggen 1978) have shown that this restriction is unnecessary, and that *uvby* indices can yield useful information for G and K stars.

Narrower filters had also restricted *uvby* studies to relatively bright stars as well; in fact, rather few open clusters with turnoffs in the F-star range, well within the original realm of calibration, were accessible with phototubes as detectors. CCD's have eliminated this restriction and the Strömgren *uvby* system has been applied to fainter, cooler and more metal-poor stars with considerable success.

At halo metallicities, a different limitation is encountered which afflicts most photometric indices of metal abundance based on weak lines, a reduction of sensitivity to declining abundances below [Fe/H] ~ -2.0. We have ameliorated this problem by the introduction of a new index, *hk*, which retains sensitivity to metallicity changes below –3.0. The index is a color difference analogous to the m_1 index, with a 90Å bandpass which covers the Calcium II H and K lines replacing the v filter where most of the iron lines are clustered. We have described the system with a set of standards (Anthony-Twarog, Laird, Payne and Twarog 1991) and demonstrated its applicability to metal-poor red giants in Twarog and Anthony-Twarog (1991)

Extension of the Strömgren system by use of CCD's is not without some pain, however. Development of suitable standards for fainter and cooler programs has lagged. Most of our applications of CCD-Strömgren photometry have exploited the large dynamic range and high internal precision afforded by advanced detectors. We have happily encountered no evident difficulty with image structure in using filters even as narrow as the Hβ narrow filter.

2. Applications to Globular Clusters

CCD technology has brought the main sequences of globulars within the reach of intermediate-band photometric systems, permitting some membership discrimination for stars which may be too faint, crowded or numerous to permit proper motion studies or radial-velocity measures. NGC 6397 has provided the first, and one of the best proving grounds for Strömgren applications. In the second CCD-Strömgren study directed at this relatively nearby cluster, Anthony-Twarog, Twarog and Suntzeff (1992) surveyed six fields in the cluster with the original intent to confirm additional photometric main sequence binary candidates suggested by Anthony-Twarog (1987). In this application, the photometric reduction code *DoPhot* was used (Mateo and Schechter 1990). For the main sequence as well as for the giant branch, the m_1 index provides sufficient discrimination of cluster members from foreground stars to permit unusually clean color-magnitude diagrams (see Figures 3, 5 and 6 of Anthony-Twarog et al. 1992). The issue of possible main sequence binaries remains unresolved, however; the statistical incidence of photometric binary candidates is too close to the expectation for chance superposition of stellar images, and these faint weak-lined stars have proven too difficult for direct radial-velocity measurement.

ω Centauri has provided a very different set of challenges to the CCD-Strömgren system. Begun several years ago as a masters' thesis, Krishna Mukherjee's study of this famously inhomogeneous cluster was designed to measure the dispersion of heavy elements among the main sequence stars. For this application, internal precision was considered more important than the accuracy of the external calibration. We followed our usual practice of **not** averaging CCD frames prior to processing or reduction. Our software uses the positional correlation between frames to match and average measurements for each star in each color before index construction. Any small zero-point shifts between similar frames can be removed this way, and the errors constructed reflect the realistic repeatability of measurement for each star. We "rediscovered" one variable dwarf Cepheid by sifting through our set of stars with anomalously large standard deviations. By comparing our assessment of measurement errors in $b-y$ and m_1 for upper main sequence stars, 0.015 and 0.017 respectively, we were able to show that the intrinsic dispersions in both these indices imply an intrinsic spread in iron-peak elements of 0.7 to 1.0 dex, entirely consistent with results from studies of the cluster's evolved stars (Mukherjee, Anthony-Twarog and Twarog 1992).

One unanticipated result from the ω Centauri study emerged from the upper main sequence surface gravity indices, which were initially scrutinized to see any discernible evidence for an age spread among the turnoff stars (none was found). Figure 6c of Mukherjee et al. (1992) shows the progression of c_1 values along the upper 1.5 mag of the nearly vertical turnoff, with the expected increase in c_1 for more evolved stars. The Strömgren system permits the determination of individual distance moduli, based on the difference δc between a star's c_1 value and the ZAMS value for its temperature; the absolute magnitude for the star's temperature is then corrected by an amount $f\delta c$. ω Cen has provided the first high-quality determination of f for metal-poor F stars, and the implied value of 8 differs from values derived in disk clusters. The temperature-dependent characterization of f derived by Nissen, Twarog and Crawford (1987) has been verified in model computations by Bell (1988) for solar and slightly metal-poor compositions, but predicts a value of 12 for F stars at the turnoff color of ω Cen. Interestingly, Allen, Schuster and Poveda (1991) noted in their analysis of photometric parallaxes for halo stars based on $uvby$Hβ data, that their derived absolute magnitudes for subdwarfs are too bright, equivalent to use of evolutionary corrections based on overly large f values.

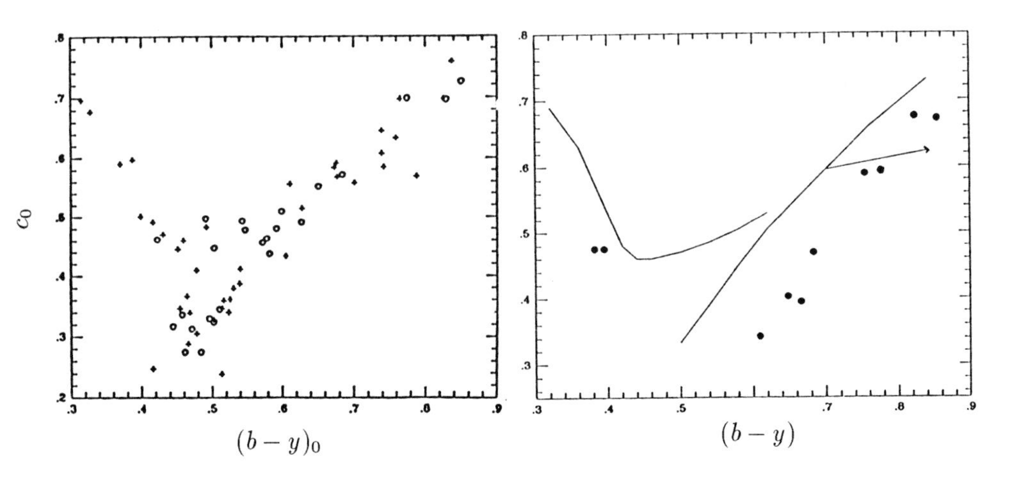

Figure 1 On the left, c_0 and $(b-y)_0$ values for field metal-poor giants. Circles indicate stars with [Fe/H] < -2.0. The right-hand panel shows a comparable diagram for photoelectric data in the cluster NGC 6397. The vector shows the shift direction for the field star sequence for $E(b-y) = 0.10$

Many of these applications in globular clusters are anchored with photoelectric photometry, and until recently, our Ca filter data was entirely photoelectric. In the course of photoelectric observations with the the $uvbyCa$ system of the widely scattered metal poor red giant sample, we noted a peculiar morphology in the $c_1, b-y$ diagram for field metal-poor giants. Figure 1a shows the reddening-corrected values for a large sample of giants, with photometric values from Twarog and Anthony-Twarog (1991). Stars with spectroscopically determined [Fe/H] values < -2.0 are noted with circles, while more metal-rich stars are noted with crosses. The implication of a metallicity-independent pseudo-HR diagram for halo giants provided one strong motivation to extend our Strömgren studies to globular clusters, with uniformly determined reddenings the expected payoff.

The typically perverse result, at least for NGC 6397 where our photoelectric indices are reliably tied to our larger field star system, is that cluster $c_1, b-y$ diagrams may well be different from the field! In Figure 1b we have reproduced the field star $c_1, b-y$ relation with the photoelectric data obtained for giants in NGC 6397. The reddening vector indicates the direction of shift for a value of $E(b-y) = 0.10$, smaller than the likely value of 0.14 for the cluster. The cluster stars evidently define a steeper relation than the field giants, and the two bluer AGB stars would lie far off the reddening-adjusted field star sequence.

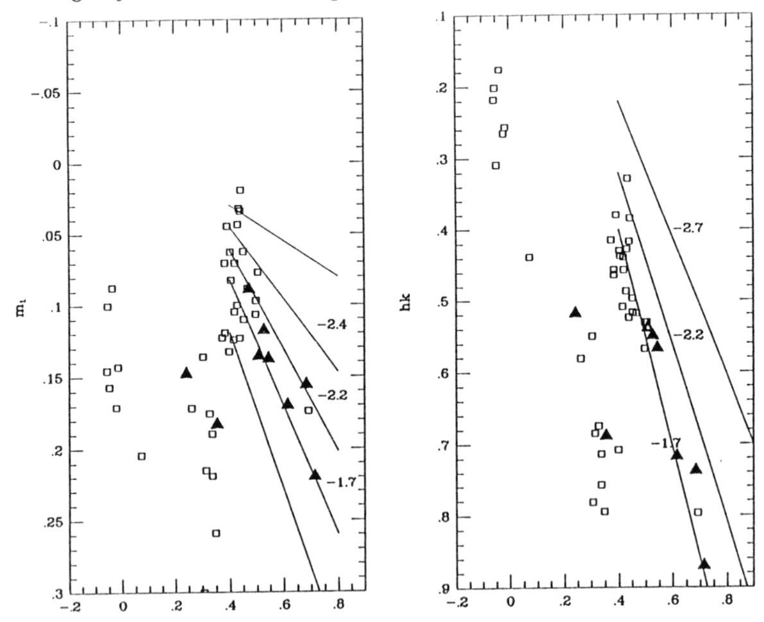

Figure 2 Metallicity indices m_1 and hk versus $b-y$ for giants in NGC 6397. Dark symbols represent results of photoelectric photometry, while the remaining points are from one CCD field in the cluster. In both diagrams, reddening corrections consistent with $E(b-y) = 0.14$ have been applied to the data. The linear sequences have been developed from observations of field stars with well-determined abundances, and are labeled with [Fe/H] values.

While we have not yet resolved this fascinating discrepancy, we have pursued CCD extensions in this and other globular clusters. Results from one field in NGC 6397 are presented here. We have found that the hk index not only provides superior metallicity information for the giants but provides cleaner membership discrimination as well. Figure 2 shows companion diagrams of the metallicity indices m_1 and hk as functions of $b-y$ color. The photoelectric data in the cluster are echoed here with darkened symbols and have been used to provide a provisional calibration of the CCD data. The linear sequences are isometallicity sequences based on data for field stars with spectroscopic abundance determinations and indicate a metallicity of -1.85 ± 0.10 for the cluster, based on the photoelectric data alone. The CCD sample is dominated by more faint giants, clustered near $(b-y)_0 \sim 0.4$. The brightest giant in this sample appears to conform to the appropriate isometallicity relation. Apart from horizontal branch stars in the upper left part of the $hk, b-y$ diagram, most of the other stars appear to be non-members.

3. Hβ Photometry in Melotte 66

The class of very old disk clusters is an exceedingly small one, and Melotte 66 has always appeared to be one of the most interesting. In photographic studies (Hawarden 1976; Anthony-Twarog, Twarog and McClure 1979) as well as more contemporary CCD surveys (Kaluzny and Shara 1988) the main sequence and giant branch display a width reminiscent of the chemically inhomogeneous globular cluster ω Centauri, although to a lesser extent. In spite a fairly large reddening (E$(B-V) \sim 0.14$ to 0.17), previous studies indicated that the breadth of the sequences is not due entirely to variable reddening. We have confirmed this is in a definitive and novel manner by one of the first large applications of Hβ photometry with CCD's.

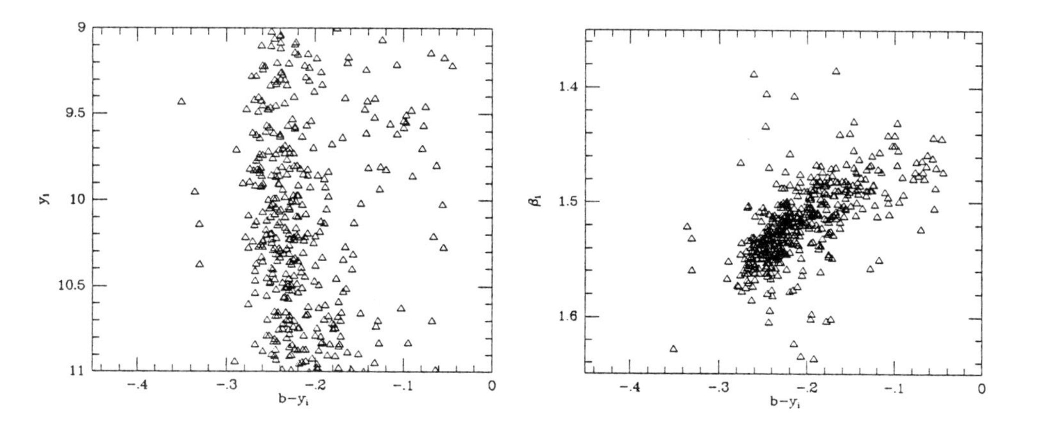

Figure 3 On the left, an instrumental color-magnitude diagram for Melotte 66 covering the turnoff region from $V \sim 16.5$ to 18.5. The right-hand panel shows the correlation between instrumental indices β_i and $b-y$ for the same sample.

Figure 3a shows a portion of the instrumental color-magnitude diagram, y_i versus $(b-y)_i$. The corresponding cmd with β as the temperature index is nearly indistinguishable over this nearly vertical stretch near the turnoff. Measurement errors in both indices for this range of magnitudes are about 0.01. There is a striking correlation between β_i and $(b-y)_i$, showing clearly that within the considerable breadth of this upper main sequence, **redder** stars are in fact **cooler** stars; if variable reddening were the cause of the main sequence width, no correlation would be expected. There is no obvious correlation between temperature indices β or $b-y$ and the metallicity index, so we cannot yet describe a cause for this sizeable temperature range at the turnoff of this very old cluster.

We want to acknowledge Stephen Shawl's collaboration on the Melotte 66 project, and to indicate our very large debt to the several students at the University of Kansas who have cheerfully assisted in data reductions over the past few years, especially Jackie Milingo, Marian Sheeran and Robert Stewart. We have depended heavily on photometric codes developed by Peter Stetson (DAOPHOT) and by Paul Schechter with Mario Mateo (DoPhot), and we can't thank them enough.

References:

Allen, C., Schuster, W.J., Poveda, A., 1991, Astron. Astrophys., **244**, 280.

Anthony-Twarog, B.J., 1987, Astron. J., **93**, 1454.

Anthony-Twarog, B.J., Laird, J. B., Payne, D.M., Twarog, B.A., 1991, Astron. J., **101**, 1902.

Anthony-Twarog, B.J., Twarog, B.A., McClure, R.D., 1979, Astrophys. J., **233**, 188.

Anthony-Twarog, B.J., Twarog, B.A., Suntzeff, N.B., 1992, Astron. J., **103**, 1264.

Bell, R. A., 1988 Astron. J., **95**, 1984.

Eggen, O.J., 1978, Astrophys. J., **221**, 881.

Gustafsson, G., Bell, R.A., 1979, Astron. Astrophys., **74**, 313.

Hawarden, T.G., 1976, MNRAS, **174**, 471.

Kaluzny, J., Shara, M.M., 1988, Astron. J., 95, 785.

Mateo, M., Schechter, P., 1990, *Proceedings of the First ESO Workshop on Data Analysis.*

Mukherjee, K.M., Anthony-Twarog, B.J., Twarog, B. A., 1992, PASP, in press.

Nissen, P.E., Twarog, B.A., Crawford, D.L., 1987, Astron. J., **93**, 634.

Twarog, B.A., Anthony-Twarog, B.J., 1991, Astron. J., **101**, 237.

Discussion

E.F. Milone: *Your colour-magnitude diagram of NGC 6397, where you showed a variable blue straggler, had a circle around two stars. Are they the two blue stragglers?*

Anthony-Twarog: We recovered one variable blue straggler, E39, an AI Velorum star, and also discovered one additional non-variable blue straggler.

W. Tobin: *Do you have any experience as to whether interference filters are sufficiently uniform for there to be no transformation problems as a function of location on the chip?*

Anthony-Twarog: I don't, and I'm not sure that it would be determinable given the standards which have been available.

T.J. Kreidl: *I've done Strömgren photometry of some open clusters and compared the photometry with published photoelectric photometry. I have seen no obvious, systematic, deviations. One should be cautious, however, that the filters are of the highest possible quality to ensure minimal variations over the field.*

A.T. Young: *I would like to make a further comment on interference filter variations. Laboratory measurements show a spatial variation of a few Å in a centimetre or so. I would recommend placing a filter at a pupil image, rather than near the field image.*

R. Florentin-Nielsen: *If you put the filter in the pupil plane, you will have a field dependance of the bandpass due to the fact the angle of incidence of light onto the filter will be different between the different position in the field. It may be safer to just have the filter a fair distance in front of the CCD and map the throughput effects.*

Anthony-Twarog: The angular effects can be kept small by using a large filter and pupil image. Conservation of throughput then ensures small angular effects.

High-Precision Photometry with CCDSs on Small Telescopes

Michael Zeilik[1]

[1] *Institute for Astrophysics, The University of New Mexico*

Abstract

From the theory of CCD operation and our experience since 1986 with a Photometrics CCD camera, we provide practical suggestions for achieving high-precision photometry on a small telescope (61-cm).

1. Introduction

Since 1986, we have employed a Photometrics CCD camera on our 61-cm telescope at Capilla Peak Observatory (Laubscher et al., 1988). One goal of this instrument has been to achieve high-precision photometry of point sources with the camera operating as a multichannel photometer. We have devised solutions that should apply to any other small observatory attempting to attain quality results. These issues include: superbias frames, flat-fielding techniques, calculations of exposure time, technique of sky subtraction, and matching of filters to CCD response functions to obtain linear transformations to standard photometric systems. With care, an observer can produce a precision of a few millimagnitudes for differential photometry.

2. CCD S/N for Sky-Subtracted Data

Newberry (1991) has provided an excellent analysis of the operation of a CCD and its use for sky-subtracted observations. In the case of a photon-limited noise (and with no "external" noise, such as scintillation), the CCD S/N equation implies the following practices for optimizing the actual S/N: (1) Increase object counts by using a CCD with high QE and insuring that the system has clean optics; (2) reduce average background count per pixel relative to object count by working at a dark site on a moonless night, choosing a CCD with low dark count, and minimizing scattered light in the system; (3) reduce number of pixels under object (but not less than 2 or will undersample) by superpixeling the CCD, changing f-ratio with reimaging optics to match the pixel size, and maintaining good focus and guiding; and (4) maximizing the number of pixels recording the sky background. Following these guidelines, even an old RCA CCD chip (which has high readout noise – about 80 electrons!) can achieve a S/N $\approx$ 1300 for a V-magnitude of 13 with a 61-cm telescope.

3. Capilla Peak Observatory

We have achieved such results with our Photometrics 815 CCD camera on a Boller and Chivens telescope at Capilla Peak Observatory in New Mexico. The site has a high elevation (2835 m) with good transparency, dark skies (V = 21.2 mag/ arcsec^2), and typical seeing of 1.3 arcsec (with occasional subarcsec seeing). Weather is photometric about 1/3 the nights; for multichannel differential photometry, we can work with light clouds and so have about half the nights available for observing. We have taken great care to keep the optics clean. The overall reflective transmission is now 95% and has stayed at that level for the past five years. When we are not observing, we run a positive-pressure ventilation system to keep dust out of the dome. The CCD camera is a dedicated instrument and rarely taken off the telescope.

From our experience observing both bright and faint point sources, we have developed the following guidelines for high-precision work: (1) Keep stars on the SAME pixels by offsetting from reference stars, maintaining good guiding (an autoguider is essential; we built our own with an intensified Fairchild CCD camera), and choose a software mask for data reduction that is about 3 times seeing for a given night; (2) try to use comparison stars in the SAME CCD field; (3) Use the shortest exposures possible to obtain the required object count, as estimated from CCD S/N equation; (4) Use the SAME (and LARGE) pixel area to measure the sky in the frame and apply a uniform software mask that avoids obvious sources.

Noise can be added in the course of CCD data reduction. To keep this noise contribution to a minimum, we have relied on the following procedures. First, we have taken a number of "superbias" frames, the mean of a few hundred exposures at the same temperature (which we keep stable to $\pm$ 0.1°C). Bias-subtracted frames contain at least the noise of the bias frame. Second, we observe the sky at twilight and at night (in blank fields) to generate our flat fields. These have been separately median-filtered to make a "master" flat field at each color for flat-field division. The S/N of the flat field must be higher than that of any pixel to which it is applied so that the S/N in the reduced object frame is not degraded.

By applying these techniques, we have achieved a precision of a 2 to 3 mmags for differential on point sources. We are now basically limited by scintillation noise. As far as accuracy is concerned, we learned a fundamental lesson when we first started observing with a KPNO CCD filter set. We found that the transformations to the Johnson-Kron-Cousins system were badly nonlinear. We therefore did synthetic photometry (Beckert and Newberry, 1989) to match the integrated response of our chip and its filters to the JKC system. Over a fairly wide color range, our custom-designed filter set is linear to $\pm$ 0.05%, so we can attain an accuracy on that order with high-precision observations.

This work was funded in part by National Science Foundation Grant AST-8903174.

References:

Beckert, D. and Newberry, M.V., 1989, Pub. Astron. Soc Pac., 101, 849.

Laubscher, B.E., Gregory, S., Bauer, T.J., Zeilik, M., and Burns, J., 1988, Pub. Astron. Soc Pac., 100, 131.

Newberry, M.V., 1991, Pub. Astron. Soc Pac., 103, 122.

Discussion

T.J. Kreidl: *Have you considered binning (2x2) to decrease the high readout noise of the RCA chip by* $\sqrt{2}$*, as well as reducing the readout time? We have found that to work very well with the old RCA system we have at Lowell Observatory.*

Zeilik: We have the ability to *superpixel* the chip. But in real-time, on-site evaluation of the data, we have found doing a superpixel or subarray more trouble than they are worth.

S.M. Howell: *I would just like to point out that the high readout noise of the RCA is likely not to be a problem for you due to your application to bright, high S/N sources. In fact, the large well depths are probably an advantage.*

Zeilik: That's correct. The high readout noise of the RCA chip does not affect high S/N observations of bright stars, especially cool ones.

Session 7

Photometry from Space

Photometry from Space

R.C. Bless and M. Taylor

Astronomy Department, University of Wisconsin-Madison, U.S.A.

Abstract

After a brief review of the traditional opportunities as well as the problems of doing photometry from space vehicles over a wide variety of wavelength regimes, several topics will be discussed. These include the factors associated with various types of spacecraft orbits (for example, near-earth, geosynchronous, planetary cruise) and with the different kinds of vehicles and platforms used (expendable, shuttle attached payloads, permanent stations, multi-purpose spacecraft, small and large projects). Since ground-based capabilities are constantly improving and expanding, it is necessary to assess those that might compete with space-based photometry. Finally, the present state of space observations as they pertain to "classical," high precision, high speed and multi-object photometry will be reviewed; possibilities for the future will be addressed briefly.

1. Introduction

This is an extremely broad topic encompassing not only a wide range of wavelengths, but a wide range of techniques as well. Even if we were not limited by the length of this paper, we would certainly be limited by our ignorance. However, we hope to address most of the broader issues of space photometry as well as a few of the more interesting and specialized developments of late.

We realize that not everyone at this meeting is an expert in space astronomy and so, we will begin with a brief review of the advantages and disadvantages of doing photometry from above the earth's atmosphere. In this context, we will also consider some of the astronomical implications associated with various types of vehicles and their orbits. The ultraviolet part of the spectrum has probably been the primary beneficiary of space photometry to date. Consequently, we will take a look at the evolution of UV photometry over the last 25 years or so. This will bring us up to the present time of the Hubble Space Telescope (HST) where three of the five instruments currently installed are capable of doing photometry: the Faint Object Camera (FOC), the High Speed Photometer (HSP), and the Wide Field Camera (WFC). We will end with a discussion of what we perceive are the primary areas of concern for space photometrists today.

2. Brief review of characteristics of space

The opportunities afforded photometry by space are well-known and are all a direct consequence of the absence of the earth's atmosphere. These advantages can be divided into three major areas:

1) Access to the entire electromagnetic spectrum: From the ground, we have access to the visible, almost all of the radio, and a few narrow regions in the infrared, primarily in the near-infrared. As is well known, the amount of absorption by the earth's atmosphere decreases with increasing altitude. In this vein, we might mention the south pole as a potentially good site for photometry. The south pole is more than 3000 m above sea level and the air there is very cold and dry opening up new "windows" in the IR part of the spectrum. These features are currently being exploited by the Center for Astrophysical Research in Antarctica (Harper 1991).

2) No scintillation and no changes in atmospheric transparency: Other papers presented at this meeting (e.g. Dravins et al. 1992) have addressed the problems associated with scintillation. However, for completeness, we will mention that the absence of the earth's atmosphere in this regard becomes very important in the search for small amplitude oscillations which occur on timescales of minutes to microseconds. Scintillation and photon noise are the two major sources that usually limit the photometric accuracy of fluctuations which occur on timescales of less than one minute while atmospheric transparency fluctuations limit the photometric detectability of variations on longer timescales (Warner 1988).

3) Reduced background: The elimination of atmospheric turbulence allows space photometrists to use smaller apertures and hence, significantly reduce the background noise. This is particularly beneficial in the infrared. From the ground, the IR sky brightness increases drastically with wavelength as a result of airglow and/or thermal emission from telescope optics and other components of the telescope (Joyce 1992). From space, the angular resolution of an object is limited only by the optics of the telescope, the properties of the detector, and the capabilities of the pointing control system. Hence, it will be much easier to discern small intrinsic variations which occur in the IR. During this meeting we have heard about the capabilities of image restoration and active optics (Shearer et al. 1992), but, at least in the case of active optics, we are limited to small fields of view.

The advantages of doing photometry from space however, do not come without a price. The difficulties include at least the following:

1) Environmental effects: These include the low temperatures in space, the temperature changes which occur over the day-night cycle, and radiation by trapped particles which can potentially damage detectors and other sensitive electronics. We will say more about this in section 4.

2) Spacecraft limitations: These limitations include restrictions on launch load which in turn affect the orbits which are achievable for a specific vehicle plus payload. Also, limitations on the power available to operate an instrument, limitations on the rate at which data can be sent to a ground-station on earth, and quality of the pointing control system are all concerns to be dealt with. On a slightly different note, scheduling space observations, especially those that are time-critical has certainly proven to be a non-trivial problem as has the determination of the precise universal time of an event which is known only as well as the calibration of the on-board clocks.

3) Political: The cost of putting an instrument into space is extremely high and consequently, politics plays a significant role. Budget problems often make it difficult to proceed on a reasonable time-scale. As a result, space astronomy suffers from long lead times and delays which can result in flying obsolescent equipment. (The spacecraft computer on HST is based on 1970's technology!). Finally, oftentimes, one has a large and cumbersome bureaucracy to deal with and, at least in the U.S., there is always the possibility of outright cancellation of a project. These aspects of space astronomy are not well suited for a university environment where one hopes to train the next generation of space astronomers.

Several other factors affect astronomical performance in space. Many of these can be categorized according to the various kinds of orbits into which a variety of vehicles can be placed. First the types of orbits:

1) Near-earth (orbital periods of roughly 100 minutes): Because of periodic earth occultations and the necessity to avoid the South Atlantic Anomaly (SAA), the observing time per orbit is severely limited. In addition, scheduling time-critical observations can be difficult because half of the time you may be on the wrong side of the earth. Commanding and data storage are more complex because one must rely on a communication satellite such as TDRSS or have more than one ground station available. Real time commanding must be scheduled in advance to insure TDRSS availability. With HST, these factors, taken together with a cumbersome ground system, have so far limited the actual (collecting photons from the program object) observing efficiency to about 10 percent, or about 10 minutes per orbit, on average.

2) Geosynchronous orbit (more generally, 24 hour orbit): More energy is required to reach the required altitude, but IUE has demonstrated the huge advantages of this type of orbit. The spacecraft can be in contact with one or another ground station for long periods of time so that data are immediately available and commanding can always be done in real time. Further, the earth subtends a much smaller angle than it does in near-earth orbits with a corresponding increase in observing efficiency. It seems to us that more serious thought should be given to the trade-off between payload size and observing efficiency than has heretofore been done, at least in the U.S.

3) Planetary cruise (for example, Voyager): Typically, these orbits offer lots of observing time, but the available power and hence, bandwidth is usually limited. The slower data rate is a concern for at least some science programs. Also, instruments for planetary work often are not optimal for stellar observations.

Let's turn next to a few considerations about launch vehicles.

1) Sounding rockets: These are relatively low-cost vehicles which provide an opportunity to test novel ideas or techniques. Although the amount of observing time is limited, sounding rockets enable the scientist to have maximum control of a project and they are an excellent way to introduce students to space science.

2) Attached payloads (i.e. space shuttle): The situation for larger vehicles is not as happy. In an ideal situation, there should exist a stable of launch vehicles capable of placing a wide variety of payloads into the orbit most appropriate for the mission. That is, the scientific and observational requirements should drive the choice of vehicle rather than the other way around. In the U.S. at least, this has not been possible since for political reasons, the orbiter was to be the all-purpose launch vehicle. The problems of this policy are many, but the chief one is that any man-involvement complicates matters enormously with little, if any, advantage. Nor is the orbiter a good platform for an attached payload requiring accurate pointings. The ASTRO payload is a good example. The pre-mission planning was an enormous task since a new timeline and associated mission products (which were numerous, in part because of the logistics involved in a manned flight) had to be developed for each new launch date which differed by more than a few weeks from the previous date. In the case of ASTRO, there were thirteen different timelines, none of which was used during the actual mission because of problems early-on. One of the chief lessons learned from that mission is that the space shuttle is not a good vehicle from which to do space astronomy when scheduling is time-dependent.

3) Expendable vehicles (for example, Titan or Delta): Even if the most expensive of these vehicles approaches the cost of a shuttle launch, they are unmanned and

consequently the logistics of these missions are considerably less complicated. NASA has long been urged to use expendable vehicles to launch scientific satellites, and it seems to be moving in that direction, albeit slowly. An interesting development which we will comment on later is the Pegasus launch vehicle.

4) Space station/Lunar base: As with any of its very large programs these mega-projects are more the product of NASA's concern with its institutional survival than with science. Furthermore, they are so far in the future, so enormously expensive, and so subject to budgetary as well as technical forces that we find it difficult to consider them as significant opportunities for anyone out of diapers.

3. Some highlights of UV photometry

Space photometry has been evolving over the last several decades with substantial efforts directed toward UV astronomy. This region of the spectrum was first explored with sounding rockets, but, it was quickly realized that longer missions were required in order to begin to understand the UV universe. In 1968, the second of the Orbiting Astronomical Observatories, OAO-2, was launched, three years after the quick failure of the first OAO spacecraft. Although characterized as an engineering flight with limited expectations, it lasted 50 months. It consisted of several 0.3 - 0.4 m telescopes with a 10' field of view (Code et al. 1970), indicative of the quality of the pointing control system. OAO-2 did broad-band filter photometry and low resolution spectroscopy with a photometric accuracy of on average, a few percent.

Among the subsequent UV missions was the Netherlands Astronomical Satellite (ANS) launched 2 years later. Again, the telescope aperture was small, 0.23 m, and the field of view, although smaller than that used in OAO-2, was still quite large, 2.5' (Wesselius et al. 1980). The broad-band photometry obtained with the ANS had a photometric accuracy of < 1 percent.

In 1978 the International Ultraviolet Explorer (IUE) was launched. Although designed to last only a few years, it has been operating for almost fifteen years! The telescope is a 0.45 m with a 10" x 20" aperture which has both a high and a low resolution spectrograph (Boggess et al. 1987). The photometric accuracy of the IUE over the entire bandpass from 1150 to 3200 Å is between 5 and 10 percent.

This brings us up to the present time of the HST. As mentioned earlier, three of the five HST instruments are capable of doing photometry. Each instrument has narrow, medium, and broad-band filters spanning the ultraviolet and optical regions of the spectrum. The FOC is capable of doing photometry at the few percent level, while the WFC I obtains optical photometry with an accuracy of 5 - 10 percent. The WFC I is not able to do photometry in the ultraviolet because of contamination build-up on the CCD. Although the HSP was originally expected to do 0.1 percent photometry on a routine basis, the spherical aberration of the HST mirror and the instability of the pointing system (jitter) have made that impossible. With 15 percent of the encircled energy contained within a 0.1" radius, compared to the original specification of 70 percent, the HSP can only do about 10 percent absolute photometry; this is the accuracy to which observations taken months apart can be intercompared. The HSP's real strength with the current state of the HST is in doing very high time resolution relative photometry.

Unfortunately, with the degraded images of the HST and the pointing instability, space photometry doesn't look much better than it did 25 years ago. Of course, there is still the possibility that the situation will improve with the second generation instruments and with the installation of COSTAR, an instrument designed to minimize image degradation resulting from spherical aberration. Unfortunately, this will be accomplished at the expense of the HSP.

4. Detectors in space

Given the many well-known advantages of CCDs over other detectors, together with the increasing read-out speeds, CCDs are becoming attractive for all but the highest speed photometry. However, there are a couple of areas of concern.

As mentioned earlier, radiation effects from trapped particles can be a serious problem. Not only can they add noise to the data, they can permanently damage portions of the detector or associated electronics. Although no permanent radiation damage has yet been detected, WFC I sees about 5000 hits per 800 x 800 CCD within 45 minutes. Generally, these cosmic rays affect individual pixels and so when half of the radiation from ST was to fall within one pixel there would have been a problem in distinguishing particle "stars" from real stars. With the spherical aberration, however, the image covers so many pixels that this effect is not seen. However, the chances of a particular star being contaminated with a cosmic ray hit is higher. In order to ease the data analysis, WFC I routinely takes two images of every field. We might just add that WFC II will use thicker CCDs than those used in WFC I, so not only will more energy be deposited by a cosmic ray, but the resulting electrons will diffuse to a larger number of pixels and could form faint blotches in an image.

Another major concern for CCDs in space has to do with obtaining flat fields. Unfortunately, there are no good, natural flat field sources. Although WFC I uses the earth as its flat field source, clouds produce streaks on the images and ocean waves produce very bright glints. In order to minimize these problems, WFC I takes a median of several flats at different orientations of the HST. WFC II will have its own internal lamps and optics to provide flat fields. However, earth flat fields will still be necessary in order to remove the signature of the optics ahead of the camera.

a. UV Photometry with CCDs

Although WFC I can do photometry in the visible at the 5-10 percent level, depending on how crowded and how large the field is (the point spread function is not constant over the entire field), because of severe contamination by an unknown source it cannot do UV photometry. Contaminants adhere to the coldest surface available, and in the case of WFC I, this is the CCD. Since it takes only a monomolecular layer of organic material to absorb UV radiation, contamination is a very serious problem at these short wavelengths. WFC I loses its UV sensitivity in a matter of hours after the CCDs are heated sufficiently to evaporate off the contaminants. It has been estimated that to work well in the UV, the rate at which contaminants are deposited on the camera will have to be at least five orders of magnitude less than the rate seen in WFC I. This means that for cold CCDs intended for use in the UV, contamination has become a major problem at a level that has not been faced before. Although the source of these contaminants is not yet known, a huge effort has been made to eliminate all potential sources of contamination for WFC II. In addition, the CCDs on WFC II have considerably lower dark currents than those on WFC I and so will be able to operate at slightly warmer temperatures, -70C rather than -90C. In any case, if CCDs are to be useful in the UV this problem must be solved.

Another problem with using CCDs in the UV has to do with the red leak in interference filters. Observing the UV radiation from a cool star, for example, is difficult because the visible light leaking through the filter, combined with the high quantum efficiency of CCDs in the red, can swamp the UV signal. To overcome this problem, the Jet Propulsion Laboratory (JPL) has undertaken a project to produce so-called Wood's filters which are based on a discovery by R.W. Wood in the 1930's that alkali metals absorb UV radiation but have a long wavelength cut-off set by the plasma frequency of free electrons. Although these characteristics have been known for more than 50 years, the challenge comes in manufacturing stable filters of this type which can survive in the space environment. Recently, JPL has been successful in producing Wood's filters which will be used for WFC II. A transmission curve for one of these filters comprised of a sodium layer 6000 Å thick deposited on MgF_2 substrates has a peak transmission of about 30 percent and is blind to radiation longward of about 2200 Å at the 10^{-5} level (Clarke et al. 1992). Despite their low transmission, these filters could be useful in UV applications and in fact, might be excellent dichroic filters.

b. IR photometry with CCDs

Compared to other wavelength regions, very little IR photometry has been done from space. The biggest problem from the ground has to do with the large backgrounds and hence the difficulty in discerning small variations. Also, IR detectors are more complex than optical CCDs (Glass 1992). Optical CCDs are capable of doing 0.1 percent photometry. Data of similar accuracy should be achievable in the IR as well. However, because of the large IR backgrounds, it is necessary to get above the earth's atmosphere before one can hope to be competitive in this regard. The Infrared Astronomical Satellite is so far the only satellite designed to do IR photometry and the accuracy obtained was only on the order of 10 percent or so.

IR detectors have seen tremendous advances over the last several years and we will no doubt continue to see them evolve. The two areas that will probably see the greatest improvement are the size of the detectors and the read noise. The largest

IR CCDs being built today are 256 x 256 pixels with typical read noise values on the order of 100 electrons. The CCDs being developed for the second generation HST instrument NICMOS will have lower read noise. These arrays will operate over a wavelength range from 0.7 - 2.5 μm with a linearity to about 150,000 electrons. The read noise of these arrays is down by a factor of 4; 25 - 30 electrons (Thompson 1992). We expect to see continued progress in this area over the next few years.

c. X-ray photometry with CCDs

We would next like to touch briefly on some interesting new developments in the x-ray regime. X-ray detectors are generally quite efficient; the problem has been in their energy resolution, that is, the width of their "filters." The workhorse detector has been the proportional counter which absorbs a large fraction of the incident photons. The amplitude of the resulting pulse is proportional to the energy of the detected photon giving an energy resolution of typically 25 percent of the energy. This corresponds roughly to optical broad-band photometry, like UBV. Below 1 keV the resolution becomes very poor, however. Here atomic absorption-edge filters–for example, beryllium, boron, and carbon–are used to isolate broad energy bands. The transmission of these filters can be measured and the counter gas stopping efficiency can be calculated to give a calibration good to about 5 percent.

A considerable step up in energy resolution is provided by CCDs specially processed for use as x-ray detectors. In order that an x-ray photon is absorbed, the depletion regions are made 10-20 times thicker than for optical CCDs. The energy resolution of such a CCD is about 100 eV at 6 keV, significantly better than that of a proportional counter. In optical terms, this corresponds very roughly to Stromgren photometry. The efficiency of such a detector is very good; in fact, from 0.5 to 8 keV, the CCD is nearly 100 percent efficient. Two 4 x 4 CCD arrays each with 512 x 512 pixels made at MIT are to be flown on the Japanese-U.S. ASTRO-D satellite in February 1993.

The resolution of a solid state detector is limited by the statistical nature of the ionization process which is only about 30 percent efficient. One technique to achieve higher energy resolution is not to measure the charge produced by an absorbed photon, but to measure the temperature rise it produces in a small (say 0.5 mm square and a few tens of microns thick) silicon element which acts like a calorimeter. For this purpose the detector must be at a very cold temperature, 50 mK. Such a detector made by McCammon (1987) at U. Wisconsin gives an energy resolution of about 1000 at 6 keV. This is narrow band photometry indeed, corresponding to narrow line filters in the optical. X-ray photometry is in the happy situation of having available extremely sensitive detectors with a wide range of resolutions.

5. Future prospects for space photometry

There are some very promising projects for the future of space photometry.

1) HIPPARCOS: Although the primary goal for HIPPARCOS is to measure accurate stellar parallaxes, a separate instrument called Tycho will be obtaining on average, 150 photometric observations of each of about 1,000,000 stars. Although the expected photometric accuracy is only on the order of 30 mmag for the brightest stars (Großmann, 1992) it will provide an enormous data base of photometric observations.

2) PRISMA: If funded, this European project will be a dedicated mission designed to exploit the long observing times available from a geosynchronous orbit. The primary objective of PRISMA is to do asteroseismology of solar-type (and later-type) stars. In addition, observations of classical pulsators, such as the δ Scuti and RR Lyrae stars, Cepheids, and rapidly oscillating Ap stars, will be obtained (Bromage 1992).

3) Pegasus: A third prospect for the future has to do with the development of a new, small, expendable launch vehicle in the U.S. called Pegasus. Pegasus is dropped out of a B-52 and then fires a two-stage rocket to reach its orbit. This is a very low-cost operation, but the project has not seen success as of yet. There have been two launch attempts, both of which have had problems. A third attempt is currently being scheduled for the end of this year to launch the X-ray satellite Alexis.

In summary, how do we see the future of photometry from space? We certainly have the technical capabilities to successfully do photometry from space: 1) we have the expertise required to build optical systems which produce very good images, 2) pointing control systems have reached a sophistication that enable the use of small apertures, and 3) a wide variety of sensitive detectors are now available or soon will be.

The cost and time required for the construction of space missions must be reduced. Long dry spells during which no new data are being acquired must be avoided. To this end, it seems to us that more emphasis should be put on several smaller, special purpose payloads rather than on a single, large, all-purpose satellite intended to do many different things. Large projects of course have their place, but they should be carefully justified as the only way to obtain crucial data. Furthermore, it should be assumed that missions will be designed for high orbits unless some compelling factors justify the serious loss of operational efficiency associated with a low orbit. Finally, greater discipline must be imposed such that a mission is descoped if cost and schedule constraints are exceeded, rather than dragging on interminably.

Even if all of these suggestions were implemented, many programs would still be very expensive. Given the forseeable fiscal constraints under which most of the nations having space programs will have to live, there will likely be a need for international cooperative efforts. Such programs can become very complex and cumbersome. We must learn how to work together effectively.

References:

Boggess, A., Wilson, R., Barker, P.J., Meredith, L.M., 1987, in: *Specific Accomplishments of the IUE*, ed. Y. Kondo, p 3.

Bromage, G.E., 1992, these proceedings.

Clarke, J.T., Burrows, C., Crisp, D., Gallagher, J.S., Griffiths, R.E., Hester, J.J., Hoessel, J.G., Holtzman, J.A., Mould, J.R., Trauger, J.T., Westphal, J.A., 1992, White Paper for WF/PC 2 Far-Ultraviolet Science.

Code, A.D., Holm, A.V., Bottemiller, R.L., 1980, Astrohpys. J. **43**, 501.

Dravins, D., Lindegren, L., Mezey, E., 1992, these proceedings.

Glass, I.S., 1992, these proceedings.

Großmann, V., 1992, these proceedings.

Harper, D.A., 1991, private communication.

Joyce, R.R., 1992 in: *Astronomical CCD Observing and Reduction Techniques*, ed. S. Howell, Astr. Soc. of Pacific Conf. Series **23**, p 258.

McCammon, D., Juda, M., Zhang, J., Holt, S.S., Kelley, R.L., Moseley, S.H., Szymkowiak, A.E., 1987, Japanese Journal of Applied Physics **26**, 2084.

Shearer, A., Redfern, R., Wouts, R., O'Kane, P., O'Byrne, C., Jordan, B.D., 1992, these proceedings.

Thompson, R., 1992, private communication.

Warner, B. 1988, in: *High speed astronomical photometry*, Cambridge University Press, Cambridge, p. 15.

Wesselius, P.R., van Duinen, R.J., Aalders, J.W.G., Kester, D., 1980, Astron. Astrophys. **85**, 221.

Discussion

D. O'Donoghue: *Will the Prisma mission be devoted to doing stellar seismology of solar-type stars?*

Taylor and Bless: Yes. the Prisma mission will do stellar seismology of solar-like stars and others. There will be a talk this afternoon when we will hear much more about the specific goals of the Prisma project.

C. Morossi: *Could you comment on damage in UV coated CCD's due to particle radiation?*

S.B. Howell: It depends on where the event hits the CCD. For example, if a radiation event destroys or damages part of the serial output register, all pixels which pass through the affected part of the output register, will be affected; possibly a large part of the array. If however a pixel in the imaging array is damaged only those pixels in the column below it are affected. Some radiation events which *damage* pixels appear to *fix* themselves on timescales of months, probably due to charge, initially deposited deep within pixels, finally let out.

Tycho Photometry: Calibration and First Results

V. Großmann

Astronomisches Institut der Universität Tübingen, Germany

Abstract

The star mappers on board the ESA Hipparcos satellite will be used as one part of the Tycho project to obtain photometric data of a quasi B and V band magnitude for the 500,000 brightest stars and one broad-band magnitude T for a further 500,000 stars. Approximately 150 million single observations of these stars will be collected during a 4.5-year mission of the satellite.

The accuracy of the photometric data gained is expected to be 0.03 mag on the average at $B = 10.5$ mag for non-variable stars. We shall get T magnitudes for stars down to a limiting magnitude of about $B = 12$ mag depending on the galactic latitude. We will give an overview of the calibration procedure and present some of the first results.

1. Introduction

The star mapper data from the ESA Hipparcos satellite provide a unique possibility to compile a large photometric catalogue covering both the southern and northern hemisphere using measurements from one single instrument.

The Hipparcos mission, especially the geometry of the star mapper grid, and the processing of the raw photon counts have been described in detail in one recent A&A issue entirely dedicated to the satellite (Astron. Astrophys. **258**, 1992). The Tycho experiment is described in detail in Høg *et al.* (1992) and references therein. Thus, we will summarize only the points important for photometry and then present an overview on the calibration procedure and some results.

The star mapper experiment determines the Tycho B_T and the V_T magnitudes for each star which is detectable by the photomultipliers. The B_T and V_T bands with effective wavelengths of $\lambda = 428$nm and $\lambda = 534$nm are not equivalent to the Johnson B and V bands: they have roughly the same limits but differ in shape (Scales *et al.* 1992). A broad-band T magnitude covering both the B_T and V_T bands is also determined. T is defined such that T=V_T for $B_T - V_T$=0. One disadvantage of the experiment is that the integration time for each star crossing the star mapper grid is the same for both bright and faint stars due to the constant scanning velocity of 168.75 arcsec/s.

2. The input data to photometry

Because the star mapper grid consists of four inclined and four vertical slits, (each 0.914″ by 2400″) each star crossing the grid will give eight peaks in the raw count data for the B_T and V_T band, respectively. These peaks may be caused by a star crossing one of the fields of view. These fields, separated by 58 degrees, are lying in the scanning plane of the satellite and therefore are called preceding and following field of view.

The compressed raw counts received from the satellite are converted to photon count rates for each band. By the use of a non-linear filtering method (cf. Bässgen *et al.* 1992) the so-called raw transits are produced.

Currently we use only a minor fraction of the raw transits including the transits of the photometric standard stars included in the Tycho input catalogue. The final photometric catalogue of the Tycho measurements however will not be based on the current input catalogue of three million stars but on the revised input catalogue of roughly one million stars derived from the data processing of the first year of the mission (Halbwachs *et al.* 1992).

In order to calibrate our observations we rely on the catalogue of photometric standard stars compiled from about 267,000 measurements (cf. Grenon *et al.* 1992) and cross-referenced with the Tycho input catalogue to give 65,058 stars (Egret *et al.* 1992). Out of this sample 21,559 stars remain with more than two sources or observations at different epochs and some of them are rejected because they are flagged as close pairs or known multiple systems in the Tycho input catalogue. Thus the working sample of photometric standards for calibration discussed below contains 13,436 stars.

3. Calibration and reduction

The task preceding the calibration is to identify the raw transits with a particular star. This is done by computing a combined probability for all transits attributed to one star transit both in time and in the standard star magnitude. Stars measured with a high background due to the elliptical orbit of the satellite are rejected. After this procedure the transits are split according to their origin: inclined/vertical slits, following/preceding field of view, upper/lower part of the slit system for the B_T, V_T, and T magnitudes, giving 24 possible combinations.

For each combination six calibration parameters are computed: offset, slit abscissa (both linear and quadratic), colour (both linear and quadratic) and a linear combination of slit abscissa and colour.

The mathematical method is discussed in Scales *et al.* (1992). After the calibration the parameters are checked for consistency (*e.g.* sudden parameter changes between two sets of observed transits). The transits are then reduced using the calculated six parameters and each transit is merged into a sorted working catalogue. At the moment this catalogue contains about 650,000 single transits.

4. Some results

We determined calibration parameters from the sample of photometric standard stars and reduced the transits of about 41,000 stars covering a time interval of about 300 days. In this period we found transits of nearly all photometric standards (96.7%).

The calibration parameters were plotted for each of the 24 combinations mentioned above and no significant sudden parameter changes could be detected over the period. However there are differences *e.g.* between inclined and vertical slit observations which show a different offset in all three bands for the upper part of the slit system.

Figure 1 compares the reduced Tycho observations with the magnitudes from the standard star catalogue for about 4250 single transits for one region of the sky. Because this region is located near a scanning node of the satellite, several transits were observed for each star in less than a year generating the stripes in the three plots. More than 98% of all transits are within the 3σ photon noise errors indicated by the solid curves. The slightly worse T magnitudes are a general phenomenon.

Especially in the B_T band, it is obvious that there is a lack of Tycho measurements for stars fainter than the 10th magnitude resulting in a shift of the mean of the distribution towards negative differences in Figure 1. This is due to the detection limit of the instrument: if the faintest detectable magnitude is $B_T = 11.7\,\mathrm{mag}$ it is not possible to get fainter measurements even though they are lying within the photon noise.

In Figure 2 we extracted the reduced observations of a 7th magnitude star which was observed for about four days. Figure 2a shows each single transit in all three bands for the inclined and the vertical slit groups respectively. In Figure 2b the same reduced measurements are averaged for time intervals of about 5.9 hours.

Going back from these plots to a bigger data sample (about 6700 single transits) we find the following mean values and 1σ errors for the differences between reduced Tycho magnitudes (single transits, vertical slits) and the ground-based measurements for the B_T band :

range	mag	transits	mean	σ
4.5	5.5	125	0.0014	0.0269
5.5	6.5	513	-0.0018	0.0336
6.5	7.5	1005	0.0027	0.0495
7.5	8.5	1083	0.0024	0.0821
8.5	9.5	1130	0.0042	0.1353
9.5	10.5	744	-0.0596	0.2313
10.5	11.5	251	-0.2610	0.3451
11.5	12.5	19	-0.6841	0.5315

The growing deviation of the mean for faint stars can be seen again in the table as discussed above. The 1σ errors are consistent with the expected pre-launch errors (*e.g.*. Grenon 1989).

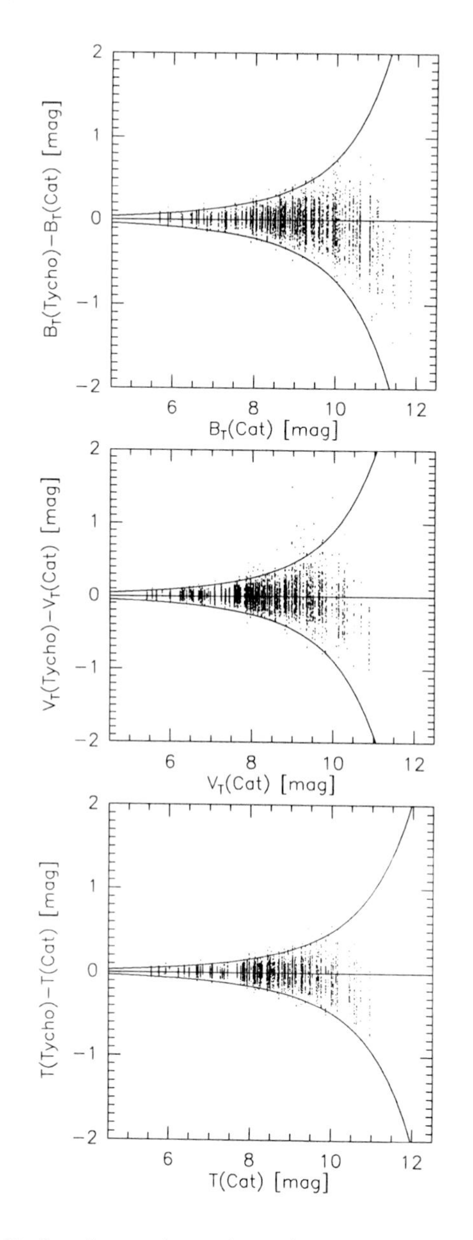

Figure 1 Reduced Tycho observations minus the catalogued values versus the catalogued values for the Tycho B_T, V_T, and T band. The solid curves indicate the 3σ limits from photon noise.

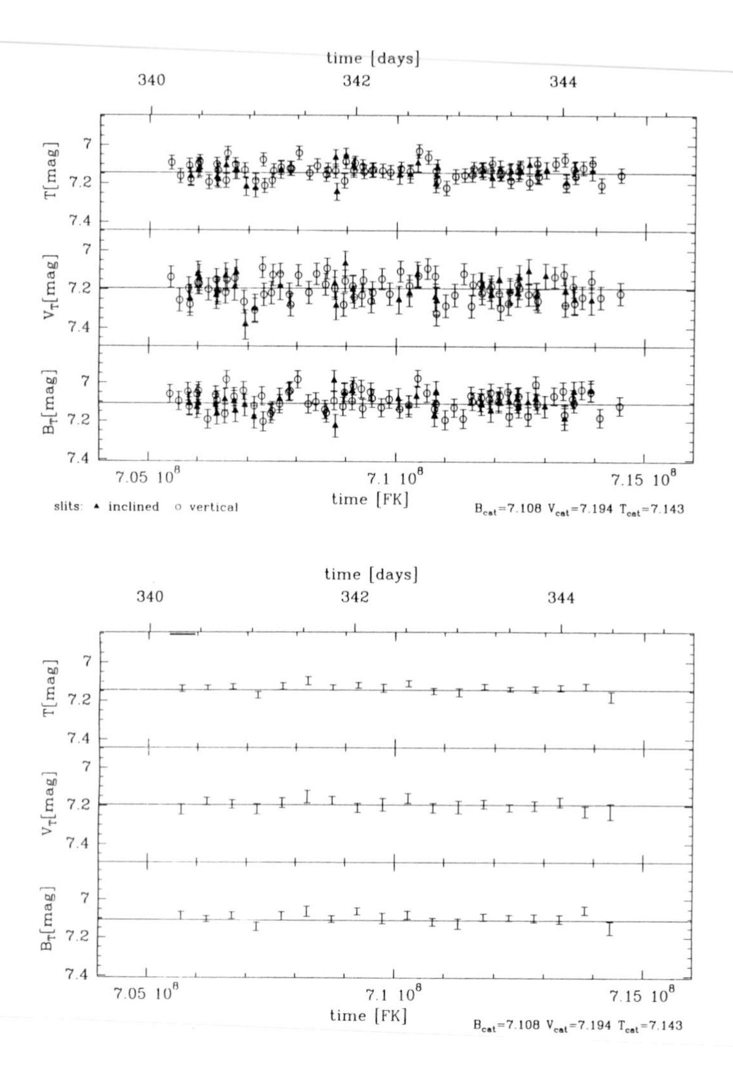

Figure 2 Example of the Tycho measurements for a 7th magnitude star in all three Tycho bands versus time. **a** the single transits across the vertical and inclined slit pairs of the Star Mapper grid. **b** average of these transits for time intervals of 5.9 hours. The lower time axis has filing keys units (one filing key is 0.042 seconds), the upper one gives the corresponding days since the launch of Hipparcos. The solid lines are the magnitudes from the standard star catalogue which are given at the lower right corner of the Figures. The 1σ errorbars include the photon noise and the error resulting from the uncertainty in the calibration parameters.

5. Current processing and outlook

On the basis of a time interval of about 300 days we presented some measurements of the star mapper photometry. The on-coming data analysis should yield better results mainly for one reason: in the final data treatment we will use only transits found on the basis of the revised Tycho input catalogue containing roughly one million stars. This will give less false identifications of transits with stars than the usage of the current input catalogue (about three million stars). A minor but important improvement is the better treatment of the background during the detection process.

Due to the decision of ESA this year, the satellite operation will be extended towards mid-1994. As a consequence we will get more (on the average about 150) observations for each star compared with the average 100 observations for a three-year mission.

Acknowledgement

This work was supported by the Bundesministerium für Forschung und Technologie under grants No. 010085029.

References:

Bässgen, G., Wicenec, A., Andreasen, G.K., Høg, E., Wagner, K., Wesselius, P.R., 1991, **258**, 186

Egret, D., Didelon, P., McLean, B.J., Russell, J.L., Turon, C., 1992 **258**, 217

Grenon, M., 1989, ESA SP-1111, Vol. II, 141

Grenon, M., Mermilliod, M., Mermilliod, J.C., 1992, Astron. Astrophys. **258**, 88

Halbwachs, J.L., Høg, E., Bastian, U., Hansen, P.C., Schwekendiek, P., Wagner, K., 1992, Astron. Astrophys. **258**, 193

Scales, D.R., Snijders, M.A.J., Andreasen, G.K., Grenon,M., Grewing, M. Høg, E., van Leeuwen, F., Lindegren, L., Mauder, H., 1992, Astron. Astrophys. **258**, 211

Discussion

W. Tobin: *When do you expect to distribute the optical disks? Also, how many times will an average star be observed in the extended mission*

Grossmann: Before the mission was prolonged by ESA we expected to distribute in 1996, now it might be 1997. On average we expect a star will be observed 150 times.

A. T. Young: *I am concerned that the standard deviations of the individual measurements are so large. It is very difficult to detect systematic errors smaller than 1/3 or 1/4 of the random errors. I worry that there may be undetected systematic effects larger than a millimagnitude.*

Faint Photometry with the HST Wide Field Camera

S.M.G. Hughes[1]

[1] *Palomar Observatory 105-24, California Institute of Technology, Pasadena CA 91125, USA*

Abstract

A series of 40 HST WFC exposures in v (F555W) and i (F785LP) have been obtained of two fields in M81, in order to search for Cepheid variables. This has provided an ideal opportunity to compare how the photometry programs DoPHOT and DAOPHOT cope with undersampled images, and present our 'best solution' to the problem of reducing WFC photometry for faint stars in crowded fields.

Although there are major obvious advantages in using WFC images (such as 0.1 arcsec resolution), measuring stellar magnitudes with WFC poses several problems in addition to those normally associated with ground-based CCDs. Not only is there a much higher cosmic ray count and the image cores are undersampled, but the aberrated optics of HST give rise to a 3 arcsec halo and a variable point spread function (PSF), and the flat fields (obtained from earth-streaked images) are not ideal. In addition, since early 1992 the chips have been seen to suffer from periodic contamination which creates a measles effect, but happily for v and i-type projects, is most pronounced at short wavelengths.

Fortunately, most of these problems should be fixed with the installation of WFPC 2 at the end of 1993, which is designed to not only correct the optical aberrations, but will have better detectors and an internal calibration ("flat-field") lamp.

However, even with the current problems, it is still possible to do faint photometry, and one of the major projects of HST, to calibrate several extragalactic distance indicators to better than 10% using Cepheids to measure distances to galaxies out to the Virgo group, is about to complete its first phase, by finding a large sample of Cepheids in M81 (Mould 1992).

Photometry programs that fit conventional Gaussian-like PSFs are inappropriate to WFC images, due to the undersampling of the diffraction-limited core spikes. Two popular photometry packages that have had their psf routines suitably modified are DAOPHOT II (Stetson 1991, which fits a Lorentz function) and DoPHOT 2.0 (based on Mateo & Schecter's 1989 DoPHOT program) with its pseudo-Gaussian routine replaced by a routine written by Abi Saha (STScI), which is designed to match the PSF shape of the WFC cores. The performance of these two packages have been compared by adding 5x100 randomly positioned pseudo-stars, covering a range of known magnitudes with photon (Poisson) noise included (using ADDSTAR in DAOPHOT II) to five identical medianed images of an M81 field (derived from 8x900 sec F555W exposures). The PSF function used by ADDSTAR to create the pseudo-stars was derived from a 7x7 grid of PSF stars generated by the TINYTIM code (Krist 1992, STScI, private communication). The differences between the measured and actual magnitudes are plotted in Figure 1, along with the standard deviation in each of several magnitude bins. Clearly, DAOPHOT does much better at most magnitudes, although this may be partly due to the use in DAOPHOT of the same PSF function that was used to create the pseudo-stars, and also because the standard version of DoPHOT 2.0 does not allow for a quadratically-varying PSF. However, at the magnitude of Cepheids (~24 at the distance of M81), each is of comparable performance, giving errors of ~0.2 mag.

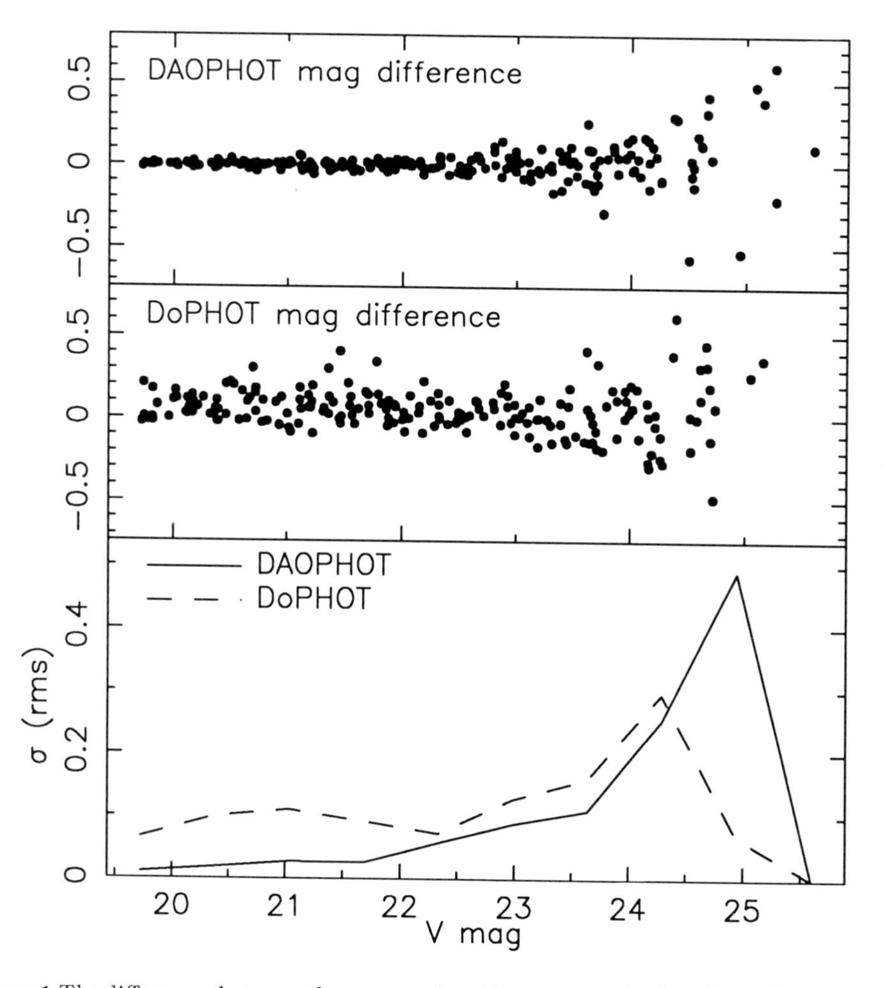

Figure 1 The differences between the measured and known magnitudes of pseudo-stars, obtained by DAOPHOT II (top panel), and DoPHOT 2.0 (middle panel). The bottom panel shows the standard deviation of these differences per magnitude bin, as a function of approximate *V* magnitude.

However, even better results are now being obtained by Stetson (1992, private communication) using ALLFRAME, a code based on DAOPHOT II, but which simultaneously analyses all common frames. Using ALLFRAME, ~30 Cepheid candidates have been found in the two M81 fields. A further 8 frames will be obtained in Cycle 2 to confirm and refine their periods.

Having obtained precise relative photometry, the data need to be calibrated. Initial calibration of the data has been made from Palomar 200-inch COSMIC frames obtained by Jeremy Mould. Here the main problem has been that about half the objects that appear stellar in the COSMIC frames are in fact extended (probably HII regions) in the WFC frames, and so all extended objects have to be eliminated before matching with the ground-based photometry. Current calibration of the zero-point has an uncertainty of 0.2 mag, but it is possible this will improve with better flat-fielded WFC data and modelling the vignetting across the WFC field (Holtzman et al 1991).

Acknowledgements

I am greatly indebted to Peter Stetson for valuable advice and for guiding me through the intricacies of DAOPHOT, to Abi Saha for sending me both the UNIX version of DoPHOT 2.0 and his WFC PSF routine, to the members of the H_0 team, in particular Jeremy Mould, Myung Lee, Barry Madore and Wendy Freedman, and to Mario Mateo for helpful advice with DoPHOT.

References:

Holtzman J.A. et al, 1991 AJ 369, L35.
Mateo M., Schecter P. 1989, in *Proc of 1st ESO ST-ECF Data Analysis Workshop*, eds. P.J.Grosbol, F.Murtagh and R.H.Warmels (Garching:ESO), p69.
Mould J.R. 1992, *Science with the HST*, ECF conference July 1992, Sardinia, in press.
Stetson P.B. 1991 in 3rd ESO/ST-ECF Data Analysis Workshop, Garching April 1992, ESO Conference and Workshop proceedings No.38, eds. P.J.Grosbol and R.H.Warmels, p187.

Discussion

M. Taylor: *I realise that using an internal calibration lamp will probably improve the flat-fielding problems significantly, but won't you still need to do external flats to determine the characteristics outside the camera optics?*

T.J. Kreidl: *Have you tried deconvolution, well aware that there may be flux-conservation problems, to see if this might help the situation?*

Hughes: Nick Weir (Caltech) has been experimenting with a new version of his Maximum Entropy code, that should be capable of conserving flux. In the meantime, we intend using deconvolved images to identify faint and crowded images, whose positions will then be used in ALLFRAME to derive photometry from the original images.

E. Budding: *Could you explain the term "internal flat-fielding"? Also, how soon is it going to be before this whole flat-fielding business can be ignored at the 0.1% level?*

Hughes: WFPC2 will have an internal calibration module that will be used to monitor the stability of the CCDs' response, but to obtain a true calibration (to remove the effects of telescope plus WFC optics plus detector response) external flat-fields will still be needed. As to when we will no longer need flat-fields, that depends not only on uniform CCD response, but also on telescope optics, which currently are certainly not uniform over large fields of view, nor stable over time, due to thermal and gravitational flexure and deteriorating optical surfaces.

R.R. Shobbrook: *In spite of the problems, you still have a distance to M81 far more accurate than before this work?*

Hughes: Yes, that is certainly true. The aim is to measure galaxy distances to better than 5%. Since we have $\sim$ 16 good Cepheid candidates from 24 epochs (includes cycle 2 time), these results show that we should be able to measure a reliable distance to an internal accuracy (ignoring reddening and P, L zero-point uncertainties) of $<$2% for M81, but eventually we intend to observe galaxies out to Virgo.

Recent Results from the High Speed Photometer

M. Taylor[1], R.C. Bless[1], M. Nelson[1], J. Percival[1], A. Bosh[2], M. Cooke[2], J. Elliot[2], W. van Citters[3], J. Dolan[4], J. Biggs[4], J. Wood[5], E. Robinson[5]

[1]*U. Wisconsin,* [2]*MIT,* [3]*NSF,* [4]*NASA/GSFC,* [5]*U. Texas*

Abstract

One of the first stellar photometry programs completed with the High Speed Photometer (HSP) on the Hubble Space Telescope (HST) was visual and ultraviolet observations of the Crab pulsar. We obtained continuous observations on four consecutive days using a visual filter (4000 - 7000 Å) and an additional observation, approximately two months later, using an ultraviolet filter (1600 - 3000 Å). Each observation has a time resolution of 10.7 μsec and spans approximately 30 minutes in duration. In addition to the observations made with the HSP, contemporaneous UBVR observations were also made at Jodrell Bank and McDonald Observatory. Some of the more prominent results include the following: 1) the main pulse arrival time is the same in the UV as it is in the optical and the radio regions of the spectrum, 2) there is essentially no difference in the shape of the optical pulse from one observation to the next, 3) the "flatness" of the peak of the main pulse suggests that the main pulse has been resolved in time, and 4) in accordance with the trend of observations from the radio to infrared wavelengths, the main pulse is slightly narrower in the UV than in the optical.

A second HSP science observing program was a long-term program to monitor the eclipsing dwarf nova, Z Chamaeleontis (P_{orb} = 107 minutes). We obtained a total of 42 observations of Z Cha in the UV (1120 - 1580 Å) each with a duration of approximately 45 minutes and separated by approximately three days. Although the majority of the observations cover the eclipse of the white dwarf and hot spot, a few observations were obtained outside-of-eclipse in order to obtain the complete light curve. During the course of this program, Z Cha underwent two "normal" outbursts in which the shape of the light curve changed dramatically. We will present a comparison of the light curve in quiescence with that during a "normal" outburst and quantify such geometrical and physical parameters as temperature and size of the white dwarf, hot spot, and accretion disk.

Discussion

D. O'Donoghue: *Would you say that there is any difference between the UV data obtained by HST and the fast photometry obtained from the ground in the optical?*

D. Dravins: *How is it possible to determine a phase shift of 45μs between the ultraviolet and visual light curves of the Crab Pulsar given that the HST onboard clock has an accuracy of only 1 millisecond?*

Taylor: We have calculated the FWHM of the UV and the V main pulse profiles, I don't think this is a function of the accuracy of the onboard clock. You will notice that the trailing edges line up very well while the leading edges of the main pulse profiles are distinctly separated so, relatively speaking, I believe it is safe to say that the UV main pulse is narrower than the V.

Stellar Photometry with the Optical Monitors

E. Antonello[1], M. Cropper[2]

[1]*Osservatorio Astronomico di Brera, Via E. Bianchi 46, 22055 Merate, Italy,*
[2]*Mullard Space Science Laboratory, University College London, Holmbury St Mary, Dorking, Surrey RH5 6NT, UK.*

Abstract

The Optical Monitors are small optical telescopes which will fly on board of the X-ray satellites SPECTRUM-X-GAMMA (JET-X experiment) and XMM. Their main scientific applications are the simultaneous observations (imaging) in the optical and UV band of the optical counterparts of X-ray sources, with limiting magnitudes of about m_V=22 (JET-X) and 24 (XMM). The OMs are developed in order to perform also serendipitous observations with high photometric precision of bright stars falling in the field of view, which should allow the detection of stellar microvariability at a level better than 10^{-5}. The main characteristics of the instruments are presented and the problems to be solved in order to reach the scientific goals are briefly discussed.

1. Introduction

The project of an Optical Monitor (OM) for X-ray satellites derives from the scientific need of having complete data coverage at various wavelengths, UV and optical, of the observed sources, from AGN to bright stars. In principle, optical observations simultaneous with X-ray ones can be performed from ground based telescopes and indeed this procedure has been adopted thus far. However, the complexity of satisfying the constraints typical of the optical telescopes (weather conditions, source observability) and of the X-ray instrumentation (e.g. orbital constraints) lead inevitably to a substantial loss of observing time, both with the optical and X-ray instrument. Therefore, the only practical way of having at the same time an optimal utilization of the time available for X-ray observations, and the wealth of simultaneous UV-optical observations, is to have a small telescope, co-aligned to the X-ray one, as part of the mission. Other advantages of the OM are the immediate identification of the optical counterparts of X-ray sources, mainly in crowded fields, and the possibility of performing a continuous monitoring for long time intervals (days). The latter opportunity along with the provided high precision of *differential* photometry measurements can be exploited for detecting in serendipity mode very small regular oscillations (microvariability) of bright stars falling in the field of view. This provides an opportunity to develop the research field of asteroseismology.

Table 1: Main characteristics of the OMs

	JET–X – OM	**XMM – OM**
Satellite	SPECTRUM–R–G (Russia), JET-X Exper. (UK, Italy, Germany, Russia, ESTEC)	XMM (ESA)
Launch date	1995	1999
Orbit	4 days	1 day
Lifetime	> 3 years	> 10 years
OM Resp.	Italy	UK, Italy, USA, Belgium
Status	Phase C	Phase B
Telescope	Ritchey–Chretien 26 cm f/10.9	Ritchey-Chretien 30 cm f/13
T3	mirror with central hole	dichroic: 150-550, 550-1100 nm
First beam		
Filter wheel	UV, V, white, stop	UV1, UV2, UV3, U, B, V, HeII, white, grism1, magnifier, grism2
SFC	8', pix. size 1.67" CCD frame transfer (UV)	22', pix. size 1" large format MCP device
Second beam		
Filter wheel	—	V, R, I, Z, Hα, white, grism1, magnifier, blur, grism2, stop
WFC	30', pix. size 6.2" CCD frame transfer	30', pix. size 1.8" large CCD, frame transfer

The present contribution contains a short description of the capability of the OM for the X-ray satellite missions SPECTRUM–X–GAMMA and XMM as regards the scientific applications related to stellar photometry.

2. The Optical Monitors

The main characteristics of the OMs are reported in Table 1 and in the schematic layout shown in Figure 1. Conceptually, the OM is a small Ritchey-Chretien telescope with two detectors: one is dedicated mainly to the imaging/photometry with good resolution for a small field (SF) of view in particular in the UV band, and the other mainly to the imaging/photometry with lower resolution for a wide (WF) field around the small one. JET–X OM will make only wide band photometry in few filters, while with XMM OM it will be possible to use several filters, grisms (for low resolution spectroscopy) and a magnifier.

In order to secure the high quality of the photometric data it is essential that the spacecraft be well stabilized, or at least a mechanism must be available for the correction of its instability. The reason for this is the nonuniformity of the CCD responsivity within each pixel; this cannot be corrected for by means of flat field techniques. In the case of JET–X OM the original project included a stabilization mechanism and a derotating system (Antonello et al., 1990) which were later removed. Since the star tracker capability of the satellite is limited, the present project foresees the use of the OM itself as an accurate star tracker for satisfying the requirements of pointing stabilization. Similarly, the high photometric precision with the CCD in XMM OM requires a high stability, and in this case the image stability against the drift of the satellite will be guaranteed by a mechanism based on a double–wedge system placed before the CCD. Both in JET–X and in XMM case, the signal needed for the stabilization is derived from the stars observed with one of the two detectors.

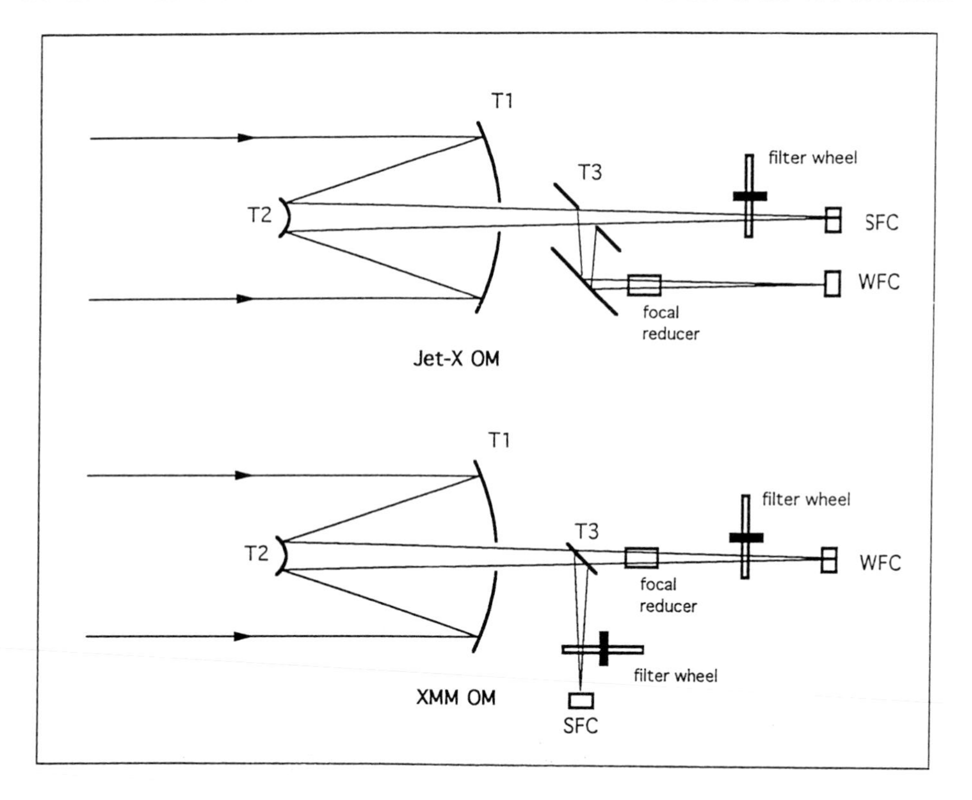

Figure 1. Schematic layout of the OMs

The main difference between JET–X and XMM OM is the detector used for observing the sources in the SF. In JET–X case it is a CCD, UV enhanced, while in XMM case it is a large format photon-counting detector based on a multi-channel plate and CCD readout system (MIC detector). In addition, the red detector in the

XMM OM is a large format CCD. In view of the large field of MIC, the term SF is somewhat of a misnomer preserved for comparison with the JET–X OM.

3. JET-X OM

In JET–X OM a drilled mirror separates the optical beam in two parts. The central part is focussed on the SF focal plane after passing through a changing filter mechanism with three positions, while the surrounding annulus is reflected and focussed on the WF focal plane by a focal reducer. The detector for the SF is a frame transfer UV enhanced CCD TH7863. The main task of the WF camera (WFC, where the detector is another frame transfer CCD TH7863 but not UV enhanced) is to perform the star tracking (with defocussed stellar images) for the stabilization. The working temperature of the two detectors is about −100°C and it is kept at that level by means of a passive radiator.

The CCD frames must be corrected for cosmic ray contamination, and the cosmic ray filtering is performed on board (see Antonello et al., 1990). With the SFC it should be possible to detect faint sources with m_V ∼23 (S/N≥3) with integration times of about a thousand seconds. Since the bit rate allotted to OM is limited (0.5 kbit s^{-1}), there are some constraints on the size and/or the number of the frames telemetred to the ground. As regards the brightest stars falling in the field of view ($m_V \geq 4$), they are observed by the WFC and used for the star tracking. A total of some tens of bright stars are selected in a threshold mode, and each stellar image is contained within few pixels (e.g. 3 x 3 pixels). A series of 1 s (or less) exposures are summed for a total of several tens of seconds of exposure time, and the subimages corresponding to the selected bright stars are telemetred.

4. XMM OM

In XMM OM the beam separator is a dichroic, because the MIC detector in the SFC is dedicated to the observation in the UV–blue band and the CCD to the observation in the red–near IR band. The blue beam is reflected and reaches the SFC after passing through the blue filter wheel, while the red beam is transmitted and reaches the WFC after passing through the image stabilization mechanism, the focal reducer and the red filter wheel. The expected angular resolution on the SFC is 2 arcsec, and it can be about 0.5 arcsec when using the magnifier placed on the filter wheel (working in white light).

With the SFC it is expected to detect a point source as faint as 24th mag with S/N=5 in 1000 s. Moreover, relatively bright stars ($m_V \geq 13$, in the V filter) can be observed with a time resolution of 0.01 second, and this capability renders this OM suitable for studying phenomena with very short time scales, such as transients. With the WFC it will be possible to study sources brighter than about $m_V = 23$ with reasonable exposure times, and it will be possible to observe stars as bright as $m_V \sim 4$ using the blurrer placed on the filter wheel for defocussing, and exposure times of few tenths of a second as in JET–X OM case.

Table 2: Limiting magnitude in white light for different source spectra for detection with XMM OM at a significance of 5σ in 1000 s; a, b and c indicate best, typical and worst background, respectively.

	blue beam			red beam		
source	a	b	c	a	b	c
B0 star	24.6	24.4	23.6	23.3	23.1	22.6
A0 star	23.6	23.4	22.5	23.6	23.4	22.6
M0 star	22.4	22.2	21.3	24.7	24.6	23.8

The allotted bit rate to this OM is 2 kbit/s, but the expected large amount of data mainly from the MIC detector requires a substantial data compression on board. This data compression will be performed partly using windows and binning techniques instead of full frame images. In view of the 10 years planned lifetime of XMM, substantial effort has been directed at the optimization of the operating modes of the instrument to maximize its capability without introducing unnecessary complexity.

5. Stellar Photometry

The expected integration times required to detect a point source using the XMM–OM are reported in Table 2 for a sample of sources and observational conditions. However, apart from the advantage over ground based instruments of observing in the UV band, here probably it is worthwhile to stress on the other advantage of observing from space, that is the possibility of avoiding the limitations given by the earth atmosphere in terms of scintillation and transparency fluctuations which are more important for the brightest stars.

The *differential* photometry precision for the bright objects is reported in Table 3. The 1σ–precision has been estimated for a source temperature of 5770 K. Owing to the brightness of the stars, the single exposures are assumed very short, 1 s or less, and the star image is spread over 4 or more pixels. In this way it is possible to observe bright stars up to the 4th magnitude without saturation and with a conversion factor from electrons to ADU close to the unity. The precisions have been estimated for a cumulative number of exposures of 30 s. In general, the effects of readout noise due to the very high cumulative number of exposures are not significant for very bright stars. As shown in Table 3, the expected minimum detectable amplitude of a sinusoidal variation of the stellar luminosity for an observing time of 3.5 days (JET–X case) can be of few 10^{-6} mag, that is close to the amplitude of oscillations of the Sun detected by ACRIM on SMM. As a comparison, we recall that the logarithm of the noise power density at a frequency of about 12 mHz of the photometric measurements is about –4.1 for a ground based 1 m telescope, –5.3 for the best CCD observations made with a ground based 4 m telescope (Frandsen, 1992, private communication) and -6.5 for a star with m_V=5 observed with the OM. The reason for the improvement

Table 3: Differential photometry precision (mag) for a cumulative number of 30 exposures of 1 s of bright stars with T = 5770 K, and minimum detectable (99% confidence level) amplitude of a sinusoidal variation of luminosity for an observing time of 3.5 days (see text).

m_V	precision (1σ)	ampl.	m_V	precision (1σ)	ampl.
4.0	6.1E–5	4.2E–6	8.0	3.9E–4	2.6E–5
5.0	9.8E–5	6.6E–6	9.0	6.2E–4	4.2E–5
6.0	1.6E–4	1.1E–5	10.0	1.0E–3	6.8E–5
7.0	2.4E–4	1.7E–5	11.0	1.7E–3	1.1E–4

which can be obtained with the OMs is the observation from space, which avoids the atmospheric scintillation and transparency changes.

The precisions reported in Table 3 have been estimated assuming a perfect stabilization of the images and of CCD temperature. Moreover, no contamination owing to cosmic rays and no degrading of the CCD performances due to radiation have been taken into account. The effects of a small wandering of the stellar images on the expected precision have been studied by means of simulations (Antonello, Poretti; this Colloquium) and will be tested in laboratory. It will be possible to correct for the cosmic ray contamination by means an on board frame comparison technique only the images of faint sources with relatively long exposure times, and it will not be possible to apply a similar correction to bright star images; therefore, we expect an incidence of about few percent of time series data contaminated by cosmic rays in the case of bright star observations.

7. CONCLUSION

In the present note we have discussed the main characteristics and expected photometric capabilities of the OMs for X–ray satellites, in particular the high precision photometry in the case of observed bright stars. While the rationale for incorporating an optical/UV telescope on the JET–X and XMM spacecraft is driven by a wide variety of needs, the unique position and capabilities of these instruments will open a new window on the study of microvariability and low level pulsations in a large number of stars. This will allow the spectrum of variability to be explored for a variety of stellar masses and evolutionary phases with the promise of a wealth of observational tests of our understanding of stellar structure.

References:

Antonello E., Citterio O., Mazzoleni F., Mariani A., Pili P., Lombardi P.: 1990, in SPIE Proceedings Conference 1235 on *Instrumentation in Astronomy VII*, 867.

High-Precision Photometry with Small Telescopes on the Moon

Michael Zeilik

Institute for Astrophysics, The University of New Mexico

Abstract

I explore the scientific gains that can be achieved by siting small (1 to 2-meter), special-purpose telescopes on the moon with the goal of high- precision observations.

1. Introduction

Robust small telescopes (1 to 2 meters) will likely pioneer astronomy on the moon. These light-weight, robotic instruments could even be soft-landed to the moon prior to the establishment of a lunar base. Each could be designed for a specialized task, such as imaging with detectors for limited ranges of the spectrum selected for specific scientific goals. Lunar APTs (LAPTs; Zeilik, 1989) would function much like those on the earth for interleaved programs of target objects. Initial instruments could operate as Lunar Transit Telescopes (LTTs). A LTT provides an unbiased, deep sample of a designated strip of sky over time. In either mode, we can imagine a network of lunar telescopes to cover the full sky (especially the southern) at a range of wavelengths to achieve a wide variety of scientific goals unattainable on earth.

The moon is by far the best site in the inner solar system for the next generation of high-performance telescopes. The main advantages of the moon relate to the fact that it is a stable, slow-rotating space platform (near the earth) on which large (and small!) structures can be built in an environment with no atmosphere. Lunar-based telescopes can make powerful utility of the lunar advantages for astronomy: no atmospheric opacity, so that all regions of the spectrum are open to view; no atmospheric seeing, so that the resolution is limited by the optics; no weather, so that long strings of precise, time-serial data are straight-forward to obtain; and no sky backgrounds to limit S/N, so even small telescopes can detect faint objects with relatively short integration times.

2. CCDs and Small Telescopes on the Moon

The Lunar advantages easily optimize the use of CCDs on small telescopes on the moon. If we apply the S/N CCD equation of Newberry (1991), we quickly see that: (1) we optimize object counts (100% transparent and constant with no air!); (2) we decrease background counts (lower sky brightness, less variation); (3) we can reduce number of pixels covered by object (no seeing, diffraction limited, match pixel size to optics); and (4) we can tailor filters to detectors and standard photometric systems as well as satisfying the sampling theorem (no second order extinction!). In conclusion: Low sky background, diffraction-limited performance optimizes the use of CCDs on small lunar telescopes to make them work effectively as large, ground-based telescopes.

In addition, small telescopes have the practical advantages that they have a low mass; are relatively inexpensive; rely on proven technology; are designed to be simple, robotic, and robust; can be soft landed; function as precursors to giant lunar telescopes to test the working conditions in the lunar environment with cosmic rays, meteoroids, large thermal variations between day and night, and charged surface dust.

3. Lunar Automatic Photometric Telescopes (LAPTs)

As a simple case, consider a 1 to 2-m, f/2 primary, f/5 to f/10 Cassegrain telescope. It will have a resolving power of 0.1 arcsec at 600 nm; a mass of some 1000 kg; and be passively cooled. As a pointed telescope (accuracy < 10 arcsec), it will have alt-azimuth mount, so that it is latitude independent. The detector would be a mosaic of CCDs (about 2000 by 2000 pixels; a 10 micron pixel size) at a focal plane, a minimum of 2 colors (with a dichroic beam splitter), and a field of view of some 10 by 10 arcmin. The CCDs would provide target acquisition: an exposure of a few seconds detects stars down 15th mag at a $S/N > 3$ (about 10 stars per frame). We would provide software masks for target identification with artificial intelligence algorithms for rapid confirmation. Once the right field is established, a 500 s exposure gives $S/N \approx 2000$ (0.05%) for a $V = 15$ point source of $S/N \approx 200$ for $V = 20$. We can rapidly attain high-precision, time-serial, continuous data.

What kinds of science could LAPTs achieve? One major target would be variable sources with multiple periodic structure, such as chromospherically active binary start systems, quasars, and AGNs. We could obtain performance impossible on the earth by collecting time-serial, continuous data of high-precision ($S/N = 100$ minimum) on selected targets. Such long data strings will allow us to extract reliably all the periods displayed by the target objects.

We can go beyond a single telescope and imagine a LAPTs network with eight telescopes, four in each hemisphere set at a 90 degree spread in longitude. A network would increase continuity of coverage, the number objects observed, and the wavelength regime for simultaneous data. Of course, it would be coordinated with any earth-based networks to maximize the scientific return.

4. Lunar Transit Telescopes (LTTs)

An even simpler design is that of a Lunar Transit Telescope (McGraw, 1992), which uses the moon's rotation as a slow and stable telescope drive. This instrument points at the meridian; the integration time is simply the time of an object's transit of the d detector at the focal plane.

Consider a 2-m telescope (mass about 1000 kg), with a FOV of about 2 degrees, and all-reflective optics. With a pin-joint structure, it would undergo passive thermal self compensation and never need to be focused. The detector would be a mosaic of UVOIR CCDs, with 0.1 arcsec pixels, and 5 bandpasses: 0.1 to 2 microns. The lunar sidereal rate gives 6.6-min integration on 2048 pixel array, so we expect limiting magnitudes of +27 at V, +25 at H, and +24.5 at K for a S/N = 10. As the sky slides by, an LTT will survey about 700 square degrees each lunar day. It will provide an unbiased, "deep" sample of homogeneous, high-precision, multicolor photometry.

An LLT will provide a vast amount of data – a strip search of the universe that will reveal millions of objects, some at physical extremes. It will be serendipity at its astronomical best, as the sample will be statistically complete down to the limiting magnitudes. For instance, we can study the morphology of distant galaxies and with multicolor imaging, provide evidence for a firm scenario of galaxian evolution. At this range of lunar astronomy, we will be constrained more by our imagination than by our data.

This work was funded in part by National Science Foundation Grant AST-8903174.

References:

Newberry, M. V., 1991, Pub. Astron. Soc. Pac., 103.122.

McGraw, J. T., 1992, in AIP Conference Proceeding 207, Astrophysics on the Moon, ed. M. J. Mumma and H. J. Smith, p. 433.

Zeilik, M., 1989, "Remote Access Lunar Photometric Telescopes" in Remote Access Automatic Telescopes, ed. D. S. Hayes and R. M. Genet (Fairborn Pres), pp. 289-298.

Discussion

A. T. Young: *I question your assumption that a telescope on the Moon is cheap, even if small. I was involved in a 1 cm telescope on Mars (Viking Lander camera), which cost about 16 million - before launch costs. Also, you have little dark time on the Moon. The Earth in the lunar sky is about 4 magnitudes brighter than the Moon in the terrestial sky.*

Zeilik: Any space mission for astronomy is expensive. In relative terms, a low-mass payload is cheaper. The phases of the Earth do make a *bright* source, especially at *full* Earth. But the Moon has no atmosphere to scatter the light, and the problem is less severe than in low-Earth orbit.

S.B. Howell: *The radiation event problem will cause real damage to the CCD. In time-series mode, these events will cause the loss of photometric information for a few frames, but most of the series will be useable.*

Zeilik: I hope that a cosmic-ray hit will not affect more than 1 or 2 pixels and can be removed in processing. However, with a flux $\sim$ 100 times that on the earth, actual damage may well result.

R.F. Garrison: *One disadvantage you didn't list, with respect to geosynchronous orbits, is that only one half of the sky is available at any given time.*

Zeilik: Yes. That's the reason I proposed a network on the Moon for *all-sky* coverage.

E. Budding: *I think that, if you want to see real data from this kind of remotely operated device, (including also the SALUTE-type instrument) before retirement, you should be pushing hard for the terrestrial-precursor version.*

Zeilik: The CCD Transit Instrument (CTI) is an already operational precursor to the Lunar Transit Telescope.

D.H.P. Jones: *I understand there are some deep craters at the poles of the Moon which are in continuous darkness.*

Zeilik: Such craters would work as excellent sites for completely-cooled infrared telescopes.

Tarmo and Silvi Oja say good-bye to Ian Elliott

A group at the prehistoric site of Navan Fort near Armagh